Clausing, 1988). Figure 2.5 shows a "House of Quality" related to the desired characteristics of an automobile door. On the left-hand side of the figure (horizontal rows) the two attributes that the customers desire are listed (i.e., easy to open and close door, and isolation). The columns (i.e., the upper right-hand part of the figure) show the FRs for engineering design. This figure is a graphic representation of a transformation matrix where the customer attributes are converted into a set of FRs.

The use of the "House of Quality" is as follows (Hauser and Clausing, 1988). Once the customer attributes are defined, engineering characteristics of a design must be listed. This process must involve both the marketing and the engineering personnel. In assessing the problems with an existing design of one's product, the customer's perceptions are also listed as shown in Fig. 2.5. In this particular example the figure states that "our door" is much more difficult to close from the outside than the competitors door. The design team then decides to investigate further because the marketing data indicate that this customer attribute is important. They identify the FRs that affect this customer attribute: energy to close door, peak closing force, and door seal resistance. The engineers decide on two FRs—the energy to close the door and the peak closing force—as good choices because they relate to the consumer attributes most closely. They then proceed to consider these two FRs (in terms of the Hauser–Clausing notation, engineering characteristics).

The next step in this approach is to identify what other FRs might be affected by changing the door-closing energy. Door-opening energy and peak closing force are positively related. Other FRs (i.e., engineering characteristics in this figure) are considered to be negatively related based on objective measures of competitors' doors, customers' perceptions, information on cost, and technical difficulties. Through this process a new door-closing target is set for the door as 7.5 ft-lb of energy. This target, noted on the very bottom of the house, establishes the FRs for the door "easiest to close."

Next consider the customer attribute "no road noise." This attribute, according to the market study, is only mildly important to the customer. Its relationship to the specification of the window is not strong, in that window design can reduce the outside noise only to a limited degree. Furthermore, in order to reduce the noise transmission through glass, it must be made heavier, which adversely affects the door-closing energy. Finally, marketing data show that customers are happy with this company's car as far as the noise is concerned. Therefore, the design team decides to leave the window's transmission of sound alone. The original constraint on noise is still accepted.

The technique of defining FRs to satisfy the perceived needs using the "House of Quality" concept is applicable in the case of an existing product. However, when a product that is new and innovative is designed for the first time, the FRs must be defined in "solution-neutral" environment without considering any physical solution in mind.

2.6 Hierarchy of FRs and DPs: Decomposition of the Design Process

There are two very important facts about design and the design process, which should be recognized by all designers:

1. FRs and DPs have hierarchies, and they can be decomposed.
2. FRs at the ith level cannot be decomposed into the next level of the FR hierarchy without first going over to the physical domain and developing a solution that satisfies the ith level FRs with all the corresponding DPs. That is, we have to travel back and forth between the functional domain and the physical domain in developing the FR and DP hierarchies.

In this section we explore these two aspects of FRs and DPs.

In considering the design of a refrigerator door in Chapter 1, we initially confined our attention to the most important problems. We considered the energy loss and the access to the food in the refrigerator. We did not concern ourselves with the specific ways in which the door was to be hung, or about the insulation technique. Only after we decide whether or not the door should be hung vertically should we worry about other details. This is always the case in all design processes. The FRs and DPs can always be decomposed into a hierarchy. Indeed, we are extremely fortunate that this is so, because we may focus on only a limited number of FRs at a time, thereby reducing the complexity of the design task immensely. The intricacy of the design task increases rapidly with the rise in the number of FRs at a given level of the FR hierarchy.

A previous discussion touched on the need to alternate between the functional domain and the physical domain in the design process. This is illustrated further here. As stated previously, the design process begins in the functional domain with the specification of the first level of FRs. Suppose we wish to design a passenger vehicle that can transport people within a city. We may specify the first level of FRs for the vehicle to be the following:

FR_1 = Ability to move forward.
FR_2 = Ability to move backward.
FR_3 = Ability to change directions.
FR_4 = Ability to stop.

Having stated these four first-level FRs, we must now conceptualize a physical vehicle that can satisfy them. Without completing this conceptualization process, we cannot decompose these first-level FRs further into lower-level FRs. We must switch to the physical domain from the functional domain, and vice versa, in order to be able to proceed with the design process. Indeed, the design process requires that we switch between the functional and the physical domains each time we move down the hierarchy. For example, in the case of the passenger vehicle, one of the

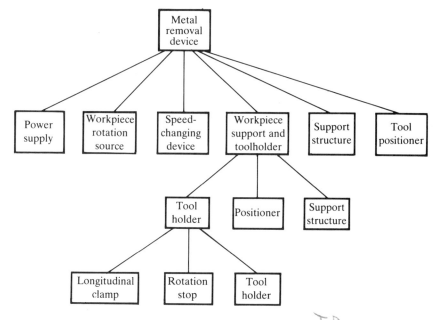

Figure 2.6. Lathe functional hierarchy.

physical solutions for satisfying FR_1 and FR_2 is to use a battery-powered d.c. electrical motor and a double-pole switch. Then FR_1 and FR_2 can be further decomposed in the context of this physcial entity.

The designer must recognize and take advantage of the existence of the functional and physical hierarchies. A good designer can identify the most important FRs at each level of the functional tree by eliminating secondary factors from consideration. Less-able designers often try to consider all the FRs of every level simultaneously, rather than making use of the hierarchical nature of FRs and DPs. Consequently, the design process becomes too complex to manage.

It is easier to illustrate the nature of the functional and physical hierarchy by analyzing an existing product. For example, the functional hierarchy and the physical hierarchy of a lathe are shown in Figs. 2.6 and 2.7, respectively. By comparing these two figures, it should be clear that we cannot simply construct the entire FR hierarchy without referring to the DP hierarchy at each corresponding level. That is, without having decided to use a *tailstock*, we could not have stated the three FRs: tool holder, positioner, and support structure.

2.7 FRs: Definition and Characteristics

In Sec. 2.6 two very important statements were made on the design and the design process. First, there is a hierarchy in both the functional space

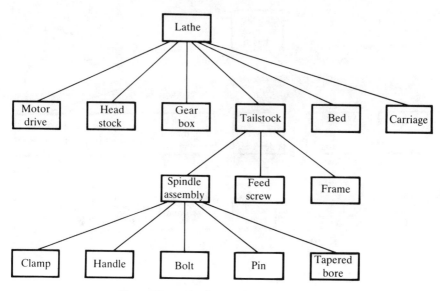

Figure 2.7. Lathe system physical hierarchy.

and the physical space. Secondly, design consists of a continuous iteration between the functional and physical spaces. By the word *function* we mean the desired output, whereas by the word *physical* we include all those things that generate the desired output. For example, an organization is a physical entity in terms of this definition. Software and hardware can be in the physical space. The function (or the desired output) may be the measurement of distance, access to the food in a refrigerator, opening of cans, mixing of liquids, etc. FRs are the designer's characterization of the perceived needs for a product, whether in terms of a device, process, software, system, or organization. The term FR has a specific meaning in the context of axiomatic design. FRs are *defined* to be the minimum set of independent requirements that completely characterize the design objective for a specific need. By definition, FRs must be independent of other FRs, and thus can be stated without considering other FRs. A set of FRs containing functions that depend on each other is called *redundant*. In this case, they can be reduced to one independent FR.

Since the designer can arbitrarily define the FRs to meet a certain perceived need, an acceptable set of FRs is not necessarily unique. For example, the FRs for the refrigerator door discussed in Chapter 1 are not unique. Another designer might have chosen a different set of FRs, depending on his or her judgement of the perceived needs. The physical solution, or the entity, will be very different depending on the particular set of FRs chosen. However, in the final analysis, the physical solution must satisfy the perceived needs, otherwise a new set of FRs must be tried. When designing a new product, it is difficult to choose the correct set of FRs on the first try. In some cases, it takes many iterations before one can be satisfied with the final set. An iteration may involve a new set of FRs or a slight modification of the old set.

Corresponding to a set of FRs, there can be many design solutions, all of which satisfy the same set of FRs. When the original set of FRs is changed, a new solution must be found. The new solution may not be derived by simply modifying the previously obtained solutions that were acceptable only for the original set of FRs. This situation is analogous to solving partial differential equations. Although the governing equation is the same, the solution may be entirely different when the boundary and/or initial conditions are changed. Many designers often make the mistake of trying to adapt or modify an existing solution when the FRs are changed by the addition of new FRs to the original ones, rather than seeking a completely new solution.

To reiterate, a good designer must have the ability to choose *a minimum number of FRs* at each hierarchical level of the FR tree. Some designers are proud that their design products can perform more functions than were originally specified. In this case, they have overdesigned the product. Consequently, it is more complex, more costly, and less reliable than is necessary. The designer who creates such a solution should go back and search for a simpler solution.

2.8 Constraints in Design

Constraints in the context of axiomatic design represent the bounds on an acceptable solution. They are of two kinds: *input constraints*, which are constraints in design specifications, and *system constraints*, which are constraints imposed by the system in which the design solution must function. The input constraints are usually expressed as bounds on size, weight, materials, and cost, whereas the system constraints are interfacial bounds such as geometric shape, capacity of machines, and even the laws of nature.

It is sometimes difficult to determine when a certain requirement should be classified as an FR or as a constraint. By definition, a constraint is different from an FR in that a constraint does not have to be independent of other constraints and FRs. In many cases cost is a constraint rather than an FR: its precise value is unimportant, as long as it does not exceed a given limit. Another distinguishing feature of constraints is that they do not normally have tolerances associated with them, whereas FRs typically have tolerances.

As we go through the design process, zig-zagging between the functional and the physical spaces, what used to be DPs at a higher level of the DP hierarchy may become constraints at a lower level of the DP hierarchy. For example, if the horizontally hung chest freezer door is accepted as the design solution that satisfies the two FRs (i.e., access to the contents and low energy loss), then the DPs of the design solution may now become *constraints* at the next level of the FR hierarchy. Therefore, when the FR associated with "access to the contents" is further decomposed into a lower level set of FRs [e.g., FR_1 = we must be able to take out the food in a chronological order (i.e., the food that went in first must come out first) and FR_2 = anyone over 5 ft 2 in. tall must able to reach any item in the refrigerator], the horizontally opening door becomes a constraint at this lower level. The design solution is now locked into concepts that use this type of door.

It becomes easier to select a set of FRs when the problem is highly constrained. The more constraints there are in a given design problem, the easier it becomes to narrow down the possible choices for FRs. In some cases the number of FRs at a given level of the FR hierarchy may be very small due to a large number of constraints, which should simplify the design process.

The probability of the same set of FRs being chosen by all designers increases as the number of constraints is increased. As the choice for FRs decreases with the increase in constraints, the number of permutations for selecting FRs decreases, which should force convergence to a similar set of FRs. This is the case when and only when the imposed constraints on all designers are exactly identical.

2.9 Design Outputs

The outputs of the design process are *information* in the form of drawings, specifications, tolerances, and other relevant knowledge required to create the physical entity. The design solution should be as simple as possible, so the design output can be conveyed with minimal effort. The total information involved should therefore be as small as possible, per Axiom 2. In Chapters 3 and 5 we formally define how we can measure the complexity using quantitative metrics. For the time being it suffices to remember: "the simpler, the better."

2.10 Design for manufacture

In the field of manufacturing, the most important words are *productivity* and *reliability*. Productivity may be defined as the *total output* value of the factory divided by the *total input*. Although such terms as *labor productivity* have been used, they are becoming useless measures in modern manufacturing operations, since the total direct labor cost is becoming a smaller fraction of the total manufacturing cost. Often the material is the dominant cost item, followed by indirect costs in discrete parts manufacturing. In view of the importance of productivity and reliability in manufacturing, we need to understand the relationship between design and manufacturing.

Design affects manufacturing in several important ways. As Example 2.1 illustrates, following the design of the product that satisfies the perceived needs, the process must also be designed. In some cases the design of products and the design of manufacturing processes must be considered at the same time. As the multilense plate case study in Chapter 6 shows, certain products cannot be manufactured unless a new process is developed. However, even in cases where the product can be manufactured by existing machines and processes, the design of the product affects the manufacturing productivity significantly. It has been stated that as much as 70–80% of manufacturing productivity is determined at the design stage. Similarly, the quality and reliability of a product also depend on both the design and the manufacturing skills.

The basic question related to *design for manufacturability* is: "How do we assure that the design decisions incorporate manufacturing concerns?" The key answer is this: when the *product and process designs* do not violate design axioms at all levels of the FR, DP, and PV hierarchies, then the product should be manufacturable. On the other hand, if the process or the product involves coupled designs, then it will be difficult to manufacture, because the manufacturing steps taken during the later stages of production can affect or undo the work of earlier stages. When the design axioms are satisfied by the product and the process, the manufacturability of the product is assured (Suh, 1985). This point is further reinforced in terms of theorems and case studies in later chapters.

The design axioms can better ensure design for manufacture if we develop specific corollaries or design rules, based on the axioms, applicable to specific instances of manufacturing. Boothroyd and Dewhurst (1983) have developed a handbook approach to assembly by giving specific rules for various classes of assembly operation. This is a useful approach for routine design operations. However, the danger of the handbook approach is that although the rules may be applicable to a given situation, they may be invoked wantonly by users even for unrelated tasks. This danger arises from the lack of a theoretical conception and scientific basis for the rules.

Stoll (1986), in his review paper, gives the following design rules for efficient manufacture, which can be easily developed for certain specific cases from the design axioms.

1. Minimize the total number of parts.
2. Develop a modular design.
3. Use standard components.
4. Design parts to be multifunctional.
5. Design parts for multi-use.
6. Design parts for ease of fabrication.
7. Avoid separate fasteners.
8. Minimize assembly directions.
9. Maximize compliance.
10. Minimize handling.

If these rules are used indiscriminately, it is obvious that the designer may act incorrectly and violate the axioms. Therefore, if rules are to be developed, their use must be constrained to specific situations. For example, the first rule may not be correct if the total number of parts is minimized at the expense of coupling FRs, since this violates Axiom 1.

2.11 Design Helix: Design, Manufacturing, and Marketing Symbiosis

In Fig. 2.3 it is shown how the "design" world works: the societal and human needs are translated into a set of FRs; a solution is synthesized through ideation to satisfy them; the solution is then analyzed for its rationality and compared with the perceived needs, to check its fidelity to the original needs. In an industrial firm the design decision on the choice of the FRs to satisfy the perceived needs are often made by the marketing people. Once the FRs are defined, then engineers are called in to develop a technical solution to fulfill the FRs. When the design is completed, the product is manufactured by designing appropriate manufacturing processes and strategies. The important point is that throughout the product design–manufacturing–marketing process, a consistent set of decisions must be made, at each stage of which one has to traverse between the functional and the physical space.

Figure 2.3 is a snapshot of a dynamic process which oversimplifies the reality in the interest of highlighting the essence of the thought process. The oversimplification has two shortcomings. First, the whole product development process does not necessarily begin from the societal and/or human needs. It can begin anywhere in the chain. For example, when a new technological breakthrough occurs, it can create a new product as well as new demand, creating markets that had not previously existed. Secondly, the product development process is not as planar as shown in the two-dimensional space in Fig. 2.3, because the process depends on time and moves forward continuously. For example, when the first round of the design process is completed, and the solution is compared with the original set of FRs and needs (which the marketing people might have developed through a "market study"), it is highly probable that the additional insight gained through the first iteration may enable the designer (or the manager) to come up with a better set of FRs, which may in turn enable the design and manufacture of more-advanced products. The availability of advanced products may then change the marketing strategy; that is, each time the design loop is criss-crossed, new information is generated which propels all related fields (e.g., engineering, marketing, etc.) to the next plateau.

This symbiotic process of the "design" world, going from "marketing" to "design" to "manufacturing" to "marketing" to "design," etc., can best be depicted by the design helix, shown in Fig. 2.9. The design helix shows

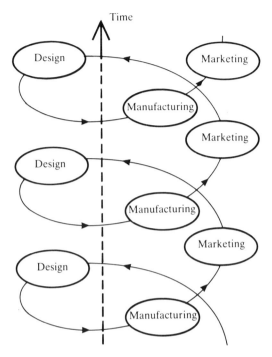

Figure 2.9 The design helix of the "design" world involves marketing, design, and manufacturing. "Marketing" in this context includes the response of "users" or customers, in addition to the selling activity.

that the product development process can begin at any stage or any place, and as a function of the time spent and the effort spent, we move to the next stage. Conversely, just because a product is designed, manufactured, and sold in the marketplace, we should not regard the process as being completed. One must identify new information from the existing product, and develop the marketing–design–manufacturing strategies for the next generation of improved products. Quotation marks are used around key words to indicate that the conventional subdivision and classification of the "design" world are somewhat arbitary.

This process of traveling up the "design"–"manufacturing"–"marketing" helix (i.e., the design helix) is sometimes random and chaotic, rather than being systematic and rational, because there is no absolute referent that provides the metric for the decisions made at each step. The design axioms given in this book provide a set of laws that provide the reference frame for the decision making.

Much more needs to be done to understand the symbiotic dynamics of the design helix, so that industrial firms can develop improved products and increase both the market size and the market share. For this to be possible, we need to deal with both micro- and macro-issues, develop models for dynamic interaction amongst the elements of the design helix, and develop most-appropiate manufacturing processes, systems, and machines. Some macro-issues clearly should involve not only engineers, but also the business enterprise, the legal profession, and the political process.

2.12 Summary

This chapter describes the design process, after defining design, as the mapping process between the functional and the physical domains. It emphasizes the notion that the first step in design is problem definition in the form of FRs and constraints, that the design process involves a hierarchy of FRs and DPs, and that the designer must switch between the functional and physical domains to carry out the design process. It defined FRs and constraints, and elaborated on the output of the design process. The concept of the design helix introduced the dynamic nature of the product development and marketing.

References

Boothroyd, G., and Dewhurst, P., *Design for Assembly—A Designers Handbook.* Department of Engineering, University of Massachusetts, Amherst, MA, 1983.

Hauser, J.R., and Clausing, D., "The House of Quality," *Harvard Business Review,* 63–73, May–June, 1988.

Jenkins, R. V. ed., "Words, Images, Artifacts and Sound: Documents for the History of Technology," *British Journal of History of Science* **20**:39–56, 1987.

Shaw, M.C., "Creative Design," *CIRP Annals* **35**(2):461–465, 1986.

Stoll, H.W., "Design for Manufacture: An Overview," *Applied Mechanics Review* **39**(9):1356–1364, 1986.

Suh, N.P., "Manufacturing and Productivity," *Proceedings of Sagamore Conference, Innovations in Materials Processing,* Bruggerman, G., and Weiss, V., eds. Plenum Publishing, NY, 1985.

Wilson, D.R., "An Exploratory Study of Complexity in Axiomatic Design," Ph.D. Thesis, MIT, August, 1980.

Wilson, H.G., MacCready, P.B., and kyle, C.R., "Lessons of *Sunraycer*," *Scientific American* **260**(3):90–97, 1989.

Problems

2.1. Develop the FR and DP hierarchies for a bicycle.

2.2. Develop the FR and DP hierarchies for color photography.

2.3. Establish the FRs for a laser printer. Are your FRs independent?

2.4. Establish the FRs for a scanning electron microscope.

2.5. Develop a design solution that can satisfy the following perceived needs. One of the requirements of computers is the ability to store information. This is sometimes done using a rapidly rotating hard disk and a read/write head that moves perpendicularly to the disk motion. The disk is made by coating a substrate with a layer of magnetic material (i.e., iron oxide powder dispersed in a polymeric matrix). The read/write head is a magnetic field sensor/generator that can either store information in the disk or read the information that is already stored in the disk. The density of the information that can be stored or read is inversely proportional to the distance between the head and the disk. As the distance between them is decreased, however, the head touches the disk and both wear, which limits the life and reliability of the device. In conventional storage devices the distance between the head and the disk is controlled by letting the head "fly" over the disk. This is done by putting a beveled edge in front of the head for aerodynamic lift as shown in the figure. The distance in commercially available storage devices is as small as 10 μm. Propose a new head design that can increase the storage density by 10-fold.

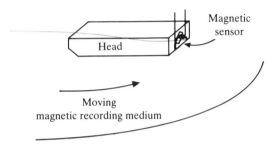

3
DESIGN AXIOMS AND COROLLARIES

3.1. Introduction

In Chapters 1 and 2 the general aspects of design are discussed, including the design process, the significance of FRs, constraints and DPs, and the hierarchical nature of FRs and DPs. Design is defined as the creation of a synthesized solution to satisfy the perceived needs through the mapping process between the FRs, which exist in the functional domain, and the DPs, which exist in the physical domain. In this chapter the axioms that apply to this mapping process are presented, and their implications are discussed. Although a few simple examples are presented in this chapter, the case studies that illustrate the use of the axioms and problem-solving techniques are discussed extensively in Chapter 6.

The basic goal of the axiomatic approach is to establish a scientific foundation for the design field, so as to provide a fundamental basis for the creation of products, processes, systems, software, and organizations. This is a significant departure from the conventional design process, which has been dominated by empiricism and intuition. Without scientific principles, the design field will never be systematized, and thus will remain a subject that is difficult to comprehend, codify, teach, and practise.

The use of the axiomatic approach to problem solving is as old as the history of science, as discussed in Chapter 1. All fields of science and technology have been transformed through increasing rigor, from the phenomenological stage to the scientific era characterized by axioms and principles. In the design area, however, it has taken longer for this idea to be applied than in other fields of science, such as mechanics and thermodynamics. One major impediment to progress in the design field has been the assumption that only subjects dealing with nature are subject to axiomatization.

As discussed in Chapter 2, the first step in design is to define a set of FRs that satisfy the perceived needs for a product (or process, systems, software, or organization). The second step is the creation of a tangible entity in the physical domain that satisfies the stated FRs with the least expenditure of resources in the form of energy, materials, and information. The physical being is characterized in terms of DPs. The central question is: "As we map DPs in the FR space, are there certain rules or axioms that are satisfied by a good design?" The uncovering of these axioms has provided significant insight to the design process itself, and forms the scientific basis for design and synthesis (Suh et al., 1978a,b).

3.2 Definition of Axioms, Corollaries, Theorems, FRs, DPs

By definition, axioms are fundamental truths that are always observed to be valid and for which there are no counterexamples or exceptions. Axioms may be hypothesized from a large number of observations by noting the common phenomena shared by all cases; they cannot be proven or derived, but they can be invalidated by counterexamples or exceptions.

A theorem is a proposition that may not be self-evident but that can be proved from accepted axioms. It is therefore equivalent to a law or principle. Consequently, a theorem is valid if its referent axioms and deductive steps are valid.

Corollaries are propositions that follow from axioms or other propositions that have been proven. Again, corollaries are shown to be valid or not valid in the same manner as theorems.

FRs are defined in Chapter 2 as a *minimum* set of *independent* requirements that completely characterize the functional needs of the product design in the functional domain. By definition, each FR is independent of other FRs. If one FR depends on another, then they are equivalent and logically identical.

DPs are the key variables that characterize the physical entity created by the design process to fulfill the FRs.

3.3 Design Axioms

There are two design axioms that govern good design. Axiom 1 deals with the relationship between functions and physical variables, and Axiom 2 deals with the complexity of design. These axioms can be stated in a variety of semantically equivalent forms. The declarative (or procedural) form of the axioms is (Suh, 1984):

        Axiom 1   *The Independence Axiom*
                        Maintain the independence of FRs.
        Axiom 2   *The Information Axiom*
                        Minimize the information content of the design.

Axiom 1 states that during the design process, as we go from the FRs in the functional domain with DPs in the physical domain, the mapping must be such that a perturbation in a particular DP must affect only its referent FR. Axiom 2 states that, among all the designs that satisfy the Independence Axiom (Axiom 1), the one with minimum information content is the best design. As discussed in Chapter 2, the term "best" is used in a relative sense, since there are potentially an infinite number of designs that can satisfy a given set of FRs. Axioms 1 and 2 may be restated as follows.

> Axiom 1 *The Independence Axiom*
> *Alternate Statement 1:* An optimal design always maintains the independence of FRs.
> *Alternate Statement 2:* In an acceptable design, the DPs and the FRs are related in such a way that specific DP can be adjusted to satisfy its corresponding FR without affecting other functional requirements.
>
> Axiom 2 *The Information Axiom*
> *Alternate Statement:* The best design is a functionally uncoupled design that has the minimum information content.

Let us now return to the refrigerator door design discussed in Chapter 1. The question is: if there are two FRs for the refrigerator door—that is, access to the stored food and minimal energy loss—is the vertically hung door a good design? We can see that the vertically hung door *violates* Axiom 1, because the two specified FRs (i.e., access to the contents and minimum energy use) are *coupled* by the proposed design. When the door is opened to take out milk, cold air in the refrigerator escapes and gives way to the warm air from outside.

What, then, is an *uncoupled* design that somehow does not couple these two FRs? One such uncoupled design of the refrigerator door is the horizontally hinged and vertically opening door used in chest-type freezers. When the door is opened to take out what is inside, the cold air does not escape since cold air is heavier than the warm air. Therefore, this type of chest freezer door does satisfy the first axiom.

We may note here that when we refer to the satisfaction of the FRs, the solution is understood to satisfy the original FRs within a certain tolerance band; that is, even in the case of the chest-type refrigerator door, there is some convective loss upon removal of the contents. However, if the FR is stated so that the energy loss is to be less than 10 calories per opening of the door, and if the energy loss associated with the chest refrigerator does satisfy this requirement, then it is an acceptable solution.

Throughout this book, design will be separated into three groups: *uncoupled, coupled,* and *decoupled* (or quasi-coupled) designs. An uncoupled design satisfies Axiom 1, whereas a coupled design renders some of the functions dependent on other functions, and thus violates Axiom 1. A coupled design may be decoupled; when the coupling is due to an insufficient number of DPs in comparison with the number of FRs that must be kept independent, we may accomplish this by adding extra components, which increases the number of DPs. A decoupled design may

Design Axioms and Corollaries 49

be inferior to an uncoupled design in the sense that it may require additional information content. The differences between these designs are discussed further in this chapter.

In order to clarify the significance of Axiom 1, let us consider two more examples qualitatively.

Example 3.1. Zero and Gain Control of a Recorder

Many recorders for capturing data during engineering experiments have two important FRs: the control of gains and the control of zero. This is typically done by turning two knobs, which represent the DPs. To set the zero and gain, one knob is turned to set the zero, and the other to control the gain. However, when the gain is set, one finds the zero has changed; therefore, one must go back to the first knob and reset the zero; this in turn changes the gain. The process has to be repeated until the zero and the gain reach the desired values. This is a typical example of a *coupled system*. There is nothing that one can do about this, short of modifying the analog electric circuit, or replacing it with a digital circuit (Wilson et al., 1979).

Example 3.2: Decoupling of a Single Screw Extruder

A single screw extruder for processing thermoplastics consists of a helical screw, housed in a cylindrical barrel which is typically warmed by electrical heaters. When the screw rotates, plastic granules are fed into the barrel section from the hopper (see Fig. 3.1). These granules are compacted in the screw channel, due to the relative motion between the plastic pellets and the barrel in the helical channel of the screw. The compacted plastic solid bed is then sheared near the solid bed–barrel interface. The solid bed melts due to the shear work, as well as to the heat transfered from the barrel to the plastic.

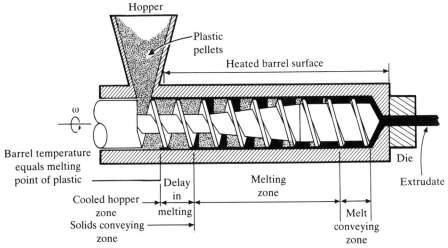

Figure 3.1. Schematic picture of a plasticating extruder. (Adapted from Tadmor and Gogos, 1979.)

The molten plastic is then pumped by the screw out of the extruder. The extrudate is then pushed through a die, which is normally attached to the extruder at its exit, to make an extrudate with a constant profile. The FRs of the extruder are the pumping rate and the temperature of the extrudate.

Is this a coupled machine? Why? If it is, can we decouple the extruder system so that the flow rate and temperature of the extrudate can be controlled independently? That is, can we create an uncoupled or decoupled extruder system?

Solution

The single screw extruder is a coupled system, since the screw speed affects both the temperature and the pumping rate of the extrudate. Furthermore, the barrel temperature affects not only the extrudate temperature, but also the pumping rate, since the viscosity of plastics also depends on temperature. (For a detailed analysis, see Tadmor and Gogos, 1979, and Appendix 3A.) Therefore, the flow rate of plastics fluctuates, and there are occasional surges that reduce the product quality and increase the waste.

The system can be made into a decoupled system by inserting a precision positive displacement pump, such as the gear pump shown in Fig. 3.2. First the speed of the gear pump is set to control the pumping rate, then the screw speed is adjusted to control the extrudate temperature. This solution has decoupled the *function* of flow rate control from the *function* of temperature control by inserting a gear pump in the physical arrangement of the hardware. (For a detailed analysis, see McKelvey, 1984; and Appendix 3B.)

Functional coupling should not be confused with *physical* coupling, which is often desirable as a consequence of Axiom 2. *Integration* of more than one function in a single part, as long as the functions remain independent, should reduce complexity. An example that illustrates the use of physical integration without compromising functional independence is the bottle/can opener design discussed in Example 3.3.

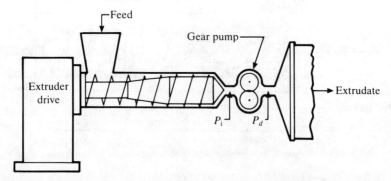

Figure 3.2. Schematic representation of pump-assisted plasticating extrusion process. (From McKelvey, 1984.)

Example 3.3: Bottle/Can Opener Design

Suppose that we are interested in designing a simple, low cost bottle/can opener that can be operated manually. The FRs of the device are

FR1: Open beverage bottles.
FR$_2$: Open beverage cans.

By definition, the two FRs are *independent,* but no mention is made of concurrency; that is, we sometimes wish to open a bottle or a can, but not both simultaneously.

Solution

A simple device that satisfies the above set of FRs is shown in Fig. 3.3. This is a very simple opener that can be made by stamping a sheet metal strip. In this design the means for achieving the two FRs independently are embodied in the same physical device rather than in two separate components. Therefore, minimal information content is required to manufacture the device.

Are the FRs coupled by the design? The design does not couple the FRs, because the act of opening cans does not interfere with or compromise the requirement of opening bottles. The FRs would be coupled only if there were an FR to open bottles and cans simultaneously, which is not the case here. The two separate functions are fulfilled by one physical piece, but without functional coupling. Physical integration without functional coupling is advantageous, since the complexity of the product is reduced, in line with Axiom 2.

The examples given in this section are rather qualitative, since so far quantitative measures of functional coupling and complexity have not been given. Chapter 4 discusses functional coupling in greater detail, and the issue of complexity is treated more extensively in Chapter 5.

Before leaving this section, it should be noted again that the identification of the *ultimate* optimal design having minimum information content cannot be guaranteed; also, there is no method for generating all potential designs. We can only identify the best design on a relative basis among those proposed, using Axioms 1 and 2. However, the ability to eliminate unacceptable or unpromising ideas in their early stages enhances the creative part of design, and reduces the cost of development and the chance for failure.

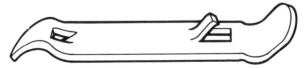

Figure 3.3. Can and bottle opener. This bottle/can opener satisfies two FRs: (1) opens cans, and (2) opens bottles. If the requirement is not to perform these two functions simultaneously, then this *physically integrated* device satisfies two *independent FRs*.

3.4 Corollaries

In the preceding section we have shown that, as a consequence of Axiom 1 and Axiom 2, physical integration is desirable when the FRs can be kept independent. This result can be used in other contexts, and therefore can be called a corollary. From the two axioms of design, many corollaries can be derived as a direct consequence of the axioms. These corollaries may be more useful in making specific design decisions, since they can be applied to actual situations more readily than can the original axioms. They may even be called design rules, and are all derived from the two basic axioms.

Corollary 1 (Decoupling of Coupled Design)
Decouple or separate parts or aspects of a solution if FRs are coupled or become interdependent in the designs proposed.

Corollary 2 (Minimization of FRs)
Minimize the number of FRs and constraints.

Corollary 3 (Integration of Physical Parts)
Integrate design features in a single physical part if FRs can be independently satisfied in the proposed solution.

Corollary 4 (Use of Standardization)
Use standardized or interchangeable parts if the use of these parts is consistent with the FRs and constraints.

Corollary 5 (Use of Symmetry)
Use symmetrical shapes and/or arrangements if they are consistent with the FRs and constraints.

Corollary 6 (Largest Tolerance)
Specify the largest allowable tolerance in stating FRs.

Corollary 7 (Uncoupled Design with Less Information)
Seek an uncoupled design that requires less information than coupled designs in satisfying a set of FRs.

Figure 3.4 shows the relationship between these corollaries and the axioms. It shows that Corollary 1 is a direct consequence of Axiom 1, whereas Corollaries 3, 6, and 7 are derived from Axioms 1 and 2. Corollaries 2, 4, and 5 are derived from Axiom 2. Some of these corollaries are self-evident, but others have much deeper implications than is apparent.

Corollary 1 states that functional independence must be ensured by *decoupling* if a proposed design couples the FRs. *Decoupling* does not necessarily imply that a part has to be broken into two or more separate physical parts, or that a new element has to be added to the existing

Design Axioms and Corollaries

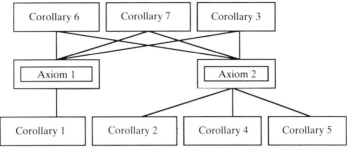

Figure 3.4. The origin of corollaries. Corollary 1 is from Axiom 1, whereas Corollaries 2, 4, and 5 are from Axiom 2. Corollaries 3, 6, and 7 are from both axioms.

design. Functional decoupling may be achieved without physical separation, although in some cases such physical decomposition may be the best way of solving the problem. Corollary 3 states that, as long as the FRs are not coupled by the physical integration of parts, the integration strategy should be followed if it reduces the information content of the design.

Corollary 2 states that as the number of FRs and constraints increases, the system becomes more complex and thus raises the information content. This implies that the conventional adage, "My design is better than yours because it does more than was intended," is misguided. A design should fulfill the precise needs defined by the FRs—nothing more and nothing less. Similarly, a process that fulfills more functions than are specified will be more difficult to operate and maintain than one that meets only the stated FRs. Reliability may also decrease when a machine fulfills more FRs than are required, because of the increased complexity.

Corollary 3 states that the number of physical parts should be reduced in order to decrease the information content, if this can be done without coupling FRs. However, mere physical integration is not desirable if it results in an increase of the information content or in the coupling of FRs. A good example of physical integration which is consistent with Corollary 3 is the bottle/can opener design discussed in Example 3.3.

Corollary 4 states a well-known design rule: use standard parts. In order to reduce inventory and minimize the information required for manufacture and assembly, special parts should not be used if standard parts can fulfill the FRs. Furthermore, the number of standard parts should be minimized so as to decrease the inventory costs and simplify inventory management, per Corollary 3. Interchangeable parts allow for the reduction of inventory, as well as the simplification of manufacturing and service operations; that is, they reduce the information content. This is even more the case if the design permits generous tolerances.

Corollary 5 is self-evident. Symmetrical parts require less information to manufacture and to orient in assembly. Not only should the shape be symmetrical whenever possible, but hole locations and other features should be placed symmetrically to minimize the information required during manufacture and use.

Corollary 6 deals with tolerances. Since it becomes increasingly difficult to manufacture a product as the tolerance is reduced, more information is required to produce parts with tight tolerances. On the other hand, if the tolerance is too large, then the error in assembly may accumulate such that FRs cannot be satisfied. Therefore, the specification of tolerances should be made as large as possible, but should remain consistent with the likelihood of producing functionally acceptable parts. The correct tolerance band is that which minimizes the overall information content. When the tolerance band is too large, the information content will increase since the subsequent manufacturing processes will require more information. Excess tolerances reduce reliability and thus increase the need for maintenance; this contributes to the increased information content.

Corollary 7 states there is always an uncoupled design that involves less information than a coupled design. This corollary is a consequence of Axioms 1 and 2. If this corollary were not true, then Axioms 1 and 2 must be invalid. The implication of this corollary is that if a designer proposes an uncoupled design which has more information content than a coupled design, then the designer should return to the "drawing board" to develop another uncoupled or decoupled design having less information content than the coupled design.

In addition to these corollaries, there can be many others. Whatever proposition can be derived from the preceding axioms and corollaries is also a corollary.

3.5 Mathematical Representation of the Independence Axiom: Design Equation

Design is defined as the mapping process between the FRs in the functional domain and the DPs in the physical domain. This relationship may be characterized mathematically. Since the characteristics of the required design are represented by a set of independent FRs, these may be treated as a vector **FR** with m components. Similarly, the DPs in the physical domain also constitute a vector **DP** with n components. The design process then involves choosing the right set of DPs to satisfy the given FRs, which may be expressed as

$$\{\mathbf{FR}\} = [\mathbf{A}]\{\mathbf{DP}\} \tag{3.1}$$

where $\{\mathbf{FR}\}$ is the functional requirement vector, $\{\mathbf{DP}\}$ is the design parameter vector, and $[\mathbf{A}]$ is the design matrix. Each line of the vector equation above may be written as

$$\mathrm{FR}_i = \sum_j A_{ij} \mathrm{DP}_j \tag{3.1a}$$

The design matrix [**A**] is of the form

$$[\mathbf{A}] = \begin{bmatrix} A_{11} & A_{12} & \cdots & A_{1n} \\ A_{21} & A_{22} & \cdots & A_{2n} \\ \vdots & \vdots & & \vdots \\ A_{m1} & A_{m2} & \cdots & A_{mn} \end{bmatrix} \quad (3.2)$$

Each element A_{ij} of the matrix relates a component of the **FR** vector to a component of the **DP** vector. In general, the element A_{ij} may be expressed as

$$A_{ij} = \partial FR_i / \partial DP_j \quad (3.3)$$

A_{ij} must be evaluated at the specific design point in the physical space unless A_{ij} is a constant. In nonlinear cases, A_{ij} varies with both FR_i and DP_j. When $m = n$, [**A**] is a square matrix. When $m = 3$ and $n = 3$, for example, [**A**] may be written as

$$[\mathbf{A}] = \begin{bmatrix} A_{11} & A_{12} & A_{13} \\ A_{21} & A_{22} & A_{23} \\ A_{31} & A_{32} & A_{33} \end{bmatrix} \quad (3.4)$$

Hereafter, Eq. 3.1 is called the *design equation*.

The left-handed side of the design equation represents "*what* we want in terms of design goals," and the right-hand side of the equation represents "*how* we hope to satisfy the DRs."

The simplest case of design occurs when all the nondiagonal elements are zero; that is, $A_{12} = A_{13} = A_{21} = A_{23} = A_{13} = A_{23} = 0$. Then Eq. 3.1 may be written for the case of $m = n = 3$ as

$$\begin{aligned} FR_1 &= A_{11} DP_1 \\ FR_2 &= A_{22} DP_2 \\ FR_3 &= A_{33} DP_3 \end{aligned} \quad (3.5)$$

A design that can be represented by Eq. 3.5 satisfies Axiom 1, since the independence of FRs is assured when each DP is changed. That is, FR_1 can be satisfied by simply changing DP_1, and similarly FR_2 and FR_3 can be changed independently without affecting any other FRs by varying DP_2 and DP_3, respectively. Therefore, the design can be represented by a diagonal matrix whose diagonal elements are only nonzero elements; that is,

$$\begin{aligned} A_{kk} &\neq 0 \\ A_{ij} &= 0 \quad \text{when} \quad i \neq j \end{aligned} \quad (3.6)$$

This satisfies the Independence Axiom and is defined as an *uncoupled design*.

The converse of an uncoupled design is the coupled design, whose design matrix consists of mostly nonzero elements. Examine the three-dimensional case where all elements of the matrix are nonzero. In this case the relationship between FRs and DPs is

$$FR_1 = A_{11}DP_1 + A_{12}DP_2 + A_{13}DP_3$$
$$FR_2 = A_{21}DP_1 + A_{22}DP_2 + A_{23}DP_3 \quad (3.7)$$
$$FR_3 = A_{31}DP_1 + A_{32}DP_2 + A_{33}DP_3$$

A change in FR_1 cannot be accomplished by simply changing DP_1, since this will also affect FR_2 and FR_3. Such a design clearly violates Axiom 1. We call this type of design a *coupled design*.

A coupled design can be *decoupled*, and this is often done in practice. Consider the special case of Eq. 3.4 where the design matrix is triangular (i.e., $A_{12} = A_{13} = A_{23} = 0$). This may be represented as

$$\begin{Bmatrix} FR_1 \\ FR_2 \\ FR_3 \end{Bmatrix} = \begin{bmatrix} A_{11} & 0 & 0 \\ A_{21} & A_{22} & 0 \\ A_{31} & A_{32} & A_{33} \end{bmatrix} \begin{Bmatrix} DP_1 \\ DP_2 \\ DP_3 \end{Bmatrix} \quad (3.8)$$

In this case the independence of the FRs can be assured if we adjust the DPs in a particular order; thus Axiom 1 is satisfied. In this situation the perturbation sequence of the DPs is the key to maintaining functional independence. If we vary DP_1 first, then the value of FR_1 can be set. Although it also affects FR_2 and FR_3, we can then change DP_2 to set the value of FR_2 without affecting FR_1. Finally, DP_3 can be changed to control FR_3 without affecting FR_1 and FR_2. If we had reversed the order and changed DP_3 first to set FR_3, and then DP_2 to set the value of FR_2, the value of FR_3 would have changed while changing DP_2. This kind of system is called a *decoupled* or *quasi-coupled* design.

The cases so far considered assumed that each element A_{ij} of the design matrix is a constant and that the design represents a linear system. In many real designs the A_{ij} are not constants; they often depict nonlinear components. In this case the design must be evaluated at a specific value of the DPs; that is, a design that is uncoupled in a certain operating range may be coupled in another operating regime if the design is nonlinear. This issue is discussed further in Chapter 4.

3.6 Other Implications of Axiom 1 for Design Practice

The definition and characteristics of uncoupled, decoupled, and coupled designs as given in Sec. 3.5 have many other implications. Some of these are discussed here.

Non-square Design Matrix Case

When $m \neq n$, the proposed design is either a coupled design or a redundant design. This can be seen by considering two simple cases. Consider first the case of $m = 3$ and $n = 2$. The relationship between FRs and DPs may be written as

$$FR_1 = A_{11}DP_1 + A_{12}DP_2$$
$$FR_2 = A_{21}DP_1 + A_{22}DP_2 \qquad (3.9)$$
$$FR_3 = A_{31}DP_1 + A_{32}DP_2$$

Suppose each element of the design matrix is zero except A_{11} and A_{22}. Then we cannot satisfy FR_3 by varying DP_1 or DP_2. On the other hand, if A_{31} or A_{32} is nonzero, then the independence of the FRs is compromised, resulting in a coupled design. This observation leads to the following two theorems:

Theorem 1 (Coupling Due to Insufficient Number of DPs)
When the number of DPs is less than the number of FRs, either a coupled design results or the FRs cannot be satisfied.

Theorem 2 (Decoupling of Coupled Design)
When a design is coupled due to the greater number of FRs than DPs (i.e., $m > n$), it may be decoupled by the addition of new DPs so as to make the number of FRs and DPs equal to each other, if a subset of the design matrix containing $n \times n$ elements constitutes a triangular matrix.

In order to prove Theorem 2, consider a coupled design which is represented by the following design equation

$$\begin{Bmatrix} FR_1 \\ FR_2 \\ FR_3 \end{Bmatrix} = \begin{bmatrix} A_{11} & 0 \\ A_{21} & A_{22} \\ A_{31} & A_{32} \end{bmatrix} \begin{Bmatrix} DP_1 \\ DP_2 \end{Bmatrix} \qquad (3.10)$$

The coupled design represented by Eq. 3.10 can be modified into a decoupled design, since the subset

$$\begin{bmatrix} A_{11} & 0 \\ A_{21} & A_{22} \end{bmatrix} \qquad (3.11)$$

is a triangular matrix. If a new design parameter DP_3 is added to the design, a new decoupled design may result which can be represented as

$$\begin{Bmatrix} FR_1 \\ FR_2 \\ FR_3 \end{Bmatrix} = \begin{bmatrix} A_{11} & 0 & 0 \\ A_{21} & A_{22} & 0 \\ A_{31} & A_{32} & A_{33} \end{bmatrix} \begin{Bmatrix} DP_1 \\ DP_2 \\ DP_3 \end{Bmatrix} \qquad (3.12)$$

In practice, many coupled designs undergo changes and become a decoupled design through a trial and error process. Coupled designs of the Eq. 3.10 type often result due to unforeseen factors, requiring extra DPs. Chapter 6 presents case studies of this type.

A different situation results when $m < n$. For example, when $m = 2$ and $n = 3$, the relationship between FRs and DPs may be written as

$$FR_1 = A_{11}DP_1 + A_{12}DP_2 + A_{13}DP_3$$
$$FR_2 = A_{21}DP_2 + A_{22}DP_2 + A_{23}DP_3 \qquad (3.13)$$

In this case, if all elements of the design matrix are equal to zero except A_{11} and A_{22}, then an uncoupled design results, and DP_3 does not affect the design. On the other hand, if $A_{12} = A_{21} = A_{23} = 0$, then a redundant design results, since FR_1 is both controllable by changing either DP_1 or DP_3 while maintaining functional independence. Furthermore, if both A_{13} and A_{23} are nonzero elements, then a coupled design results. Thus, Theorems 3 and 4 may be stated as:

Theorem 3 (Redundant Design)
When there are more DPs than FRs, the design is either a redundant design or a coupled desgin.

Theorem 4 (Ideal Design)
In an ideal design, the number of DPs is equal to the number of FRs.

When the design is a redundant design because there are more DPs than FRs, it may be possible to combine some of the DPs that affect a given FR to form a "dimensionless" group. For example, in the case of $A_{12} = A_{21} = A_{23} = 0$ discussed earlier in relation to Eq. 3.13, DP_1 and DP_3 may be so combined (e.g., by taking their quotient DP_1/DP_3) that the effect of changing DP_1 is tantamount to changing DP_3. Another possibility is simply to fix DP_3 at a constant value, and vary only DP_1. In some special cases redundancy may be required for increased reliability of particular subsystems, and thereby increased safety of the entire system. In some cases, as shown in Sec. 7.7, redundancy can be used to minimize the information content and improve manufacturability of a design.

Many designs exhibit the deficiency of having either $m > n$ or $m < n$. The single screw extruder example given in Example 3.2 illustrated the case where $m > n$, resulting in a coupled design. Another example is that of the reaction injection molding (RIM) machine.

Example 3.4: Reaction Injection Molding (RIM) Machine

Automotive parts, such as fenders, are made of polyurethane. Polyurethane parts are manufactured in a cyclic process by carefully metering and then mixing two liquids: polyol and diisocyanate. The mixing is done by impinging two liquid streams against each other in a mixing chamber. The mixing chamber is a small, cylindrical open space into which two liquid jets are injected at high speeds, which collide and mix in the chamber. The mixing mechanism of the impingement mixing is as follows.

As the liquid stream jet flows into the mixing chamber, the liquid undergoes laminar–turbulent flow transition at a low Reynolds number of

Design Axioms and Corollaries

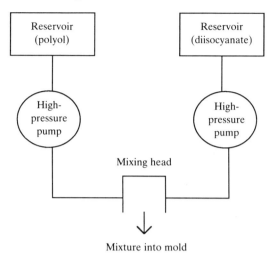

Figure 3.5. Schematic diagram of a commercial RIM machine. High-flow rate, high-precision, and high-pressure pumps impinge viscous liquid streams in the mixing chamber for mixing and delivery into a mold.

about 50.* The turbulent eddy size depends on the Reynolds number; the larger the Reynolds number is, the smaller the turbulent eddy size. The eddies from two different streams mix in the mixing chamber before the liquids go into the mold. The liquid mixture is delivered quickly to a mold in which the liquids react rapidly to form solid polyurethane before the mixture solidifies in the flow channel. A commercial RIM machine consists of two high-accuracy, high-flow rate, high-pressure pumps to deliver the liquids to the impingement mixing head (see Fig. 3.5).

The FRs of the RIM machine are

FR_1 = Deliver liquid at high flow rate (Q).

FR_2 = Deliver an adequately mixed liquid (X).

FR_3 = Deliver properly metered liquids (M).

The process is constrained to use an impingement mixing head. In order to manufacture RIM products successfully, the pump speed (W) is adjusted and the impingement nozzle size (D) is varied. Therefore, the DPs are the pump speed (W) and the nozzle size (D). Is this a good design? How would you improve upon it?

Solution

This original design has many problems. When the flow rate is high, there is a large pressure drop across the pumps; the leakage flow exits through small gaps between the pump rotor and the pump housing, which in turn

---

* The Reynolds number is defined as $Re = \rho VD/\mu$, where ρ is the mass density of the liquid, V the speed of the stream, D the diameter of the nozzle, and μ is the viscosity of the liquid. It is a dimensionless number.

contributes to the error in the metering ratio and the flow rates. The expansion of the delivery tubes (or hoses) when the pressure is suddenly changed also contributes to the metering ratio error. These problems have been analyzed by Tucker and Suh (1977), and the essence of the analysis is given in Appendix 3C. The power of the axiomatic approach is that preliminary design decisions can often be made without the exhaustive analysis of the details of the original design, although the quantitative analysis renders greater confidence in making design decisions.

The design equation may be written as

$$\begin{Bmatrix} FR_1 \\ FR_2 \\ FR_3 \end{Bmatrix} = [A] \begin{Bmatrix} DP_1 \\ DP_2 \end{Bmatrix}$$

$$\begin{Bmatrix} Q \\ X \\ M \end{Bmatrix} = [A] \begin{Bmatrix} W \\ D \end{Bmatrix}$$

(a)

where [A] is the design matrix, or the coupling matrix. The design matrix shows the change in FRs in response to changes in the DPs as

$$[A] = \begin{bmatrix} \partial Q/\partial W & \partial Q/\partial D \\ \partial X/\partial W & \partial X/\partial D \\ \partial M/\partial W & \partial M/\partial D \end{bmatrix} = \begin{bmatrix} \times & 0 \\ \times & \times \\ \times & \times \end{bmatrix}$$

(b)

where the ×'s indicate nonzero values, and hence a dependence between an FR and a DP.

In Eq. b, $\partial Q/\partial W$ represents the effect of the pump speed on the high flow rate required of the machine; it clearly is not negligible. The next element, $\partial Q/\partial D$ signifies the effect of the nozzle diameter on the high flow rate. Since the high flow rate is dictated largely by the rotational speed of the positive displacement pump, $\partial Q/\partial D$ may be negligible and hence is set to zero. The element $\partial X/\partial W$ is the term that measures the change in the quality of mixing due to the variation in the rotational speed of the pump. In view of the dependence of the eddy size ι on the Reynolds number as $\iota/D = \mathrm{Re}^{-3/4}$, it cannot be ignored. Similarly, $\partial X/\partial D$ is not negligible. The effect of the rotational speed on the metering ratio is very sensitive to the rotational speed, and thus × is designated for $\partial M/\partial W$. The effect of the nozzle diameter on the metering $\partial M/\partial D$ is not very large but is still not negligible. When the diameter gets smaller, the pressure drop across the gear pump increases and the leakage flow across the pump increases, which in turn affects the metering ratio. The relative ratio of two chemicals is very important, and even a small variation can adversely affect the mechanical properties of the final product.

Equation 3.13 shows that it is not possible to control the three FRs separately, since only two DPs can be adjusted. The design has clearly rendered the FRs dependent, and violates the Axiom 1, per Theorem 1 (coupling due to insufficient number of DPs). It is clear that we need to add one more DP, per Theorem 2 (decoupling of coupled design).

This problem can be solved by decoupling the function of pumping the liquids (which is done by the original large pumps) from the metering function (which was also done by the pumps). A modification of the physical system that eliminates the metering difficulty consists of adding two small hydraulic pump–motor elements that are mechanically coupled (as shown in Fig. 3.6) to the original system with the large pumps. This mechanically coupled pump–motor arrangement enables one of the pump–motors to act as a motor when the relative flow rate of this liquid component is greater than the set ratio, and the other to act as a pump when the relative flow rate of this liquid stream is less than the set value (Suh and Tucker, 1979).

The leakage flow through these metering pumps is small, since the pressure drop across these metering pumps is small, being proportional to the difference between the actual and the set relative flow rates. (For a detailed analysis of a similar problem, see Suh and Tucker, 1979; Tucker and Suh, 1977; Appendix 3C). If G is the gear ratio between pumps, then the design equation may be written for differential changes in FRs as functions of the change in DPs:

$$\begin{Bmatrix} dQ \\ dx \\ dM \end{Bmatrix} = \begin{bmatrix} \partial Q/\partial W & \partial Q/\partial D & \partial Q/\partial G \\ \partial W/\partial W & \partial X/\partial D & \partial X/\partial G \\ \partial M/\partial W & \partial M/\partial D & \partial M/\partial G \end{bmatrix} \begin{Bmatrix} dW \\ dD \\ dG \end{Bmatrix}$$

$$= \begin{bmatrix} \times & 0 & 0 \\ \times & \times & 0 \\ \times & \times & \times \end{bmatrix} \begin{Bmatrix} dW \\ dD \\ dG \end{Bmatrix} \quad (c)$$

The differential form of the design equation is useful when the relationship between FRs and DPs is nonlinear. For the linear case, the elements of the design matrix are constant and Eq. c can be readily integrated.

Equation c shows that the modified design, with the coupled mechanical pumps inserted between the large pumps and the mix chamber, is a decoupled design. According to Eq. c, if we set the flow rates of the large pumps first, followed by the determination of the nozzle diameter, and finally set the gear ratio of the metering pumps, then we can maintain the independence of the three FRs (i.e., Q, X, and M).

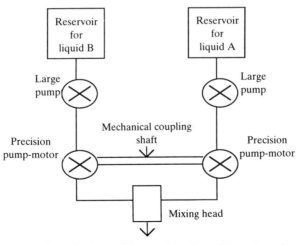

Figure 3.6. Reaction injection molding machine for making polyurethane parts.

So far we have discussed various design cases where the number and specific set of FRs remained the same; only the number of DPs were changed to satisfy Axiom 1. An equally important question is: what is the effect on the design when a set of FRs is changed by: (1) the addition of a new FR; (2) the substitution of an FR with another; or (3) the selection of a completely different set of FRs? Consider the case of an uncoupled design given by

$$\begin{Bmatrix} FR_1 \\ FR_2 \end{Bmatrix} = \begin{bmatrix} A & 0 \\ 0 & B \end{bmatrix} \begin{Bmatrix} DP_1 \\ DP_2 \end{Bmatrix} \quad (3.14)$$

If the designer decides to add one more FR (i.e., FR_3) to the set given in Eq. 3.14, then we must also add a new DP (i.e., DP_3), per Theorems 1 (coupling due to insufficient number of DPs) and 4 (ideal design). However, there is no prior assurance that DP_3 will not affect other FRs (i.e., FR_1 and FR_2), in which case a coupled design results. Obviously, a good designer will try to find a new DP that will affect only the newly added FR. When a given set of FRs is changed by substitution of one of its FRs with a new one, a new set of DPs that satisfies Axiom 1 must be chosen, although on rare occasions the old set of DPs may still satisfy the new FRs.

Based on these observations, Theorem 5 may be stated as follows.

> Theorem 5   (Need for New Design)
> When a given set of FRs is changed by the addition of a new FR, or substitution of one of the FRs by a new one, or by selection of a completely different set of FRs, the design solution given by the original DPs cannot satisfy the new set of FRs. Consequently, a new design solution must be sought.

Theorem 5 is also a very important design principle that is frequently overlooked by many designers, causing many problems in terms of cost over-run, less than optimum design, and marginal products. Instead of searching for a completely new design when the set of original FRs is changed by addition or deletion of the FRs, many designers simply change the existing design to satisfy the new set of FRs. This can be a serious mistake; although in some cases (when the original design is an uncoupled design, for example) such a modification may be made, it is more prudent to go back to the beginning and reconceptualize the solution. If such a reconceptualization leads to a simple modification of the original design, so much the better. However, one should not assume a priori that a modified design solution would satisfy the new set of FRs.

Design Procedure

Axiom 1 also has an implication for the design procedure, providing for the quick development of an uncoupled design. When we analyze an existing design, we must determine the design matrix to determine whether the design violates Axiom 1 (the Independence Axiom). However, when we do not have a pre-existing design ready for analysis, but must develop a completely new design, we can make use of Axiom 1 effectively during the concept development stage. In this case we know the FRs before the design process begins, since we arbitrarily choose a set of FRs that we want to satisfy. Furthermore, we know that the number of DPs must be equal to the number of FRs, per Theorem 4 (ideal design). Hence, we should not randomly come up with a design and see whether it satisfies Axiom 1.

A better strategy is to look for a set of DPs that will yield a diagonal or triangular matrix; that is, during the design process, the designer must constantly ask, "Which is the DP that can satisfy this particular FR and yield a diagonal or traingular matrix?" For example, the solution in Example 3.4 was motivated by knowing a priori that we needed one more DP, and that it must be chosen so that the design matrix must be either a diagonal or a triangular matrix. Comprehensive understanding of Axiom 1 can be a powerful tool for designers.

The design matrix also provides a tool for developing a design or research and development (R&D) strategy. For example, suppose the design matrix is of the form

$$\begin{bmatrix} A_{11} & A_{12} & A_{13} \\ A_{21} & A_{22} & A_{23} \\ A_{31} & A_{32} & A_{33} \end{bmatrix} = \begin{bmatrix} \times & 0 & ? \\ ? & \times & 0 \\ \times & 0 & \times \end{bmatrix}$$

That is, we are not certain whether A_{13} and A_{21} are 0 or $\times$. However, the known values of A_{ij} indicate that the best we can hope for is a decoupled design, since $A_{31} \neq 0$. Therefore, the most critical piece of information needed is whether or not $A_{13} = 0$. If it is zero we will have a decoupled design, regardless of the specific value of A_{21}. Similarly, if $A_{13} \neq 0$ the design is coupled, regardless of the value of A_{21}. Consequently, the design or R&D effort should be concentrated on the determination of A_{13} rather than A_{21}.

In some cases, as we search for a set of DPs that can satisfy the FRs in accordance with the Independence Axiom, we may find that the original set of FRs chosen earlier does not represent the perceived needs with fidelity. This situation arises during the early phase of the design process for two reasons. First, FRs are often chosen in the absence of specific design solution in mind at the onset of the design process. As we develop specific design solutions, shortcomings of the original set of FRs become more apparent. Secondly, the design process must proceed in the absence of complete information, often based only on qualitative reasoning, or the laws of nature, and factual information. The designer should always feel free to choose an improved set of FRs that may represent the societal need with a greater degree of fidelity.

Designers are sometimes forced to changed the FRs, because they are unable to conceive a design that can satisfy the original set of FRs without violating Axiom 1. In this case the designer may end up modifying FRs, dropping some of the FRs, or choosing a complete new set of FRs. In this case, when the design process is completed, the final solution must be compared with the original perceived needs, to determine whether the solution is acceptable.

Throughout the design process the design axioms, corollaries and theorems must always be kept in mind and applied whenever applicable. The importance of looking for DPs that will yield either a diagonal or a triangular design matrix cannot be overemphasized.

3.7 Information in Axiomatic Design

So far we have discussed the Independence Axiom (Axiom 1), which deals with FRs and their relationship to DPs in design. Equally important, however, is the Information Axiom (Axiom 2), which deals with complexity. The basic concept involved in Axiom 2, including the definition of information content, is discussed in this section.

The final output of the design process is a block of information for use in subsequent manufacturing or other operations. The information can be in the form of drawings, equations, material specifications, operational instructions, software, etc. Depending on the nature of the information generated, and on the interaction of this information with information available at later stages of implementation, the subsequent operations can be more difficult and costly, or less. For example, it is easy to draw a shaft and specify its dimensions and tolerances, but in order to manufacture the shaft we have to choose a set of machines and manufacturing processes. If we select a wrong machine, the machining task can be quite complex, and thereby entail a great deal of information. On the other hand, we all know that if we have the correct tool, the job can be done much more easily. This raises several questions that must be answered before we can deal with information in axiomatic design: What is a wrong machine? How do we measure quantitatively the information content in a design? How can we select the best design based on this information measure? How does design affect manufacturing?

It is relatively easy to think of complexity in a qualitative manner, but complexity is rather difficult to measure quantitatively. Complexity is associated with the difficulty of achieving a task. Going back to the shaft problem, for example, what would you answer to the question, "How much information is associated with the shaft?" If you consider the atomic arrangement and impurity content in the material used for the shaft, the geometry, the color, etc., there can be untold amounts of information. However, much of that information may be useless information in making design decisions, if the only things we need to know are the length and the accuracy of the shaft. Suppose the length of the shaft is 4 m. How much information is there? Is there more information in a 4-m shaft than in a 2-m shaft? We cannot answer these questions unless we know more about the *utility* of that information. That implies that *information can be defined only in the context of FRs*. If the length of the shaft must be known to be 4 ± 0.1 m, then we *know* how accurately we have to measure the length of the shaft to satisfy the FRs; that is, we now have information of the FR of the shaft. In this case the information is related to the *probability* of measuring the length of the shaft within the specified tolerance.

We can generalize this observation concerning the information content of the shaft as follows. *Information is the measure of knowledge required to satisfy a given FR at a given level of the FR hierarchy*. This statement may sound too abstract to be of any value. Many examples and case studies are given elsewhere (especially in Chapter 8) to clarify further the meaning of this statement.

Definition of Information

To be able to complete a given task, information is required; that is, in the absence of the requisite information, we do not have the required knowledge to execute the task. In the case of the task of measuring the length of the shaft, we needed the information on the accuracy (or the tolerance) of the shaft. However, that may be necessary but not sufficient information. Depending on what kind of implement we use, we may or may not measure the length of the shaft to the desired accuracy. For example, if the shaft has to be measured to be $4 \pm 0.000,001$ m (or 1 μm), then an ordinary measuring tape may not be used with any degree of confidence. We can use the tape measure, but the probability of measuring it to within the accuracy specified by the FR may be very low. In order to increase the probability, we will need more-accurate devices. This discussion suggests that *the notion of information is very closely related to the probability of achieving the FR*.

The information content of a design may be defined quantitatively as the logarithm of the probability of fulfilling the specified FR. If the FR is to have a shaft length of 4 ± 0.1 m, then the probability of being within the tolerance defines the information. If we *assume* a uniform probability density along the length of the shaft (which is unlikely to be the case in real manufacturing operations), the probability, p, of producing an acceptable shaft is given by the ratio of tolerance to the dimension

$$p = \frac{2(0.1 \text{ m})}{4 \text{ m}} = \frac{1}{20} \qquad (3.15)$$

The factor of 2 in the numerator signifies the tolerance can be either greater or less than 4 m by 0.1 m. Information content I is *defined* in terms of probability as

$$\text{Information} = I = \log_2(1/p) = \log_2(20) = 4.32 \text{ bits}$$

or, more generally,

$$I = \log_2(L/2\,\Delta L) = \log_2(1/p) \tag{3.16}$$

where L is the length, ΔL is the allowable tolerance on the length, and p is the probability.

The definition of information given by Eq. 3.16 is the same as that used in information theory (Shannon and Weaver, 1949; Brillouin, 1962). The logarithmic definition for information is convenient to use. When a given design involves several tasks with associated probabilities, the overall probability is the product of probabilities of all associated events. That is, the overall probability of independent events is multiplicative. Therefore, a logarithmic definition is used so that the information measure is additive, rather than multiplicative, even when the overall probability of independent events is the product of the probabilities of each event. The exact form of the definition of information content is not important, as long as it is an accurate predictor of *relative* complexity, and we use a consistent definition.

By generalizing Eq. 3.16, the information content associated with a given FR may be thought of as a logarithm of the ratio range/tolerance, assuming that the tolerance is uniformly distributed throughout the range. Equation 3.16 may be rewritten as (Wilson, 1980)

$$I = \log_2(\text{range/tolerance}) \tag{3.17}$$

where the ratio (tolerance/range) defines the probability of success. This dimensionless definition of information is particularly useful in design and manufacturing, since we have to deal with many different variables and dimensional quantities (e.g., length, hardness, cost, temperature, voltage, current, and velocity). In the domain of probability, all variables, quantities, and parameters can be treated alike, regardless of their physical origin.

In some cases, instead of the logarithm to the base 2, the natural logarithm ln (i.e., $\ln = \log_e$) is sometimes used. When the natural logarithm is used, the resultant information is measured in the unit of nats rather than in bits. 1 nat is approximately equal to 1.443 bits. In this book, both of these measures for information will be used.

Information Content and Functional Independence

Information content associated with the FRs of an uncoupled design can be obtained by simply adding the information associated with each of the FRs at each level of the FR hierarchy. However, in the case of a coupled design, any one DP can affect all other FRs. Therefore, the information content cannot be defined a priori since the information content depends on the particular path followed in varying the DPs. Consequently, the information associated with a coupled process is greater than that of a decoupled process, which in turn is greater than that of an uncoupled process.

Based on these observations. Theorems 6 and 7 may be stated as follows.

> Theorem 6 (Path Independence of Uncoupled Design)
> The information content of uncoupled design is independent of the sequence by which the DPs are changed to satisfy the given set of FRs.
>
> Theorem 7 (Path Dependence of Coupled and Decoupled Designs)
> The information contents of coupled and decoupled designs depends on the sequence by which the DPs are changed and on the specific paths of change of these DPs.

3.8 Relationship Between Axiom 1 and Axiom 2

Since Corollary 7 states that there is an uncoupled design whose information content is less than a given coupled design, it may appear that Axioms 1 and 2 are interrelated, thereby representing only one axiom. For example, one might think that Axiom 1 is a consequence of Axiom 2, since an uncoupled design that obeys Axiom 1 has the minimum information content, compared with the coupled and decoupled designs. However, a closer examination reveals that it is wiser to keep these axioms as two independent propositions. For example, in the absence of Axiom 1 one might choose a coupled design that happened to have less information than a particular uncoupled design, rather than to seek another uncoupled design having less information content (see Corollary 7). Conversely, in the absence of Axiom 2, there is no way in which we can choose the best design among uncoupled designs satisfying Axiom 1.

In an actual design process, one always starts out with Axiom 1 and seeks an uncoupled design. Only after several designs that satisfy Axiom 1 are proposed can we apply Axiom 2 to determine which is the best among those proposed. The ability to use Axiom 1 effectively is a hallmark of the creative designer.

3.9 Summary

In this chapter the axioms, corollaries, and theorems were stated and their implications were discussed. In addition, definitions were given to ensure uniformity in the use of axiomatic reasoning in design. The axioms and corollaries are meaningful only in terms of the definitions adopted for FRs, DPs, and constraints. Based on the Independence Axiom, designs may be classified into three kinds: uncoupled, decoupled, and coupled designs. A redundant or coupled design results when there are more DPs than FRs; a coupled or an infeasible design results from the converse situation.

Information content was defined in terms of the probability of attaining a set of FRs. Information is meaningful only in the context of trying to satisfy the FRs.

The design axioms are

> Axiom 1   *The Independence Axiom*
> Maintain the independence of FRs.
>
> Axiom 2   *The Information Axiom*
> Minimize the information content of the design.

The corollaries are as follows:

> Corollary 1   (Decoupling of Coupled Designs)
> Decouple or separate parts or aspects of a solution if FRs are coupled or become interdependent in the designs proposed.
>
> Corollary 2   (Minimization of FRs)
> Minimize the number of FRs and constraints.
>
> Corollary 3   (Integration of Physical Parts)
> Integrate design features in a single physical part if FRs can be independently satisfied in the proposed solution.
>
> Corollary 4   (Use of Standardization)
> Use standardized or interchangeable parts if the use of these parts is consistent with FRs and constraints.
>
> Corollary 5   (Use of Symmetry)
> Use symmetrical shapes and/or components if they are consistent with the FRs and constraints.
>
> Corollary 6   (Largest Tolerance)
> Specify the largest allowable tolerance in stating FRs.
>
> Corollary 7   (Uncoupled Design with Less Information)
> Seek an uncoupled design that requires less information than coupled designs in satisfying a set of FRs.

The theorems presented in this chapter are

> Theorem 1   (Coupling Due to Insufficient Number of DPs)
> When the number of DPs is less than the number of FRs, either a coupled design results, or the FRs cannot be satisfied.
>
> Theorem 2   (Decoupling of Coupled Design)
> When a design is coupled due to the greater number of FRs than DPs (i.e., $m > n$), it may be decoupled

by the addition of new DPs so as to make the number of FRs and DPs equal to each other, if a subset of the design matrix containing $n \times n$ elements constitutes a triangular matrix.

Theorem 3 (Redundant Design)
when there are more DPs than FRs, the design is either a redundant design or a coupled design.

Theorem 4 (Ideal Design)
In an ideal design, the number of DPs is equal to the number of FRs.

Theorem 5 (Need for New Design)
When a given set of FRs is changed by the addition of a new FR, or substitution of one of the FRs with a new one, or by selection of a completely different set of FRs, the design solution given by the original DPs cannot satisfy the new set of FRs. Consequently, a new design solution must be sought.

Theorem 6 (Path Independence of Uncoupled Design):
The information content of an uncoupled design is independent of the sequence by which the DPs are changed to satisfy the given set of FRs.

Theorem 7 (Path Dependence of Coupled and Uncoupled Designs):
The information contents of coupled and decoupled designs depend on the sequence by which the DPs are changed and on the specific paths of the changes of these DPs.

References

Brillouin, L., *Science and Information Theory*. Academic Press, NY, 1962.
McKelvey, J.M., "Performance of Gear Pumps in Polymer Processing," *Polymer Engineering and Science* **24**(6):398–402, 1984.
Shannon, C.E., and Weaver, W., *The Mathematical Theory of Communication*. University of Illinois Press, Urbana, IL, 1949.
Suh, N.P., "Development of the Science Base for the Manufacturing Field Through the Axiomatic Approach", *Robotics and Computer Integrated Manufacturing* **1**(3/4):397–415, 1984.
Suh, N. P., and Tucker III, C.L., U.S. Patent 4,170,319, 1979.
Suh, N.P., Bell, A.C., and Gossard, D., "On an Axiomatic Approach to Manufacturing Systems," *Journal of Engineering and Industry, Transactions of A.S.M.E.* **100**:127–130, 1978a.
Suh, N.P., Kim, S.H., Bell, A.C., Wilson, D.R., Cook, N.H., and Lapidot, N.,

"Optimization of Manufacturing Systems through Axiomatics", *Annals of CIRP* **27,** 1978b.

Tadmor, Z., and Gogos, C.G., *Principles of Polymer Processing.* John Wiley, NY, 1979.

Tucker III, C.L., and Suh, N.P., "Fluid Delivery and Metering for Reaction-injection Molding," *Journal of Engineering for Industry* **99**(3):678–681, 1977.

Wilson, D.R., "An Exploratory Study of Complexity in Axiomatic Design," Ph.D. thesis, MIT, 1980.

Wilson, D.R., Bell, A.C., Suh, N.P., and Van Dyck, F., "Manufacturing Axioms and their Corollaries," *Proceeding of North American Manufacturing Research Conference,* Ann Arbor, MI, Vol. 7, 1979.

Problems

Some of the problems given here are not fully specified. Furthermore, some of the problems deal with technologies that are not specifically addressed in the text. Students are expected to conduct background study as part of their assignment, something that a designer has to do frequently.

3.1. State a corollary other than those given in Chapter 3.

3.2. Define a theorem other than those given in this book.

3.3. Prove Theorem 7.

3.4. Analyze the design of the Honda Accord (only operational features) in terms of the axioms, corollaries, and theorems. Perform a similar analysis of the General Motors Oldsmobile Cutlass.

3.5. Design a magnetic thin-film head for a "write/read" Winchester drive used in digital information storage devices that can increase the information storage density by an order of magnitude. State the FRs and constraints first, and propose a design. Write the design equation for your proposed design. Is it a coupled design?

3.6. Design a 35-mm camera that can automatically focus and set the correct exposure.

3.7. Design an "educational system" that can provide lifelong learning (sometimes called continuing education) to practising engineers.

3.8. Decouple the conventional single screw extruder discussed in this chapter without using a precision gear pump.

3.9. Design an ideal international trade system that satisfies at least the following FRs:

FR_1 = No country can have a current surplus that is greater than 10% of its GNP.

FR_2 = No country can incur a foreign debt that requires annual interest payments of more than 10% of its GNP.

FR_3 = No country should have less per capita income (including gifts) than 5% of the average per capita income of OECD nations.

3.10. Consider the following design matrices.

(a) $\begin{bmatrix} A_{11} & A_{12} & 0 \\ A_{21} & 0 & A_{23} \\ A_{31} & A_{32} & A_{33} \end{bmatrix}$ (b) $\begin{bmatrix} A_{11} & A_{12} & A_{13} \\ 0 & A_{22} & A_{23} \\ 0 & 0 & A_{23} \end{bmatrix}$

(c) $\begin{bmatrix} A_{11} & 0 & A_{13} \\ 0 & A_{22} & 0 \\ A_{31} & 0 & A_{33} \end{bmatrix}$ (d) $\begin{bmatrix} A_{11} & A_{12} & 0 \\ 0 & A_{22} & A_{23} \\ 0 & A_{32} & A_{33} \end{bmatrix}$

(e) $\begin{bmatrix} A_{11} & A_{12} & 0 \\ 0 & A_{22} & A_{23} \\ 0 & A_{32} & 0 \end{bmatrix}$ (f) $\begin{bmatrix} 0 & A_{12} & A_{13} \\ 0 & A_{22} & A_{23} \\ A_{31} & 0 & A_{33} \end{bmatrix}$

What kinds of designs do these matrices represent? How should the design be modified to satisfy Axiom 1?

3.11. When a design equation is given as

$$\begin{Bmatrix} FR_1 \\ FR_2 \\ FR_3 \end{Bmatrix} = \begin{bmatrix} A_{11} & 0 & A_{13} \\ A_{21} & A_{22} & 0 \\ A_{31} & A_{32} & A_{33} \end{bmatrix} \begin{Bmatrix} DP_1 \\ DP_2 \\ DP_3 \end{Bmatrix} \quad (a)$$

Axiom 1 states that DP_3 selected to satisfy FR_3 is a wrong choice since it also affects FR_1, and that we should search for a new DP_3 that can satisfy FR_3 without affecting FR_1. Also, if such a solution is not available, then we have to search for a DP_1 that will not affect FR_2 and FR_3, and for a DP_2 that will not affect FR_1 and FR_3.

However, Eq. a can be rewritten as:

$$\begin{Bmatrix} FR_2 \\ FR_1 \\ FR_3 \end{Bmatrix} = \begin{bmatrix} A_{21} & 0 & A_{22} \\ A_{11} & A_{13} & 0 \\ A_{31} & A_{33} & A_{32} \end{bmatrix} \begin{Bmatrix} DP_1 \\ DP_3 \\ DP_2 \end{Bmatrix} \quad (b)$$

Based on this equation, we may say that we have alternative ways of satisfying
the Independence Axiom; that is, to select a DP_2 that can satisfy FR_3 without affecting FR_2, etc.
What is wrong with this argument?

3A: ANALYSIS OF SINGLE SCREW PLASTICATING EXTRUDER

3A.1 Introduction

Appendices in this chapter are presented to provide an analytical basis for the design decisions made in the text. Although they are not essential in understanding the materials presented in the text, they are given to show that in design, synthesis, and analysis must be part of the feedback process illustrated in Fig. 1.1.

Single screw plasticating extrusion is a basic polymer processing technique used in the polymer industry. This technique is widely adopted to process many different kinds of materials; it is primarily used to transport, melt, and mix thermoplastics that melt at high temperatures. The material is supplied in granular form and is extruded into sheets, tubes, and various other profiles.

3A.2 Description of Plasticating Extrusion and Extruder

The cross section of an extruder is shown in Fig. 3A.1. It consists of a screw with a helical channel section inside a barrel that is typically heated with electric band heaters. The granular plastic pellets are fed into the extruder from a hopper by gravity. The granular plastic is transported forward by the frictional force between the plastic pellets and the barrel. The solid plastic granules pack together in the channel of the screw a short distance into the barrel, forming a solid bed (see Fig. 3A.2).

Because of the mechanical shear work done to the plastic at the barrel–solid bed interface (due to the relative motion between the solid bed and the barrel), and because of heat transfered from the barrel to the plastic (this occurs at low rotational speeds of the screw when the plastic temperature is lower than the barrel temperature, due to limited shear work), the plastic near the barrel surface melts. This thin molten plastic layer is transported to the front side of the channel, due to the wiping action of the barrel, forming a molten pool as shown in Fig. 3A.2.

The thickness of the molten plastic layer near the barrel–solid bed interface remains nearly constant throughout the melting zone, but the size of the molten pool increases and the width of the solid bed decreases as

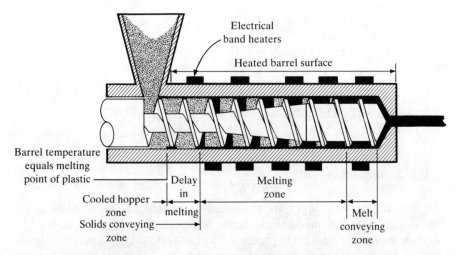

Figure 3A.1. Schematic picture of a plasticating extruder. (From Tadmor and Klein, 1970.)

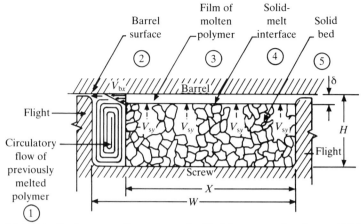

Figure 3A.2. Idealized cross section compared with (a) cross section from a PVC cooling experiment and (b) cross section from a low-density polyethylene cooling experiment. (Courtesy of Tadmor and Klein, 1970).

the solid bed continues to melt. The fluid particles in the molten pool undergo a toroidal vortex motion, which causes mixing of the molten plastic by laminar mixing. Laminar mixing takes place by stretching of the interface area between two or more different phases of the plastic.

Eventually, when the solid bed is completely molten, which normally occurs at about two-thirds of the way down the extruder, the molten plastic is further mixed and pumped through a die attached to the extruder.

3A.3 Analysis of Plasticating Extrusion

As shown in Fig. 3A.1, the extruder may be divided into three zones: (a) solids-conveying zone; (b) melting and mixing zone; and (c) pumping and metering (also known as melt conveying) zone. Each of these zones will be analyzed. Readers who are interested in details of the extrusion process should consult Tadmor and Klein (1970).

The Solids-conveying Zone

Figure 3A.3 shows a section of the solid bed in the screw channel. The screw with a right-hand helical channel rotates counterclockwise, and the solid plug moves down the channel section with speed V_{pl}, due to the frictional force between the solid plug and the barrel surface. The frictional force opposes the downward motion of the solid bed. The solid plug also has a constant angular velocity component $V_{p\theta}$. The overall mass flow rate G may be expressed in terms of V_{pl}, and the geometry of the

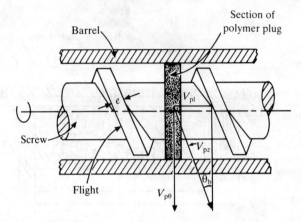

Figure 3A.3. Section of the solid plug. Velocities are measured relative to the screw. $V_{p\theta}$ and V_{pz} are measured at the barrel surface. V_{pl} is independent of the channel depth. (From Tadmor and Klein, 1970).

screw may be expressed as

$$G = V_{pl}\rho_b \left(\frac{\pi}{4}(D_b^2 - D_s^2) - \frac{eH}{\sin \bar{\theta}} \right) \tag{3A.1}$$

where D_b is the inside diameter of the barrel, $D_s = D_b - 2H$, H is the channel depth, e is the flight width, and $\bar{\theta}$ is the mean helix angle. V_{pl} may be expressed in terms of the tangential speed of the barrel surface V_b, with

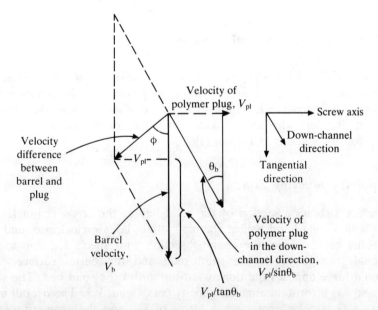

Figure 3A.4. Velocity vector diagram for calculating the velocity difference between barrel and solid plug. (From Tadmor and Klein, 1970.)

respect to the coordinate axis that rotates with the screw as

$$V_{pl} = V_b = \frac{\tan \phi \tan \theta_b}{\tan \phi + \tan \theta} \tag{3A.2}$$

where ϕ is the "friction angle" formed between the velocities of the solids and the barrel surface (see Fig. 3A.4). From Eqs. 3A.1 and 3A.2, the "friction angle" ϕ can be found, given the mass flow rate through the extruder or vice versa.

However, Eq. 3A.1 will be useful only if ϕ can be determined independently, based on other considerations. For this purpose, we consider the force and torque balance (i.e., the equilibrium condition) by applying Newton's laws. Figure 3A.5 shows the forces acting on the solid plug. From the considerations of the equilibrium of forces along the axial and the tangential directions, we obtain (Tadmor and Gogos, 1979)

$$\cos \phi = K \sin \phi + M \tag{3A.3}$$

where

$$K = \frac{\bar{D} \sin \bar{\theta} + \mu_s \cos \bar{\theta}}{D_b \cos \bar{\theta} - \mu_s \sin \bar{\theta}} \tag{3A.4}$$

$$M = 2 \frac{H}{W_b} \frac{\mu_s}{\mu_b} \sin \theta_b \left(K + \frac{\bar{D}}{D_b} \cot \bar{\theta} \right)$$

$$+ \frac{W_s}{W_b} \frac{\mu_s}{\mu_b} \sin \theta_b \left(K + \frac{D_s}{D_b} \cot \theta_s \right)$$

$$+ \frac{\bar{W}}{W_b} \frac{H}{Z_b} \frac{1}{\mu_b} \left(K + \frac{\bar{D}}{D_b} \cot \bar{\theta} \right) \ln \frac{P_2}{P_1} \tag{3A.5}$$

where W is the width of the channel, P_1 is the initial pressure at $z = 0$, P_2 is the pressure at any down-channel distance z_b, subscripts s and b refer the screw and barrel, respectively, and an overbar denotes the mean value over the channel depth.

If we know the pressure rise, then M can be computed from Eq. 3A.5, and then ϕ can be computed from Eq. 3A.3. The mass flow rate is then

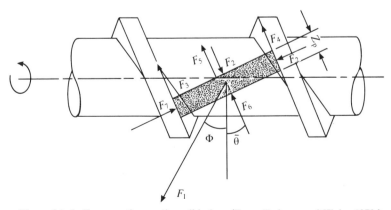

Figure 3A.5. Forces acting on the solid plug. (From Tadmor and Klein, 1970.)

computed from Eq. 3A.1. The power consumed in the solids-conveying zone is

$$\mathbb{P}_w = \pi N D_b W_b Z_b \mu_b \cos \mu \frac{P_2 - P_1}{\ln(P_2/P_1)} \tag{3A.6}$$

where N is the rotational speed of the screw, equal to $V_b/\pi D_b$.

The Melting Zone

Several basic assumptions are made in order to analyze the melting process in the melting zone. As stated earlier, the melting of the solid bed occurs at the solid bed–barrel interface, where a thin molten layer exists. The molten plastic flows into the molten pool, due to the wiping action of the barrel, which drags the plastic into the pool. The whole process is assumed to occur at a steady state. The solid bed is assumed to be homogenous, deformable, and continuous, and all physical properties are assumed to be constant. As the solid bed melts at the interface, the width of the solid bed decreases gradually. The analysis presented here follows the work of Tadmor and Klein (1970).

Figure 3A.6 shows a small differential element of the solid bed and the molten film near the barrel surface. The z-axis is parallel to the axis of the extruder, the y-axis is perpendicular to the barrel surface, and ϕ is the

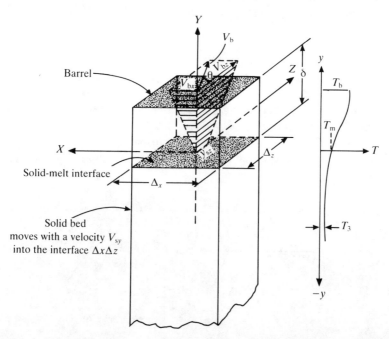

Figure 3A.6. Differential volume perpendicular to the solid melt interface. Velocity profiles and temperature profiles in the film and solid bed are shown. (From Tadmor and Klein, 1970.)

helix angle of the channel. The solid bed has a local down-channel velocity $\mathbf{V}_{sz}$ and a local velocity component into the melt film of $\mathbf{V}_{sy}$. $\mathbf{V}_b$ is the barrel surface velocity with component $\mathbf{V}_{bz}$ and $\mathbf{V}_{bx}$. The shear work done on the film is due to the velocity difference between the barrel surface and the solid bed, which is

$$\mathbf{V}_j = \mathbf{V}_b - \mathbf{V}_{sz} \tag{3A.7}$$

$$|V_j| = (V_b^2 + V_{sz}^2 - 2V_b V_{sz} \cos \theta)^{1/2} \tag{3A.7a}$$

The shear rate is V_j/δ, where δ is the thickness of the film. The cross-channel velocity V_{bx} determines the rate of melt removal.

The heat transferred from the barrel and the viscous work done on the film is used to raise the temperature of the solid bed and supply the heat of fusion, λ, of the material undergoing melting at the film–solid bed interface. Equating these energy terms, the rate of melting per unit down-channel distance ω is obtained as

$$\omega(z) = \left(\frac{V_{bx}\rho_m[k_m(T_b - T_m) + (\mu/2)V_j^2]X}{2[\lambda + C_s(T_m - T_{s0})]} \right)^{1/2} \tag{3A.8}$$

The width of the solid bed, X, changes as a function of z, T_{s0} is the temperature of the solid bed far away from the interface, k_m is the thermal conductivity of the melt, μ is the viscosity of the polymer, C_s is the heat capacity of the solid bed, and ρ_m is the density of the melt.

Using this melting rate given by Eq. 3A.8, the change in the width of the solid bed can be obtained from a mass balance. When the solid bed velocity is constant and the physical properties do not vary as functions of temperature, the solid bed width, X, can be expressed as

$$\frac{X}{W} = \left(1 - \frac{\Psi}{2}\frac{Z}{H}\right)^2$$

where

$$\Psi = \Phi\sqrt{W}\,H/G \tag{3A.9}$$

$$\Phi = \frac{V_{bx}\rho_m\{k_m(T_b - T_m) + (\mu/2)V_j^2\}}{2[\lambda + C_s(T_m - T_{s0})]}$$

The Pumping and Metering Zone

The flow in the pumping and metering zone, where the plastic is completely molten, is controlled by the drag of the plastic by the barrel surface and the backward flow due to pressure increase. For isothermal flow of an incompressible Newtonian liquid in a screw of constant channel depth, the flow rate in an extruder is given by

$$Q = \tfrac{1}{2}\pi N D_b \cos \theta_b \, W H F_d - \frac{W H^3}{12\mu}\frac{\Delta P_s}{L}\sin \bar{\theta}\,(1 + f_L)F_p$$

where

$$f_L = \frac{1}{\tan^2 \theta} - \frac{6\mu V_{bz} H}{H^3(\Delta P/L)}$$

$$F_d = \frac{16W}{\pi^3 h} \sum_{i \text{ odd}}^{\infty} \frac{1}{i^3} \tanh\left(\frac{i\pi H}{2W}\right) \tag{3A.10}$$

$$F_p = 1 - \frac{192H}{\pi^2 W} \sum_{i \text{ odd}}^{\infty} \frac{1}{i^5} \tanh\left(\frac{i\pi W}{ZH}\right)$$

and N is the frequency of the rotation of the screw.

3A.4 Concluding Remarks

Equations 3A.1–3A.5 and 3A.8–3A.10 show that the melting rate, the flow rate, and the pressure rise in the extruder are affected and controlled by the rotational speed of the screw. Therefore, the single screw plasticating extruder is a coupled system.

References

Tadmor, Z., and Gogos, C.G., *Principles of Polymer Processing*. John Wiley, NY, 1979.

Tadmor, Z., and Klein, I., *Engineering Principles of Plasticating Extrusion*. Van Nostrand Reinhold, NY, 1970.

3B; PERFORMANCE OF GEAR PUMP-ASSISTED EXTRUDER*

3B.1 Introduction

The purpose of this appendix is twofold: (1) to emphasize that after the design concept is developed and checked in terms of the axioms, it must be optimized through analysis; and (2) to show that the gear pump can, indeed, isolate the function of metering the plastic flow rate out of the extruder from the function of controlling temperature. The performance of the gear pump depends on the leakage flow that exists in these pumps, which was investigated by McKelvey (1984). In Example 3.2, Sec. 3.2, the design analysis was done while assuming that there was no leakage flow, as a first approximation to the problem, which is the proper step to take.

---

*Condensed from the paper "Performance of Gear Pumps in Polymer Processing," by James McKelvey, *Polymer Engineering and Science* **24**(6):398–402, April, 1984.

Analysis such as that given here can check the validity of this assumption. The purpose of providing a detailed analysis here is to point out that both synthesis and analysis are welded in design as illustrated in Fig. 1.1.

The gear-pump-assisted plastication extrusion process is shown schematically in Fig. 3.2. The gear pump is added to a plasticating extruder to decouple the FR associated with the flow rate from the FR of the temperature control of the extrudate. Such an arrangement improves the uniformity of flow from the extruder by reducing the costly surges and fluctuations in extrusion rate that are present in almost all extruders (McKelvey and Rice, 1982; 1983).

In this appendix, the performance characteristics of the gear pump itself are analyzed. The characteristics of interest are the *volumetric efficiency* and the *energy efficiency* of the gear pump (McKelvey, 1983). These quantities are defined, and semi-empirical equations, which relate them to the polymer properties and the operating variables of the process, are given.

3B.2 Volumetric Relationships

Figure 3B.1, which is a schematic representation of a gear pump, shows the diameters D and D_0 and the width W of the gear wheels. The liquid at the inlet side of the pump fills the cavities of the gear wheels and is transported to the high-pressure side of the pump through the rotation of the gear wheels. The upper half of a tooth on one wheel fits in the lower half of the cavity on the other wheel at the nip, thereby displacing the liquid and sealing the outlet side from the inlet side.

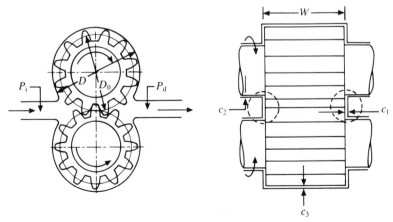

Figure 3B.1. Diagram of gear pump, showing the principal dimensions. The encircled area is the probable site for leakage flow. (From McKelvey, 1984.)

To calculate the volumetric displacement rate, the flow is treated as a plug flow, having an average speed V passing through an annulus having a rectangular cross section of width w and height H, where

$$V = \pi N[D - \tfrac{1}{2}(D - D_0)] = \frac{\pi DN}{2}\left[1 + \frac{D_0}{D}\right] \quad (3B.1)$$

$$H = \tfrac{1}{2}(D - D_0) = \frac{D}{2}\left(1 - \frac{D_0}{D}\right) \quad (3B.2)$$

Hence, the volumetric displacement rate Q_D is given by

$$Q_D = 2VWH \quad (3b.3)$$

where the factor 2 is introduced because there are two gear wheels. Introducing Eqs. 3B.1 and 3B.2 into Eq. 3B.3 yields

$$Q_D = \alpha N \quad (3B.4)$$

where

$$\alpha = \frac{\pi D^2 W}{2}\left[1 - \left(\frac{D_0}{D}\right)^2\right] \quad (3B.5)$$

Equation 3A.4 is valid only when there is no leakage flow from the high-pressure to the low-pressure side of the pump. Because all pumps have finite clearances, they exhibit leakage. From the axiomatic design point of view, this leakage determines whether or not functional independence is maintained. When this leakage flow is less than the allowable tolerance in the FRs, the effect of leakage can be ignored. The flow equation for the pump, when the leakage flow is taken into account, may be written as

$$Q = \alpha N - Q_L \quad (3B.6)$$

where Q_L is the volumetric rate of leakage.

The volumetric efficiency of the pump, ε_v, is defined as

$$\varepsilon_v = \frac{Q}{\alpha N} = 1 - \frac{Q_L}{\alpha N} \quad (3B.7)$$

The effectiveness of a pump as a surge-suppression device is directly related to its volumetric efficiency.

The primary leakage occurs at the gap between the ends of the gear wheel and the housing. In the right-hand diagram of Fig. 3B.1 this space is encircled. The leakage flow occurs under the influence of the pressure differential ΔP, where ΔP is the difference between the discharge and inlet sides of the pump.

Assuming that the molten polymer behaves as a power-law fluid, the flow rate is related to the pressure drop as

$$Q_L \propto \Delta P^{1/n} \quad (3B.8)$$

where n is the exponent in the power law. The leakage flow is dependent on the temperature and can be expressed by

$$\exp(bt) \tag{3B.9}$$

In order to take the effect of the shear developed in the leakage areas by the rotation of the gear wheels into consideration, it is assumed that the rate of shear is proportional to the rotational frequency, N, and that the effect of shear rate on the flow is given by

$$(\dot{\gamma})^{1-n} = (N)^{1-n} \tag{3B.10}$$

Combining Eqs. 3B.8–3B.10, the following empirical expression for leakage in gear pumps is obtained:

$$Q_L = K_L \left(\frac{\Delta P}{\Delta P°}\right)^{1/n} \left(\frac{N}{N°}\right)^{1/n} \exp[b(T - T°)] \tag{3B.11}$$

where the quantities with the superscript ° are arbitrary constants that scale and non-dimensionalize the variables. They have the values: $\Delta P° = 1{,}000$ p.s.i., $T° = 500°F$, and $N° = 1$ r.p.m. The coefficient K_L is the leakage rate that would occur under conditions such that $\Delta P = \Delta P°$, $N = N°$, and $T = T°$.

3B.3 Power Requirements

The mechanical power p imparted to the pump drive shaft has two components: a portion p_w used for flow work, and the remainder dissipated in fluid friction in the pump p_f. The total mechanical power is then

$$p = p_w + p_f \tag{3b.12}$$

The power required for the flow work is given by

$$p_w = Q \, \Delta P \tag{3B.13}$$

The energy efficiency ε_E for the pump is defined by

$$\varepsilon_E = \frac{p_w}{p} = \frac{1}{1 + (p_f/p_w)} \tag{3B.14}$$

The power dissipated as fluid friction is obviously a function of the viscosity and the applicable shear rate, which may be written as

$$p_f \propto \tau \dot{\gamma} \propto \eta \dot{\gamma}^2 \tag{3B.15}$$

Assuming the same temperature and shear dependence as before, Eq. 3B.15 becomes

$$r_f \propto \dot{\gamma}^{n+1} \exp(-bT) \tag{3B.16}$$

and noting that the shear rate should be proportional to the speed N, the power dissipation equation can be written

$$p_f = K_F(N/N°)^{n+1} \exp[b(T° - T)] \tag{3B.17}$$

where the coefficient K_F has units of power. K_F is the frictional power dissipated in the pump when $N = N°$ and $T = T°$.

3B.4 Experiment

The experimental apparatus is shown schematically in Fig. 3B.2. It consists of a 2.5-in. NRM "Spacemaker" extruder having a 25-HP drive, which supplies molten polymer to the pump. The pump is a Luwa 36×36 "Cinox," having helical gears with the following dimensions: $W = 36$ mm, $D = 36$ mm, and the volumetric displacement $\alpha = 25.6$ cm^3/rev.

The pump is driven by a 5-HP variable-speed drive. The drive shaft is connected to the pump shaft through a torque transducer to measure instantaneous mechanical power input to the pump.

A flow control device is attached to the pump outlet to regulate the discharge pressures. Pressure transducers are located in the flanges at the pump inlet and discharge. A thermocouple projecting into the melt stream leaving the pump is installed in the flow control device.

The data collected during an experiment consist of the mass flow rate G, the inlet and discharge pressures P_i and P_d, the pump speed N, the torque applied to the pump drive shaft Ω, and the temperature of the polymer being discharged from the pump. The pressures were recorded on a strip chart, as a function of time. Three polymers were tested. Polymers 1 and 2 are thermoplastic elastomers that differ primarily in viscosity. Polymer 3 is a polymethyl methacrylate (PMMA) material.

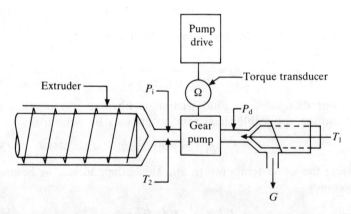

Figure 3B.2. Schematic representation of the experimental apparatus. (From McKelvey, 1984.)

Equation 3B.6 can be written

$$\left(\frac{Q}{N}\right) = \alpha - \left(\frac{Q_L}{N}\right) \qquad (3B.18)$$

and upon introducing Eq. 3B.11 it becomes

$$\left(\frac{Q}{N}\right) = \alpha - K_L\left[\left(\frac{\Delta P}{\Delta P^\circ}\right)^{1/n}\left(\frac{N}{N^\circ}\right)^n \exp[b(T - T^\circ)]\right] \qquad (3b.19)$$

Equation 3B.19 shows that a plot of (Q/N) versus the quantity in the large brackets should be a straight line with slope $-K_L$, which is shown in Fig. 3B.3 for Polymer 2. The value of the exponent n used in this correlation is 0.35. This provides the best least-square fit of the data to Eq. 3B.19.

The frictional power p_f plotted against the right-hand side of Eq. 3B.17 should be a straight line with slope K_f. Figure 3B.4 shows the plot of the power data. The frictional power loss, p_f, was found by subtracting the power, p_w, for flow work from the total power input to the pump, p. The value of the exponent n is determined from the leakage flow correlation.

Table 3B.1 summarizes the experimentally determined parameters for the three polymers and the coefficient of correlation for the data. Material 1 is a thermoplastic elastomer with the following properties: temperature coefficient of viscosity $(1/b) = 111°F$, and density given by the equation $\rho = 1/(1.15 + 0.000,25T)$. Polymer 2 is a thermoplastic elastomer similar

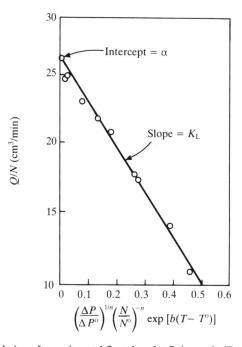

Figure 3B.3. Correlation of experimental flow data for Polymer 2. (From McKelvey, 1984.)

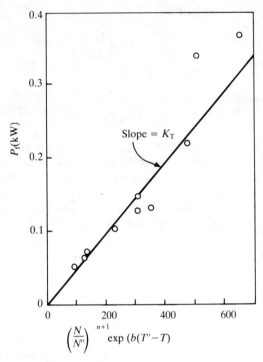

Figure 3B.4. Correlation of experimental power data for Polymer 2. (From McKelvey, 1984.)

to Polymer 1 but with lower viscosity. The values for density and temperature coefficient used for Polymer 1 were used for this polymer. Polymer 3 is a PMMA material. Its temperature coefficient was taken as $(1/b) = 45°F$ and its density was calculated from the equation $\rho = 1/(0.83 + 0.000,35T)$.

It is possible to examine the effect of the variables T, ΔP, and n over the operating range of interest by using the definition of volumetric efficiency (Eq. 3B.7), and the numerical values of the parameters K_L and n of Eq. 3B.11. Figure 3B.5 shows plots of ε_v versus ΔP for Polymer 1 at temperatures of 375 and 425°F. The range of speeds and the maximum pressure differentials covered in these plots are 50–200 r.p.m. and

TABLE 3B.1 Summary of Experimentally Determined Parameters*

Polymer	n	K_L (cm^3/min)	$K_F \times 10^5$ (kW)	Number of Data Points	Correlation Coefficient for Leakage	Correlation Coefficient for Power
1	0.47	7.5	59	49	0.98	0.84
2	0.35	30.9	49	10	0.99	0.95
3	0.40	3.0	220	15	0.96	0.95

* The numerical values of the parameters pertain to the case where the arbitrary constants of Eq. 3B.11 and 3B.17 have the following values: $\Delta P° = 1,000$ p.s.i., $N° = 1$ r.p.m., $T° = 500°F$.

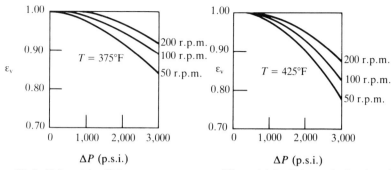

Figure 3B.5. Volumetric efficiency versus pressure differential for Polymer 1, showing effect of temperature and pump speed. (From McKelvey, 1984.)

3,000 p.s.i., respectively. The decrease in ε_v as ΔP increases is caused by the displacement flow increasing directly with speed, whereas the leakage flow is affected only indirectly by speed, through its effect on viscosity. These results indicate the conditions under which the gear pump cannot fulfill the FR, due to an excessive leakage flow.

3B.5 Characterization of Performance

The pump performance can be characterized by specifying the pumping rate, the required power, and the uniformity of delivery. In order to calculate these performance parameters, the values of the three experimentally determined pump parameters (K_L, K_F, and n) as well as the polymer density at the pumping temperature must be determined at a given inlet and outlet pressure.

The effect of the variation in the flow rate can be determined by considering a sudden change in the inlet pressure. If the inlet pressure changes from P_i to P'_i, thus changing the pressure differential from ΔP to $\Delta P'$, we obtain from Eq. 3B.11

$$Q - Q' = K_L \left(\frac{\Delta P}{\Delta P^\circ}\right)^{1/n} \left(\frac{N}{N^\circ}\right)^{1-n} \exp[b(T - T^\circ)]\left[1 - \left(\frac{\Delta P'}{\Delta P}\right)^{1/n}\right] \quad (3B.20)$$

where Q' and $\Delta P'$ are the flow rate and pressure differential obtained after the change in inlet pressure. Equation 3B.20 can be written

$$\delta Q = (\alpha N - Q)\left[1 - \left(\frac{\Delta P'}{\Delta P}\right)^{1/n}\right] \quad (3B.21)$$

From the definition of volumetric efficiency (Eq. 3B.7), we can write

$$\frac{\delta Q}{Q} = \left(\frac{1}{\varepsilon_v} - 1\right)\left[1 - \left(\frac{\Delta P}{\Delta P'}\right)^{1/n}\right] \quad (3b.22)$$

TABLE 3B.2 Characterization of Pump Performance for Polymer 1 (from McKelvey, 1984)

G (lbs/h)	T (°F)	P_d (p.s.i.)	P_i (p.s.i.)	δP_i (p.s.i.)	N (r.p.m.)	δP_d (p.s.i.)	G (%)	p (kW)
100	450	1,000	200	50	30.3	2	0.4	0.21
100	450	2,000	200	50	33.8	7	0.7	0.32
100	450	3,000	200	50	41.6	16	1.1	0.47
100	500	1,000	200	50	30.7	3	0.6	0.16
100	500	1,000	200	50	36.4	11	1.2	0.27
100	500	1,000	200	50	50.5	22	1.6	0.43
200	450	1,000	200	50	60.2	2	0.4	0.52
200	450	2,000	200	50	65.0	5	0.5	0.74
200	450	3,000	200	50	75.7	12	0.9	1.02
200	500	1,000	200	50	60.7	2	0.4	0.39
200	500	2,000	200	50	68.8	8	0.9	0.61
200	500	3,000	200	50	87.1	17	1.2	0.91

The flow rate from the pump, for power-law fluids, is governed by a relationship of the form

$$Q \propto P_d^{1/n} \tag{3B.24}$$

from which we can write

$$\frac{\delta Q}{Q} = \left(\frac{P_d'}{P_d}\right)^{1/n} - 1 \tag{3B.25}$$

where P_d' is the discharge side pressure after the drop in inlet pressure. Comparing Eqs. 3B.22 and 3B.25, we can write

$$\left(\frac{P_d'}{P_d}\right)^{1/n} - 1 = \left(\frac{1}{\varepsilon_v} - 1\right)\left[1 - \left(\frac{P_d' - P_i'}{P_d - P_i}\right)^{1/n}\right] \tag{3B.26}$$

Equation 3B.26 provides the means of calculating P_d'. Having calculated P_d', the change in the flow rate can be calculated from Eq. 3B.25.

A pump performance chart calculated for Polymer 1 is shown in Table 3B.2, for two levels of temperature and pumping rate, and for three levels of pressure differential. The effect of a surge of 50 p.s.i. in the inlet pressure on the pumping rate is given in the table.

References

McKelvey, J.M., "Energy Utilization in Extrusion," *SPE Technical Papers* **24**:921–923, 1983.

McKelvey, J.M., "Performance of Gear Pumps in Polymer Processing," *Polymer Engineering and Science* **24**(6):398–402, April, 1984.

McKelvey, J.M., and Rice, W.T., "Gear-pump Assisted Plasticating Extrusion, Part 1: Laboratory Studies," *Advances in Polymer Technology* **15**:10–35, Winter, 1982.

McKelvey, J.M., and Rice, W.T., "Retrofitting Plasticating Extruders with Gear Pumps," *Chemical Engineering* **90**:89–94, January 24, 1983.

3C: CONTROL OF METERING RATIO USING MECHANICALLY LINKED GEAR PUMPS–MOTORS IN REACTION INJECTION MOLDING*

3C.1 Introduction

The purpose of this appendix is to show how the design concept that satisfies Axiom 1 must be analyzed further to complete and optimize design. Design involves both analysis and synthesis. The analytical tools for synthesis are the design axioms, which must be employed to check the acceptability of the design. Once the physical entity is created, it must be dissected and analyzed, using the laws of nature to generate the best optimal design.

Reaction injection molding (RIM) is a process in which two or more liquid resin components are rapidly mixed and injected into a mold where they react and cure to form a finished solid part. Typically, a polyol is reacted with an isocyanate to make polyurethane. This process has attracted considerable attention recently, because it permits the making of large parts with a relatively inexpensive machine. At present, RIM is used widely in making automotive parts such as fascia and in manufacturing shoe soles.

Metering is defined as the problem of accurately controlling the flow of fluids into the mixing head, and thus metering here will encompass portions of the system that are not always thought of as metering equipment. The word system denotes a collection of mechanical, hydraulic, and electrical components, and should not be confused with the resin combinations, which are often also called systems.

The resins used in RIM are viscous liquids (0.2–$2\,N\,s/m^2$ or 200–$2,000\,cP\,s$) and react quickly, some having a gel time as brief as a few seconds. Mix ratios range from $3:1$ to $1:1$, and flow rates are usually 0.5–$5\,kg/s$ (1–$10\,lb/s$), although some larger machines exist. A typical RIM cycle might consist of an injection time of 2–5 seconds and a reaction and cure time of 60 seconds.

Mixing of the viscous resins is achieved by impinging the fluid streams at high velocity in a small mixing chamber that connects directly to the mold cavity. Pressures of 7–$20\,MPa$ ($1,000$–$3,000$ p.s.i.) are required to force the fluids through the small nozzles, so impingement mixing is often called high-pressure mixing. It offers the advantages of small mixing chamber volume, easy cleanout, and good mixing at high flow rates; but it requires that the metering system deliver the fluids in the proper ratio at all times. Most applications require 1% metering accuracy.

---

* Adapted from Tucker and Suh (1977).

3C.2 Conventional RIM

Most conventional RIM equipment use high flow-rate, high-pressure pumps to deliver each component from its reservoir to the mixing head. Usually the pumps have adjustable displacement and are driven by constant-speed motors. To make a shot, the motors are started and allowed to come to the steady state, while the fluids recirculate. Valves in the mixing head are then opened to direct the fluid into the mixing chamber and start the shot. To end the shot, the valves are switched back to the recirculating position, a ram cleans the mixed fluid out of the mixing chamber, and the pumps may be shut down to await another shot. The different machines may vary in detail, but almost all of them follow this general pattern.

The power requirement for this type of machine is fairly large. If we consider a machine delivering 2 kg/s (4.4 lb/s) at 14 MPa (2,000 p.s.i.), the actual power input to the fluid is given by

$$P = PQ \tag{3C.1}$$

For a fluid with a specific gravity of 1.1, we find that 25.5 kW (34 HP) are input to the fluid. When the efficiency of the motors is accounted for, it is easy to see why two 40-kW (50-HP) motors may be required for even a moderately sized RIM machine.

Metering errors may be broken into steady-state and dynamic errors, the former being those that are present during the entire shot, the latter occurring only for the short periods during the shot. The primary source of steady-state error is nonideal behavior of the pump. While an ideal positive-displacement device has an output proportional to the shaft speed,

$$Q = \omega d \tag{3C.2}$$

there is always some slip or leakage flow. A better model for the pump is (Blackburn, et al., 1960)

$$Q = \omega d - \frac{\Delta P}{R_L} \tag{3C.3}$$

where the second term represents the leakage flow, ΔP is the difference between inlet and outlet pressures, and R_L characterizes the resistance of the pump to leakage. As an example, consider a pump putting out 0.002 m^3/s (~122 in.^{3}/s) against a pressure of 14 MPa (~2,000 p.s.i.). If the pump has a resistance to leakage of 1×10^8 kPa s/m^3, then the leakage flow will introduce a 4.7% error in the flow rate. This amount of error is unacceptable, and in practice it is eliminated by adjusting the pump displacement until the desired output is obtained.

Some control must also be exercised over the fluid temperature, since R_L will change as the viscosity changes. Manufacturers of RIM equipment have expended considerable effort to make pumps with large R_L. By decreasing the magnitude of the leakage flow, the amount of drift due to temperature variations or pump wear is also reduced. These pumps are expensive and critical components of the machine.

The elasticity of the hoses connecting the pump to the mixing head is an important source of dynamic error. If the pressure in the line required to force the fluid around the recirculating loop is not the same as that required to force the fluid into the mixing chamber, then there will be some volume change in the hose when the shot is started. Since this volume change occurs between the metering point and the delivery point, there will be an error in fluid delivery at the beginning of the shot. It will appear as if one component begins flowing into the mixing head before the other component; this type of error is often called lead-lag. However, the hoses cannot be replaced by a more rigid conduit, since the mixing head is mounted directly on the mold, and the molds must often be tilted to get optimum filling.

The order of magnitude of this error can be estimated by forming a simple dynamic model of the situation [see, for example, Shearer et al. (1967) for more detail on this type of modeling and analysis]. The model that will be used in this case contains only three elements. The pump acts as a flow source, supplying a flow rate of magnitude Q_p regardless of what occurs downstream. The elasticity of the hoses is modeled as a fluid capacitance, and the equation describing the relationship between flow into the capacitance, Q_c, and the pressure in the line, P, is

$$Q_C = C \frac{dP}{dt} \qquad (3C.4)$$

where the quantity C is the fluid capacitance of the hose. The last element models the pressure drop required to make the fluid flow at a given rate. It is assumed that most of this resistance to flow is concentrated in the nozzle in the mixing head, and that a simple linear relationship

$$P = RQ \qquad (3C.5)$$

relates the pressure to the flow into the mixing chamber. These elements may now be combined to find one differential equation that describes the system:

$$\frac{dP}{dt} = \frac{1}{C}\left(Q_P - \frac{P}{R}\right) \qquad (3C.6)$$

Initially the system is at pressure P_R, the recirculating pressure, and at time zero the valves are opened to start the shot. Equation 3C.6 can be solved to give the time-history of the pressure

$$P = P_R + \left(\frac{Q_P}{R} - P_R\right)(1 - e^{-t/RC}) \qquad (3C.7)$$

The steady-state value of pressure is

$$P_s = Q_P R \qquad (3C.8)$$

which is the pressure required to supply a flow of Q_P to the mixing chamber. The flow rate into the mixing head will be

$$Q = \frac{P_R}{R} + \frac{(P_S - P_R)}{R}(1 - e^{-t/RC}) \qquad (3C.9)$$

The error in flow rate will be Q_c, the flow rate that goes into the fluid capacitance; that is,

$$Q_c = Q_p - Q \qquad (3C.10)$$

This error flow is given by

$$Q_c = \left(\frac{P_S - P_R}{R}\right)e^{-t/RC} \qquad (3C.11)$$

Figure 3C.1 shows Q and Q_c as given by this equation. The characteristic time τ over which this error occurs is $\tau = RC$. We can integrate the error in flow rate to find the error in the volume of fluid delivered:

$$V_E = \int_0^\infty Q_c \, dt \qquad (3C.13)$$

Substituting and performing the integration,

$$V_E = (P_S - P_R)C \qquad (3C.14)$$

As an example, suppose that $(P_S - P_R)$ equals 700 kPa (100 p.s.i.), and c equals 2.4×10^{-11} m^5/N (0.01 in.^{5}/lbf), based on a 20-ft length of 1-in. hydraulic hose. If 14 MPa (2,000 p.s.i.) is required to deliver 2×10^{-3} m^3/s (122 in.^{3}/s), then $R = 7 \times 10^9$ N s/m^5 (16.4 lbf s/in.^{5}) and the characteristic time τ is 0.17 second. The volume error, V_E, would be about 1.7×10^{-5} m^3 (about 1 in.).

Although this is the volume missing from the material that has been delivered, the amount of material delivered off-ratio will be even greater. If we assume for the moment that the dynamic error for the other component does not take place in ratio, then an off-ratio condition will

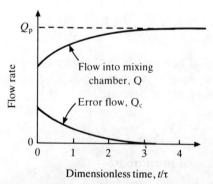

Figure 3C.1. Flow into the mixing chamber and error flow with a dynamic metering error.

exist until the error flow Q_c dies away to zero. All of the fluid delivered until that time will be off-ratio. This volume will be of the order of the desired flow rate Q_P multiplied by the characteristic time τ, or about 3.4×10^{-4} m³ (21 in.³) for this example.

Equations 3C.9, 3C.11, and 3C.14 show that one way of eliminating this error is to make P_s and P_R equal. In fact, this is a solution that is used. Adjustable orifices or needle valves in the mixing head are set so as to make the recirculating and delivery pressures equal. There is no change in the volume of the hose, and thus no dynamic error.

Another solution would be to allow the dynamic errors to take place, but to ensure that the flow rates are always in the desired ratio. If β is the volumetric mixing ratio, and we use asterisks to denote the quantities associated with the second component, then we see that since

$$Q_P = \beta Q_P^* \tag{3C.15}$$

then the requirement that

$$Q = \beta Q^* \tag{3C.16}$$

for all time is equivalent to saying

$$Q_c = \beta Q_c^* \tag{3C.17}$$

If we expand this expression, we get

$$\frac{(P_s - P_R)}{R} e^{-t/RC} = \beta \frac{(P_s^* - P_R^*)}{R^*} e^{-t/R^*C^*} \tag{3C.18}$$

Thus, the two flows will always be in ratio if

$$(P_s - P_R)/R = \beta(P_s^* - P_R^*)/R^* \tag{3C.19}$$

and

$$RC = R^*C^* \tag{3C.20}$$

To eliminate lead-lag ratio errors while using this approach would require both an adjustable orifice in the mixing head and an adjustable capacitance (perhaps a hydraulic accumulator) in the delivery line. The complexity of the mechanical requirements, and the difficulty in finding the proper adjustments, make it apparent why this solution to the lead-lag problem is not popular.

Returning to the first means of eliminating lead-lag, we note that the pressure drop across the adjustable orifices will also vary with changing fluid viscosity. This is another reason why temperature control is important to metering accuracy.

A few RIM systems deliver the fluids by forcing them out of hydraulic cylinders, and control the flow rate and ratio by the controlling the speed and position of the driving pistons. Although this type of device is not subject to the same sort of steady-state error encountered with pumps, it should be noted that the same power requirements and dynamic-error problems will be present in the piston-type systems.

3C.3 MIT RIM System

The basic concept of the MIT fluid-delivery and metering system is the creation of an uncoupled system. This is done by maintaining the independence of the FRs which are the high pressure, high flow rate, and high accuracy. In conventional equipment these FRs are usually performed by a single pump. Hydraulic accumulators are used to store fluid energy and to minimize instantaneous power requirements. As shown in Fig. 3C.2, low flow rate, high-pressure pumps are used to charge each accumulator with a resin component. A check valve then seals each accumulator off from its charging system, and the accumulators are discharged simultaneously through the metering system and into the mixing head. The fluid that has been put into the accumulators over a period of perhaps 1 minute is discharged in a few seconds. Recharging of the accumulators can be done while the previous part is cured and demolded.

If it is desired to have a shot to last for 3 seconds deliver 6 kg (13.2 lb) at 14 MPa (2,000 p.s.i.), and if 60 seconds are available in which to charge the accumulators, the flow rate through the pumps need be only 9.09×10^{-5} m^3/s (1.44 gal/min), and the power input to the system is 1.27 kW (1.70 HP). This is quite a reduction compared with the 25.5 kW (34 HP) input to a conventional machine for the same pressure and output flow rate. The same amount of energy is required to push a given volume

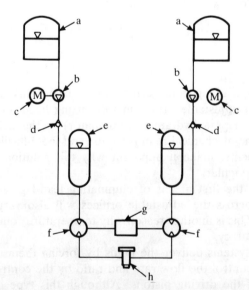

Figure 3C.2. Schematic of the MIT RIM system showing: (a) reservoirs, (b) high-pressure pumps, (c) motors, (d) check valves, (e) accumulators, (f) metering pumps, (g) mechanical device linking the metering pumps, and (h) impingement mixing head.

of the fluids into the mixing chamber at the high flow rate, but the MIT system inputs this energy over a much longer period, thus decreasing the power requirement and allowing the machine to be much smaller and lighter.

The laboratory device that was built to test this concept cost less than $8,000 in parts (1976 price) and it occupies only about 6 m^2 (65 ft^2) of floor area; yet it has produced 6.8 kg (15 lb) parts in 1 second and is capable of flow rates up to 11.3 kg/s (25 lb/s) at 1:1 ratio. The cycle time of this type of machine is limited by the time required to recharge the accumulators, whereas conventional machines may have cycles only slightly longer than the pour time. This tradeoff is discussed later.

The metering system requires no additional power inputs; it consists of a positive-displacement device (a pump or hydraulic motor) mounted in each line between the accumulator and the mixing head, the two devices being linked mechanically to turn in the proper ratio (for instance, by gearing the shafts together). This metering scheme was chosen because it makes use of the system's ability to run in ratio without any metering device. The flow rate into the mixing head depends on the pressure-flow characteristic, or fluid resistance, of the nozzle and connecting lines, and on the driving pressure supplied by the accumulator. By choosing the pressure to which the accumulators are charged, both the flow rate and the mixing ratio can be chosen. The accumulator pressure decay during the course of the shot, and the flow rate will also decay somewhat.

However, by choosing the amount of gas in the bladders of the accumulators, one may change the pressure–volume characteristic of the accumulators, and to some extent maintain the ratio during the entire shot. Malguarnera (1975) (see also Malguarnera and Suh, 1976) who designed and built the accumulator delivery system, actually did make urethane parts without any metering system, merely by the proper choice of the gas bag and charging pressures. Unfortunately, the accumulator system alone cannot be relied on to maintain accurate ratios over a long period of use, so the addition of some sort of metering device is necessary.

As noted previously, the accuracy of a positive-displacement device depends on the pressure difference between the inlet and outlet ports. With the metering system as designed, if one component tends to run too quickly, then its metering pump acts as a hydraulic motor, slowing down the fluid stream, and transmitting torque to the other meter, which acts as a pump and speeds up the other fluid stream. By arranging the system initially so that the two fluids run very close to the correct ratio, the torque transmitted by the mechanical linkage is minimized. Since the torque on a positive-displacement device is proportional to the pressure difference across it, this also has the effect of minimizing the pressure drop across the metering pump, greatly increasing the accuracy.

As an example, we use the same flow rate and resistance to leakage for the pump as in the previous example. This time the pressure drop across the pump should be of the order of 350 kPa (50 p.s.i.). Thus, the leakage flow now causes an error of only 0.12%, compared with 4.7% in the previous example. If overall errors of 1% are acceptable, then the metering pumps can be treated as ideal positive-displacement devices, and an allowance for leakage flow need not be made. This means that, once the volumetric mix ratio is known, the ratio of the pump outputs may be set once and need not be adjusted further.

As sketched in Fig. 3C.2, the only elements of the metering system are the two positive-displacement pumps or motors and the mechanical device that links their two shafts. In the experimental apparatus built to test this concept (Tucker, 1976), two timing belts and an additional shaft were used. This arrangement required four geared pulleys for the belts; with four gear choices available, the ratio could be set as accurately as it could be specified by the resin manufacturer. Note that since we require only that the rotational speeds of the two shafts be in ratio, any backlash present can cause only a small, one-time error. If the gearing is at all precise, this error is not significant.

Since the metering unit transmits very little power, and may be small and lightly constructed, and since no external power input is required, it may be installed directly onto the mixing head. This places all of the fluid capacitance effects of the connecting hoses upstream of the metering unit, and eliminates their contribution to the lead-lag problem entirely. The effect is the same as setting $C = 0$ in Eq. 3C.14, which causes the error in the volume of fluid delivered to go to zero. Tests with an experimental machine have proven that this metering concept can be quite effective.

It should be apparent that the metering system allows the use of less-critical components to obtain accuracy comparable with that of conventional equipment, or to increase accuracy by using comparable components. The accumulator delivery method allows large parts to be made quickly, using a small, inexpensive machine. As mentioned previously, the cycle time of such a machine is limited by the time required to recharge the accumulators. If the accumulators were recharged very quickly (perhaps in a few seconds), then virtually all of the improvements in power consumption would be lost. The design is best utilized when one delivery and metering device serves one mold press. Conventional machines, which input the required power continuously, can serve a number of presses when installed with a suitable switching system. This route has been taken to reduce the cost of the metering machine per press.

If the MIT system can make the use of one metering machine per mold economical, then a much greater degree of flexibility would be introduced into the manufacturing situation. Not only would a manufacturer no longer have to use a large number of molds to keep the operation economical, but different molds could be run with different materials and cycle times. Although a large number of factors are important in deciding which approach is best for a particular situation, this new design offers some clear opportunities for improvements.

At the time this work was done, the MIT RIM system was thought to be a sound design, as the foregoing statements attest. However, when the design was analyzed later using Axiom 1 (see Sect. 6.9), it was found that the design had basic flaws. It is a coupled design. To decouple it, either the FR for high flow rate or the FR for mixing must be eliminated.

Nomenclature

C = fluid capacitance, m^5/N (in.^{5}/lbf)
d = pump displacement, m^3/rad (in.^{3}/rad)
P = fluid pressure, kPa (p.s.i.)
P_R = recirculating pressure
P_S = steady-state pressure during the shot
ΔP = pressure difference between the inlet and outlet ports of pump
P = actual input power to the fluid system, kW (HP)
Q = fluid volume flow rate, m^3/s (in.^{3}/s or gal/min)
Q_c = flow into the fluid capacitance; also the error in flow to the mixing head
Q_P = pump output flow and desired flow into mixing head
R = fluid resistance, kPa s/m^3 (p.s.i./gal min)
R_L = resistance to leakage flow across a pump
V_E = error in volume of fluid delivered
β = volumetric mixing ratio
τ = time constant characterizing dynamic errors
ω = angular velocity of pump shaft, rad/s

References

Blackburn, J.F., Reethof, G., and Shearer, J.L., eds,., "Characteristics of Positive-displacement Pumps and Motors," *Fluid Power Control,* 96–101. MIT Press, Cambridge, MA, 1960.

Malguarnera, S.C., "Analysis, Design, and Construction of a High Flow Rate Polyurethane Mixer," SM Thesis, Department of Mechanical Engineering, MIT, Cambridge, MA, 1975.

Malguarnera, S.C., and Suh, N.P., "M.I.T. Liquid Injection Molding System, Part II", *Society of Plastics Engineers 34th Annual Technical Conference,* Atlantic City, NJ, 214–216, 1976.

Shearer, J.L., Murphy, A.T., and Richardson, H.H., *Introduction to System Dynamics.* Addison-Wesley, Reading, MA, 1967.

Tucker, C.L., "A Metering System for Reaction Injection Molding," SM Thesis, Department of Mechanical Engineering, MIT, Cambridge, MA, 1976.

Tucker III, C.L., and Suh, N.P., "Fluid Delivery and Metering for Reaction-injection Molding," *Journal of Engineering for Industry,* **99**(3):678–681, 1977.

4
THE INDEPENDENCE AXIOM AND ITS IMPLICATIONS

4.1. Introduction

As discussed in previous chapters, the first of the two design axioms is called the Independence Axiom. The Independence Axiom states that the independence of FRs must be maintained during the design process. This means that the choice of a physical embodiment must be made so as not to *couple* FRs when the DPs are changed. In the case of the refrigerator door, the two FRs (i.e., access to food and energy loss) were coupled when the door was opened to remove the food. Therefore (since Axiom 1 was violated), the design was not acceptable. In contrast, the chest-type freezer did not violate Axiom 1, as a slight loss of cold air during the opening and closing of the door could be tolerated. In this chapter we explore further the implications and significance of this first axiom.

There are many questions related to Axiom 1 that must be an answered. For example:

1. Since everything in the universe is ultimately related to everything else, to what extent can FRs overlap without violating the independence requirement as given by Axiom 1?
2. How can we measure functional independence quantitatively?
3. How do we represent Axiom 1 mathematically?
4. How do we actually apply Axiom 1 to real problems?
5. How do we distinguish between FRs and constraints?
6. When the relationship between FRs and DPs is not linear, how do we determine the elements of the design matrix for this nonlinear case?
7. How do we derive corollaries and theorems from the axioms? How do we show that they are, indeed, correct corollaries and theorems?

The material presented in this chapter should provide a basis for answering these questions.

It is again important to remind the reader that the first step in the design process is *problem definition*; that is, defining the FRs that characterize the need. Once this is done, a physical solution that satisfies the need (i.e., FRs) can be sought or created. During the search for an acceptable solution, many plausible solutions are often synthesized, then evaluated. Axiom 1 plays a key role at this stage, by providing an analytical criterion for evaluating the proposed design solutions. Many design decisions can be made by using Axiom 1 qualitatively, without resorting to numerical analysis, when the Independence Axiom is flagrantly violated. This is similar to using the second law of thermodynamics qualitatively to weed out perpetual-motion machines.

However, when the violation is subtle, quantitative measures for functional independence are required. This is particularly the case when we are dealing with a nonlinear design. When the relationship between FRs and DPs is nonlinear, the design may satisfy Axiom 1 only in a very narrow window defined by specific values of DPs. That is, Axiom 1 may be satisfied only in a specific design span (i.e., specific values of DPs) when the FRs are nonlinearly dependent on the DPs.

In this chapter the concept of functional coupling is clarified further. Sections 4.2 and 4.3 describe the graphical representation of two-dimensional design space, and illustrate the concept of functional coupling and the relationship between the functional and physical domain pictorially. Section 4.4 presents the nonlinear design matrix and its significance. For multidimensional design we must have quantitative measures for functional independence, which is the subject of Sec. 4.5. Since the design matrix must be insensitive to the specific metric used, Sec. 4.6 discusses the scaling of FRs and DPs. The concept of functional coupling cannot be divorced from the tolerance within which the FRs may be taken to be acceptable. This issue is discussed in Sec. 4.7.

4.2 Graphical Representation of the Mapping Process

In Chapter 3 the design process is represented as a mapping operation, moving from the functional domain to the physical domain. The nature of mapping between a given **FR** vector and a **DP** vector having design matrix [**A**] was given by the design equation as:

$$\{\mathbf{FR}\} = [\mathbf{A}]\{\mathbf{DP}\} \tag{3.1}$$

For the two-dimensional case, [**A**] may be represented as

$$[\mathbf{A}] = \begin{bmatrix} A_{11} & A_{12} \\ A_{21} & A_{22} \end{bmatrix} \tag{4.1}$$

This relationship can be represented graphically to provide further insight into the concept of functional independence (Rinderle, 1982; Suh and Rinderle, 1982; Rinderle and Suh, 1982). What we want to demonstrate is how the magnitudes of the elements of the design matrix affect the relationship between DPs and FRs, and how a DP can couple FRs by graphic means to aid the visualization of functional coupling.

By definition, FRs are independent of each other. Therefore, the **FR** vector for a case with two degrees of freedom may be represented in a functional space (i.e., functional domain) defined by two axes, FR_1 and

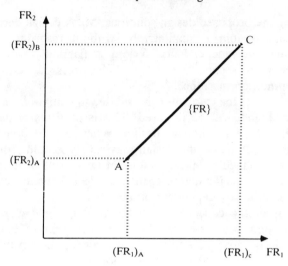

Figure 4.1. The functional requirement vector {**FR**} in a functional domain defined by the axes FR_1 and FR_2. FR_1 and FR_2 axes are orthogonal to each other, since FRs are independent.

FR_2, as shown in Fig. 4.1. The axes FR_1 and FR_2 define the functional domain. The FR_1-axis is orthogonal to the FR_2-axis, since FR_1 is independent of FR_2, by definition. In order to map the physical domain graphically onto the functional domain, the two axes that define the physical domain, the DP_1- and DP_2-axes, may be superimposed on the (FR_1–FR_2) domain as shown in Fig. 4.2. Unlike the FR_1- and FR_2-axes, the axes DP_1 and DP_2 are not necessarily orthogonal to each other, since they are chosen for convenience, without regard to the independence of DPs from each other. What we will show through this graphic representation is that Axiom 1 requires that DPs and FRs be parallel to each other, and the DPs must therefore be orthogonal to each other since the FRs are.

The relationship between the physical domain and the functional domain is governed by the angles α_1 and α_2, which represent the angle between the FR_1-axis and the DP_1-axis, and the angle between the FR_2-axis and the DP_2-axis, respectively. When α_1 and α_2 are zero, the DP_1- and DP_2-axes are parallel to the FR_1- and FR_2-axes, respectively. The case is illustrated in Fig. 4.3a. Since the corresponding FR- and DP-axes are parallel to each other, and since the DPs are orthogonal to each other, FR_1 is dependent only on DP_1, and FR_2 is dependent only on DP_2. Therefore, the case shown in Fig. 4.3a is an *uncoupled* design.

Changing FRs from State A to State C of the uncoupled design can be done by simply changing DP_1 from A to D, and DP_2 from A to B. In the uncoupled case the order of change does not affect the FRs. Thus, we may either change DP_2 before changing DP_1, or change them simultaneously. The design is totally *path independent* when the independence axiom is not

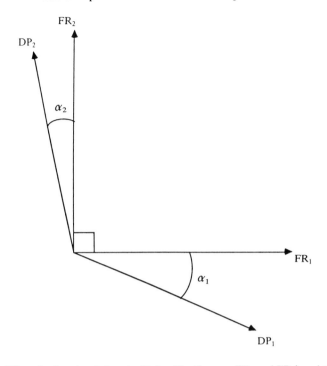

Figure 4.2. When the functional domain (defined by the axes FR_1 and FR_2) and the physical domain (defined by the DP_1 and DP_2 axes) are superimposed, the relationship between the two domains is defined by two angles, α_1 and α_2.

violated; that is, when the design is an *uncoupled* design. What this means is that, when the design is uncoupled, we may change any FR without worrying about the effect of the change on the other FRs. The uncoupled case illustrated by Fig. 4.3a is a special case of Eq. 3.1, given by

$$\begin{Bmatrix} FR_1 \\ FR_2 \end{Bmatrix} = \begin{bmatrix} A_{11} & 0 \\ 0 & A_{22} \end{bmatrix} \begin{Bmatrix} DP_1 \\ DP_2 \end{Bmatrix} \quad (4.2)$$

Chapter 3 points out that, in the case of decoupled design, the order of change of DP must be fixed in order to satisfy Axiom 1. This can be illustrated graphically using Fig. 4.3. If one of the angles, say α_1, is equal to zero and the other is not, then the design is decoupled, as shown in Fig. 4.3. In this figure the DP_1- and the DP_2-axes are not orthogonal to each other, and the DP_2-axis is not parallel to the FR_2-axis, although the DP_1-axis is parallel to the FR_1-axis. Because of the nonorthogonal, nonparallel relationship between these axes, State A in the functional domain is represented by $(FR_1)_A$ and $(FR_2)_A$, whereas State A in the physical domain is represented by $(DP_1)_A$ and $(DP_2)_A$. State E has the same DP_1 value as State A. Because the DP_2- and the FR_2-axes are not parallel, the value of FR_2 cannot be altered by changing the value of DP_2 without affecting the value of FR_1. However, changing the value of DP_1 in order to change the value of FR_1 does not affect the value of FR_2.

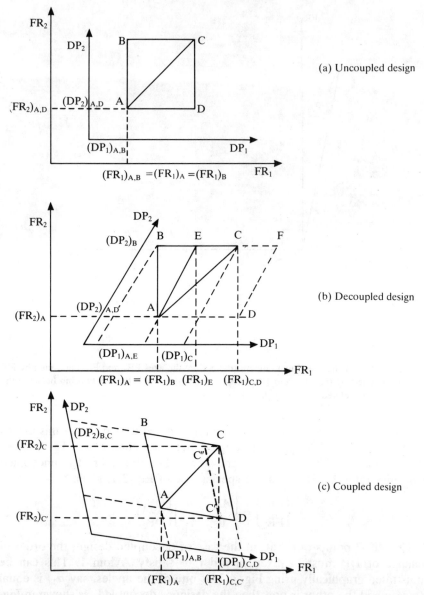

Figure 4.3. Graphical representation of the design mapping procedure from the F- to the P-Domain. $(FR_1)_{A,B}$ implies that $(FR_1)_A$ and $(FR_1)_B$ are equal to each other.

The best way to change the state of FRs from State A to State C of the decoupled design is to change DP_2 from $(DP_2)_A$ to $(DP_2)_B$ [note that $(DP_2)_B = (DP_2)_E = (DP_2)_C = (DP_2)_F$], and then to change DP_1 from $(DP_1)_E$ to $(DP_1)_C$. If we had reversed the order of change, changing DP_1 first, then the only way in which we could have moved from A to C would have been via A to D to F to C; this is much more complex. This is because each time we change DP_2, the value of FR_1 is affected, although this is not intended. The initial value of DP_1 chosen [i.e., $(DP_1)_D$] to set the value of FR_1 at $(FR_1)_C$ did not change, whereas the value of DP_2 is changed to set the value of FR_2. However, the initial value of FR_1 changed from $(FR_1)_C$ to $(FR_1)_F$.

As these examples illustrate, the decoupled (or quasi-coupled) design is *path dependent*. The preferred path, namely A to E to C, which is the first case cited, can be represented as

$$\begin{Bmatrix} FR_1 \\ FR_2 \end{Bmatrix} = \begin{bmatrix} A_{11} & A_{12} \\ 0 & A_{22} \end{bmatrix} \begin{Bmatrix} DP_1 \\ DP_2 \end{Bmatrix} \qquad (4.3)$$

This design matrix is triangular, and represents the decoupled design. In this case the sequence of change of DPs is very important, because, if it is altered, then Axiom 1 may be violated. The gear pump-assisted single screw extruder shown in Fig. 3.2 is a good example of a decoupled design as illustrated in Example 4.1. Many engineering designs are decoupled designs, because most iterative design processes lead to decoupled designs.

In Chapter 3 it is stated that a coupled design cannot satisfy Axiom 1. This can be illustrated graphically, using Fig. 4.3c. When the angles α_1 and α_2 are nonzero, and $\alpha_1 \neq \alpha_2$, the case is *coupled*, in contrast to either the uncoupled or the decoupled design, which is illustrated in Fig. 4.3c. In this figure the DP_1- and DP_2-axes are neither orthogonal to each other nor parallel to the FR_1- and FR_2-axes, respectively. Thus, in this case we must follow a very complex path to change FR_1 from $(FR_1)_A$ to $(FR_1)_C$, and FR_2 from $(FR_2)_A$ to $(FR_2)_C$.

It is difficult to reach State C from State A of the coupled design (shown in Fig. 4.3c) by simply changing DP_1 from $(DP_1)_A$ to $(DP_1)_C$, and then DP_2 from $(DP_2)_A$ to $(DP_2)_C$, since any change in DP_1 affects the values of both FR_1 and FR_2, and so does DP_2. When the state is changed from Point A to Point C' so as to obtain the correct value of FR_1, the value of FR_2 is also changed from $(FR_2)_A$ to $(FR_2)_{C'}$. Furthermore, as FR_2 is changed from $(FR_2)_{C'}$ to $(FR_2)_C$, FR_1 is also shifted to $(FR_1)_{C''}$. Subsequently, as the value of FR_1 is readjusted from $(FR_1)_{C''}$ to $(FR_1)_C$, FR_2 again assumes a new value. In order to go from State A to State C in the functional domain, both FR_1 and FR_2 must therefore be iterated continuously until they converge at Point C. In some nonlinear cases they may never converge.

In a coupled design it should again be noted that each change in DP_1 or DP_2 affects both FR_1 and FR_2. A characteristic of coupled design is that to change the state of the FRs the values of all of the elements of the design matrix must be known and specified for every state, and that it is not possible to change the value of any one FR without affecting all others. Only unique sets of values of FRs are allowable in a coupled design. Such a design lacks flexibility and requires tight tolerance. Clearly, the coupled design violates Axiom 1 and is *path dependent*.

Example 4.1: Gear Pump-assisted Single Screw Extruder

Question: Describe, in terms of the design equation, how the single screw extruder of Example 3.2, which is a coupled design, is decoupled by adding a positive-displacement gear pump between the exit of the extruder and the die.

Solution

The FRs of the single screw extruder are the pumping rate Q and the tempeature of the extrudate T. These FRs are satisfied in a single screw extruder by changing the screw speed, which affects both the extrusion rate and the extrudate temperature. The barrel temperature also affects the temperature of the extrudate, but in most cases it has less of an effect than the screw speed. Therefore, in the single screw extruder, the relationship between the FRs and the DPs is

$$\left\{\begin{array}{c}\text{Pumping rate}\\ \text{Extrudate temperature}\end{array}\right\} = \begin{bmatrix} \times \\ \times \end{bmatrix} \{\text{Screw speed}\} \tag{a}$$

The system described by Eq. a is a coupled system, since the number of DPs is less than the number of FRs, per Theorem 1 discussed in Sec. 3.6.

When the positive-displacement gear pump is attached to the extruder, the relationship between the FRs and the DPs changes to

$$\left\{\begin{array}{c}\text{Pumpimg rate}\\ \text{Extrudate temperature}\end{array}\right\} = \begin{bmatrix} \times & 0 \\ \times & \times \end{bmatrix} \left\{\begin{array}{c}\text{Gear pump speed}\\ \text{Screw speed}\end{array}\right\} \tag{b}$$

The gear pump speed affects the extrudate temperature by changing the residence time of plastics in the extruder (i.e., the longer the transit time is, the higher the extrudate temperature, due to greater work being done on the plastic). The design matrix of Eq. b is triangular, thus representing a decoupled (or quasi-coupled) design. Note that in this design the gear pump speed must be changed before the screw speed; otherwise Axiom 1 would be violated. This is characteristic of decoupled designs.

4.3 Further Discussion of FR Isograms and DP Isograms

Suppose we came up with a design. When we wrote the design equation for the design relating FRs to DPs, the design matrix turned out to be that represented by Eq. 4.4, which is for a linear coupled design:

$$\left\{\begin{array}{c}FR_1\\ FR_2\end{array}\right\} = \begin{bmatrix} A_{11} & A_{12} \\ A_{21} & A_{22} \end{bmatrix} \left\{\begin{array}{c}DP_1\\ DP_2\end{array}\right\} \tag{4.4}$$

where

$$\begin{bmatrix} A_{11} & A_{12} \\ A_{21} & A_{22} \end{bmatrix} = \begin{bmatrix} 0.87 & 0.20 \\ -0.50 & 0.98 \end{bmatrix} \tag{4.4a}$$

DP isograms can be plotted in the function space of nonsingular, linear designs, as shown in Fig. 4.4 for the case given by Eq. 4.4a. The DP

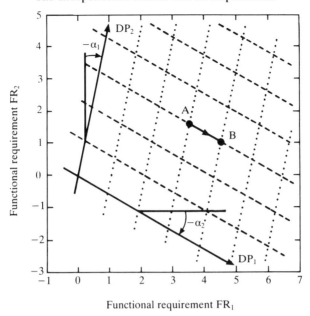

Figure 4.4. DP isograms plotted in the function space for the coupled design described by Eq. 4.4a. (From Rinderle, 1982.)

isograms intersect the function coordinates at angles of α_1 and α_2, given by

$$\alpha_1 = \tan^{-1}(-A_{12}/A_{22})$$
$$\alpha_2 = \tan^{-1}(A_{21}/A_{11})$$
(4.5)

The effect of changing a single design parameter is clearly illustrated in Fig. 4.4. DP_1 varies along an isogram of DP_2 (e.g., as we move from A to B in Fig. 4.4). In this case FR_2 changes by 0.58 units for each unit change in FR_1, and neither of the FRs can be varied without affecting the other when one of the DPs is changed.

When the design matrix given is such that $A_{11} = A_{22} = -A_{12} = A_{21}$ or $A_{11} = A_{22} = A_{12} = -A_{21}$, we have an interesting case. Here the DPs affect the FRs with equal impact. For example, when $A_{11} = A_{22} = 0.8$ and $A_{12} = -A_{21} = 0.6$, the DP isograms are as shown in Fig. 4.5. In this case the DP isograms are mutually orthogonal in the function space. The α are equal.

In some situations it is easier to visualize the effect of coupling if we express the DP vector as a function of the FR vector:

$$\{\mathbf{DP}\} = [\mathbf{A}]^{-1}\{\mathbf{FR}\}$$
(4.6)

where $[\mathbf{A}^{-1}]$ is the inverse matrix of $[\mathbf{A}]$. The inverse matrix exists if the design matrix $[\mathbf{A}]$ is a square matrix whose determinant is non-zero and whose elements are constants (i.e., the relationship between $\{\mathbf{FR}\}$ and

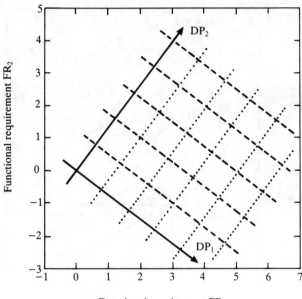

Figure 4.5. DP isograms plotted in the function space for the design described by Eq. 4.4, where $A_{11} = A_{22}$ and $A_{12} = -A_{21} = 0.6$. Note that the isograms are orthogonal. (From Rinderle, 1982.)

{**DP**} is linear). The inverse matrix $[\mathbf{A}]^{-1}$ is given by (Hildebrand, 1952)

$$[\mathbf{A}]^{-1} = \frac{1}{|\mathbf{A}|} \text{adj}[\mathbf{A}] \tag{4.7}$$

where $|\mathbf{A}|$ is the determinant of the design matrix $[\mathbf{A}]$ and its adjoint matrix (adj $[\mathbf{A}]$) is defined as

$$\text{adj}[\mathbf{A}] = \begin{bmatrix} a_{11} & a_{21} & \dots & a_{n1} \\ a_{12} & a_{22} & \dots & a_{n2} \\ \vdots & & & \vdots \\ a_{1n} & & \dots & a_{nn} \end{bmatrix} = [a_{ji}] \tag{4.8}$$

and where

$$a_{ij} = (-1)^{i+j} M_{ij}$$

Here M_{ij} is the minor of A_{ij}, obtained as the determinant of the remaining square array after the row and column containing element A_{ij} have been deleted.

The inverse matrix obtained from Eqs. 4.4 and 4.4a is represented in Fig. 4.6, which show FR isograms plotted in the DP space. For convenience, the DP space is represented by two orthogonal axes DP_1 and DP_2. Therefore, although by definition FR_1 and FR_2 are orthogonal, they are not perpendicular to each other. The isograms of FR_1 intersect lines of

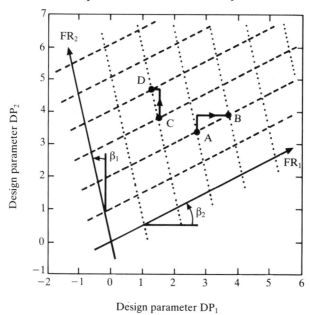

Figure 4.6. FR isograms plotted in the design space for the coupled design described by the inverse transform of Eqs. 4.4 and 4.4a. (From Rinderle, 1982.)

constant DP at an angle β_1, whereas FR_2 isograms intersect lines of constant DP_2 at angle β_2 (Rinderle, 1982). The angles β_1 and β_2 are given by

$$\beta_1 = \tan^{-1}(A_{12}/A_{11})$$
$$\beta_2 = \tan^{-1}(-A_{21}/A_{22})$$
(4.9)

In this representation the FR isograms of coupled and decoupled designs are not perpendicular to each other, even though, by definition, FRs are orthogonal to each other. This is a result of arbitrarily representing DPs as being orthogonal in the physical domain although they may not be orthogonal.

The foregoing discussion on graphical representation of the mapping process going from FRs to DPs illustrated the following points:

1. In an uncoupled design the DPs are also orthogonal to each other and each DP controls only one specific FR. Stated in another way, even when DPs are orthogonal to each other, if DP axes are not parallel to their corresponding FR axes, the design cannot be an uncoupled design. (Note that by definition, all FRs are orthogonal to each other.)
2. When DPs are dependent on each other, the design may be either coupled or decoupled, but cannot be an uncoupled design. In a decoupled two-dimensional design, one of the DPs must be parallel to one of the FRs.

Example 4.2: Passive Filter Design Problem*

In order to illustrate how the design equation can be used in practice, let us consider the two electric circuits shown in Figs a and b, which are two different designs proposed for a passive filter by an engineer without any background in axiomatic design. This filter is to be used in a simple instrumentation system (consisting of a transducer, demodulator, filter, and recorder), designed to record the displacement (such as shown in Fig. c) of a mechanical system. Two types of displacement transducers are to be evaluated: a Pickering precision LVDT model DTM-5 and a displacement transducer based on a four-active-arm strain-gage bridge. The a.c.-excited transducers produce an amplitude-modulated displacement signal. (Figure d shows an example of an amplitude-modulated displacement signal. The carrier frequency is 40 Hz in this plot. See Appendix 4A for analysis of these

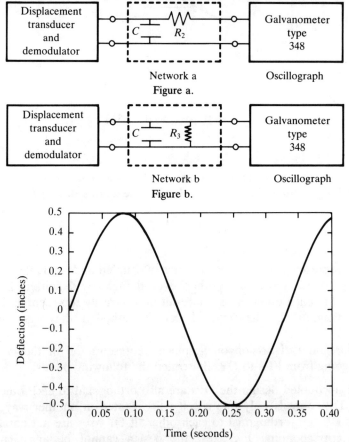

Figure a.

Figure b.

Figure c. A 3 Hz sinusoidal displacement to be recorded.

Figure 4.2. When the functional domain (defined by the axes FR_1 and FR_2) and the physical domain (defined by the DP_1 and DP_2 axes) are superimposed, the relationship between the two domains is defined by two angles, α_1 and α_2.

---
* Adapted from Rinderle (1982).

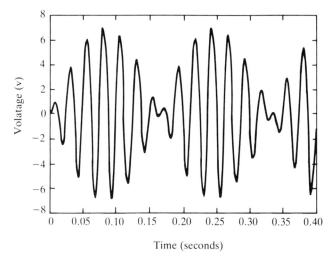

Figure d. The displacement signal shown in Fig. c amplitude modulated at 40 Hz. (From Rinderle, 1982).

waveforms.) The transducer output passes through a full-wave, phase-sensitive demodulator (see Fig. e). The demodulator output drives the networks that being considered in this example.

The networks filter and attenuate the signal, which then drives the galvanometer in order to make a record of the deflection of the mechanical system. A simulated deflection record is shown in Fig. f. The simulated deflection (see Fig. f) shows a significant ripple on the deflection record, as a result of incomplete filtering of the carrier. The signal is recorded using a light-beam oscillograph equipped with two types of magnetically damped galvanometers. The specifications for the displacement transducers, demodulator and galvanometers are given in Table a.

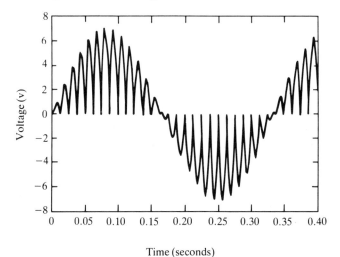

Figure e. Full-wave, phase-sensitive demodulation of the signal shown in Fig. d. (From Rinderle, 1982.)

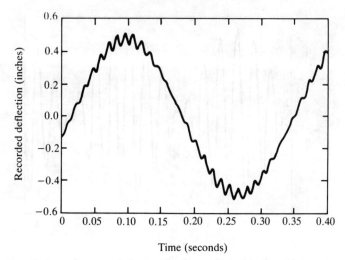

Figure f. The displacement record obtained by passing the signal shown in Fig. e thorugh a low-pass filter with a single pole at 10 Hz. A comparison with Fig. c shows the errors introduced by nonideal filtering. (From Rinderle, 1982.)

The displacement signal to be recorded has a spectral content in the range of 0–2 Hz. The displacement transducers are excited at 60 Hz. The demodulated output of the transducers consists of the desired signal, the rectified carrier near 120 Hz, and higher harmonics of the carrier. The carrier magnitude is two-thirds that of the signal, as shown by a spectral analysis of the demodulated transducer output. (See Appendix 4A for details of the analysis.)

The design task is to specify a network to match the demodulated transducer output to the galvanometer in the light-beam oscillograph. The network must suppress the carrier frequency while passing the undistorted displacement signal of interest. The network must also attenuate the signal so that the deflection record is properly scaled. The specific task lies in identifying the better filter between those shown in Figs. a and b.

Solution to Passive Filter Design Problem

The FRs may be stated as

FR_1 = Suppress the carrier without distorting the displacement signal.

FR_2 = Attenuate the signal to obtain the proper scale.

The specifications for the filter characteristics are based on the degree of carrier suppression and the fidelity of the displacement signal. Both residual carrier and signal distortion contribute to the error in the recorded signal. To determine the cutoff frequency or pole position that will minimize this error, an error analysis must be done and a criterion for the pole position must be developed. The analysis in Appendix 4B shows that the desired pole position of the filter is 6.84 Hz. The desired full-scale magnitude of the deflection record was 6 in., which is roughly equal to the full scale in the light-beam oscillograph.

TABLE a

(A) *Pickering LVDT Model DTM-5 Specifications*

Device characteristics	
Linear range	+0.50 in.
Linearity	+0.5%
Sensitivity	1.5 V/V in.
Output impedance	1860 Ω
Nominal input voltage	6 V @ 60 Hz
Output voltage	4.5 V
Primary resistance	72 Ω
Secondary resistance	240 Ω
In situ performance	
Full-scale output	+4.5 V
Full-scale deflection	+0.5 in.
Input voltage	6 V @ 60 Hz

(B) *Strain-gage Bridge Displacement Transducer Characteristics*

Device characteristics	
Bridge characteristics	4-active-arm
Gage type	Resistance
Gage resistance	120.0 Ω + 0.15%
Gage factor	2.03 + 0.5%
Output impedance	120 Ω
In situ performance	
Full displacement strain	0.001,16
Bridge excitation	10 V @ 60 Hz
Full-scale bridge output	0.015 V (after demod.)

(C) *Galvanometer Characteristics*

Ext. damping resistance	120 Ω
Undamped natural frequency	100 Hz
Flat frequency range (5%)	0–60 Hz
Terminal resistance	98 Ω
Sensitivity	6.71 μA/in.
Maximum safe current	10 mA
Viscous damping	0.13
Nominal magnetic damping	0.51

The FRs can be specified more quantitatively as

FR_1 = Design a low-pass filter network with a filter pole at 6.84 Hz.

FR_2 = Obtain d.c. gain such that the full-scale deflection results in ±3 in. light-beam deflection.

The two DPs are capacitance C and resistance R (i.e., R_2 for Network a and R_3 for Network b).

In order to investigate which network (Fig. a or Fig. b) better meets these two FRs, one can analyze the electrical response of the circuits shown in Figs. a and b. The details of the displacement transducer and the demodulator are given in Fig. g for the LVDT and the strain-gage transducer.

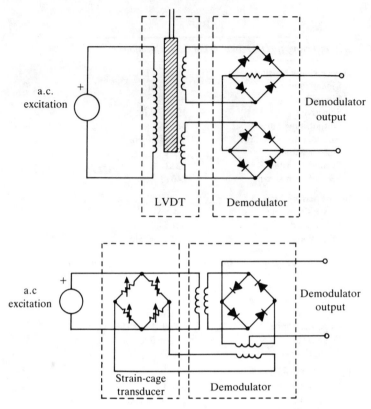

Figure g. Demodulator configurations for the LVDT and strain-gage displacement transducers.

The passive filter networks shown in Figs. a and b can be generalized as shown in Fig. h, where the displacement transducer is replaced by an output impedance R_s and a demodulated signal excitation V_s, and the galvanometer is shown by the impedance R_g. The network shown in Fig. h can be analyzed using the Kirchhoff current law to obtain the transfer function

$$\frac{V_0}{V_s} = \frac{R_g R_3}{(R_2 + R_g)(R_s + R_3) + R_3 R_s + (R_2 + R_g)R_3 R_s Cs} \tag{a}$$

Figure h.

where s is the Laplace variable. Equation (a) is in the form of $(V_0/C_s) = 1/(\tau s + 1)$, which is typical for a first-order system. From this transfer function, expressions for the filter cutoff frequency ω_c and the galvanometer full-scale deflection D are derived as

$$\omega_c = \frac{2\pi(R_2 R_s + R_2 R_3 + R_g R_s + R_g R_3 + R_3 R_s)}{(R_2 + R_g)R_3 R_s C} \quad \text{(b)}$$

$$D = \frac{|V_0|}{G_{\text{sen}}} = \frac{|V_{\text{fs}}|R_g R_3}{G_{\text{sen}}(R_2 R_s + R_2 R_3 + R_g R_s + R_g R_3 + R_3 R_s)} \quad \text{(c)}$$

where V_{fs} is the full-scale transducer output after demodulation, which is $(V_s)_{\text{max}}$.

The elements of the design matrix may be written as

$$\begin{aligned} A_{11} &= \left\{\frac{\partial \omega_c}{\partial C}\right\} \bigg/ \omega_c \\ A_{12} &= \left\{\frac{\partial \omega_c}{\partial R}\right\} \bigg/ \omega_c \\ A_{21} &= \left\{\frac{\partial D}{\partial C}\right\} \bigg/ D \\ A_{22} &= \left\{\frac{\partial D}{\partial R}\right\} \bigg/ D \end{aligned} \quad \text{(d)}$$

The elements of the design matrix are as follows (see Appendix 4A):

For Design a

$$[\text{DM}] = \begin{bmatrix} -\dfrac{1}{C} & -\dfrac{R_s}{(R_s + R_g + R_2)(R_2 + R_g)} \\ 0 & -\dfrac{1}{(R_s + R_g + R_2)} \end{bmatrix} \quad \text{(e)}$$

For Design b

$$[\text{DM}] = \begin{bmatrix} -\dfrac{1}{C} & -\dfrac{R_g R_s}{R_3(R_3 R_s + R_g R_s + R_3 R_g)} \\ 0 & \dfrac{R_g R_s}{R_3(R_3 R_s + R_g R_s + R_3 R_g)} \end{bmatrix} \quad \text{(f)}$$

where R_s is the transducer source impedance and R_g is the galvanometer impedance. The design matrix can be obtained by evaluating Eqs. e and f at the design point. This evaluation involves determining the values of the DPs required to obtain the desired function, and then substituting the DP values into the design matrix given in Eqs. e and f.

The results are shown as a series of isograms in Figs. i–m. Figure i shows FR isograms plotted in design space for the LVDT–Network a case. The isograms are nearly aligned with the coordinates of the design space, which indicates that this is a nearly uncoupled case. Figure j shows FR isograms for the strain-gage–Network a case. The isograms of filter cutoff frequency are

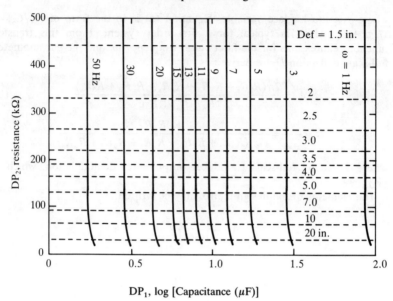

Figure i. FR isograms plotted in the design space for the LVDT–Network a–348 galvanometer case.

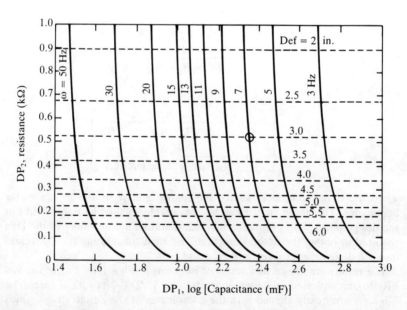

Figure j. FR isograms plotted in the design space for the strain gage–Network a–348 galvanometer case.

The Independence Axiom and its Implications   113

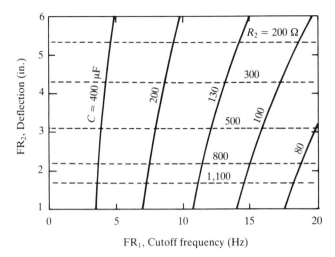

Figure k. DP isograms plotted in the function space for the strain gage–Network a–348 galvanometer case.

not aligned with the R_2 coordinate in the design space. (A logarithmic scale is used in these figures. Thus, the angles between isograms and coordinates in the figure do not correspond to the true degree of coupling.)

Figure k shows the DP isograms in the function space for the case shown in Fig. j. It shows that, whereas a change in capacities affects only the filter cutoff frequency, a change in resistance R_2 affects both the deflection and the cutoff frequency. Therefore, the design represented by the strain-gage–Network a case is a decoupled (or quasi-coupled) design.

Figures l and m are the DP isograms for cases using the network shown in

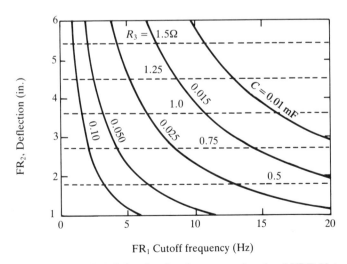

Figure l. DP isograms plotted in the function space for the LVDT–Network b–348 galvanometer case.

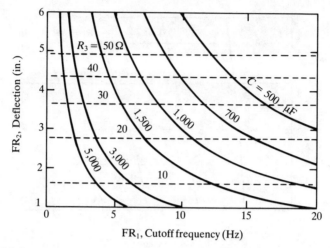

Figure m. DP isograms plotted in the function space for the strain gage–Network b–348 galvanometer case.

Fig. b. Both of these designs are quasi-coupled (or decoupled) ones. The FRs in the LVDT–Network b case are very sensitive to small changes in the DPs. In this case R_3 must be set at 0.823 Ω. If R_3 is increased by 0.18 Ω, then the deflection and the filter pole position will both change by 22%.

4.4 Nonlinear Design Matrix

In many designs the relationship between the **FR** vector and the **DP** vector is nonlinear. That is, as the value of DP changes, the value of FR increases or decreases nonlinearly. For this case Eqs. 3.1 and 3.2 may be written more generally as

$$\{\mathbf{FR}\} = [\mathbf{A}]\{\mathbf{DP}\} \tag{3.1}$$

where the elements of the design matrix [**A**] are given by

$$A_{ij} = \frac{\partial(\mathrm{FR}_i)}{\partial(\mathrm{DP}_j)} = f(\text{DPs and other parameters}) \tag{3.2}$$

For this nonlinear case the design can be evaluated only at a given design point; that is, even an *uncoupled* design at a given operating point may become a *coupled* design far away from this design point. For example, in the case of the gear pump-assisted single screw extruder (see Example 3.2) and Appendix 3B), if the gear pump is set at a very low speed, then the pressure built up by the screw of the extruder behind the gear pump is so great that the plastic may be compressed and the mass flow rate may change, even though the gear pump speed has not been altered. Conversely, if the gear pump speed is set so high that the extruder can barely supply sufficient material, then the screw speed may have a significant effect on the flow rate. In both of these instances, because of the nonlinearity of the system, a decoupled design becomes a coupled design under extreme operating conditions.

4.5 Quantitative Measures for Functional Independence

In Secs. 4.2 and 4.3 the relationship between the FRs in the functional space and the DPs in the physical space is presented graphically for a two-dimensional case, to illustrate two important factors that determine the independence of the FRs. First, it was shown that the angle (or orthogonality) between the DPs determines the sensitivity of FRs to changes in DPs. Secondly, the alignment of the DP-axes with the corresponding FR-axes determines the independence of FRs. In this section this discussion is further generalized in the n-dimensional case, so as to develop general qualitative *measures* for functional independence. The quantitative measures are called *reangularity*, R, and *semangularity*, S. We again begin the discussion with the two-dimensional case.

The design equation for the two-dimensional case may be written as

$$\begin{Bmatrix} FR_1 \\ FR_2 \end{Bmatrix} = \begin{bmatrix} A_{11} & A_{12} \\ A_{21} & A_{22} \end{bmatrix} \begin{Bmatrix} DP_1 \\ DP_2 \end{Bmatrix} = \begin{Bmatrix} A_{11} \\ A_{21} \end{Bmatrix} DP_1 + \begin{Bmatrix} A_{12} \\ A_{22} \end{Bmatrix} DP_2 \quad (4.10)$$

or

$$\mathbf{FR} = \mathbf{C}_1 DP_1 + \mathbf{C}_2 DP_2 = FR_1 \mathbf{i} + FR_2 \mathbf{j} \quad (4.11)$$

where the vectors $\mathbf{C}_k$ are

$$\mathbf{C}_1 = \begin{Bmatrix} A_{11} \\ A_{21} \end{Bmatrix} \quad \mathbf{C}_2 = \begin{Bmatrix} A_{12} \\ A_{22} \end{Bmatrix}$$

In the functional space, Eq. 4.11 may be represented graphically as shown in Fig. 4.7. The angular relationship, θ, between $\mathbf{C}_1$ and $\mathbf{C}_2$ (this is

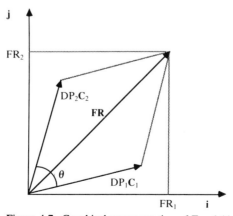

Figure 4.7. Graphical representation of Eq. 4.11.

the angle between DPs) can be obtained by taking the dot product and normalizing:

$$\cos\theta = \frac{\mathbf{C}_1 \cdot \mathbf{C}_2}{|\mathbf{C}_1| \cdot |\mathbf{C}_2|} \quad (4.12)$$

If we define Reangularity as $R = \sin\theta$, R may be expressed as (Rinderle, 1982; Rinderle and Suh, 1982; Suh and Rinderle, 1982)

$$R = \sin\theta = (1 - \cos^2\theta)^{1/2} \quad (4.13)$$

For the n-dimensional case, Eq. 4.13 may be written as

$$R = \prod_{\substack{i=1,\,n-1 \\ j=1+i,\,n}} \left(1 - \frac{(\sum_{k=1}^{n} A_{ki}A_{kj})^2}{(\sum_{k=1}^{n} A_{ki}^2)(\sum_{k=1}^{n} A_{kj}^2)}\right)^{1/2} \quad (4.14)$$

where A_{ij} is an element of the design matrix.

The summations in the denominator of Eq. 4.14 are the squared magnitudes of the columns in the design matrix. These terms are used to normalize the columns of the matrix. When the columns are already normalized, the summations are equal to unity. The normalization is necessary in order to make the measure insensitive to changes in the scale of any of the DPs.

The summation in the numerator is the dot product of the ith and jth columns of the design matrix. This summation, when properly normalized by the denominator, can be thought of as the cosine of the angle between the vectors in the ith and jth columns of the design matrix. Since this cosine term is squared and subtracted from unity, the argument of the product operator is the square of the sine of the same angle. The reangularity is the absolute value of the product of the sines of all the angles between pairs of DP isograms in the function space. The maximum value of reangularity occurs when the isograms are mutually orthogonal. As the degree of coupling increases, so the value of R decreases. In the limit when R is zero, two or more of the isograms are parallel and the design is completely coupled. Figure 4.8 shows contours of reangularity for the two-dimensional case, as a function α_1 and α_2. The contours are straight lines since in the two-dimensional case $R = \cos(\alpha_1 - \alpha_2)$.

As stated earlier, the reangularity measures the orthogonality between the DPs. However, the functional independence cannot be characterized by R alone, since we need to know the angular relaionship between the corresponding axes of DPs and FRs. Therefore, in addition to R, one more measure is required to characterize the functional independence fully. The measure for this correspondence among pairs of DP and FR variables is called semangularity, and may be written as

$$S = \prod_{j=1}^{n} \left(\frac{|A_{jj}|}{(\sum_{k=1}^{n} A_{kj}^2)^{1/2}}\right) \quad (4.15)$$

The Independence Axiom and its Implications 117

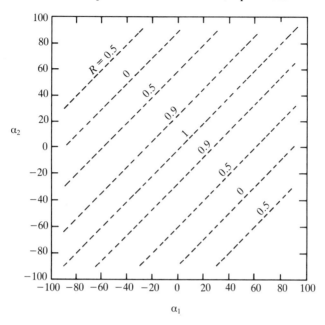

Figure 4.8. Contours of reangularity for 2-dimensional designs as a function of α_1 and α_2.

The denominator is a normalizing factor that is equal to unity if the columns of the design matrix are normalized. When the design matrix is normalized, semangularity is simply the product of the absolute values of the diagonal elements of the design matrix. If the design matrix is properly normalized, then the diagonals are unity when all off-diagonal elements are zero. When the semangularity is unity, the DP isograms are parallel to the coordinates in the function space, and the FR isograms are parallel to the coordinates of the design space. "Semangularity" is derived from Latin words meaning "same angle quality."

Contours of semangularity are shown as solid lines in Fig. 4.9 for the two-dimensional case. When $S = 1$ the design is uncoupled, provided that R is also equal to unity. When the design is in the shaded region, DPs are not related to proper FRs. The important point to note again is that the functional independence of FRs is known only when the angular relationship between the DPs and the alignment of DPs with their corresponding FRs are characterized. R measures the former, and S measures the latter.

Appendix 4D provides theorems on measures of functional independence and the modularity of independence measures. These theorems are useful in understanding the characteristics of the design matrix, but are not essential in following the main discussion in this chapter.

Example 4.3: Determination of the Reangularity and Semangularity
Consider the two designs represented by the two design matrices a and b. Determine the reangularity and the semangularity. What is the difference

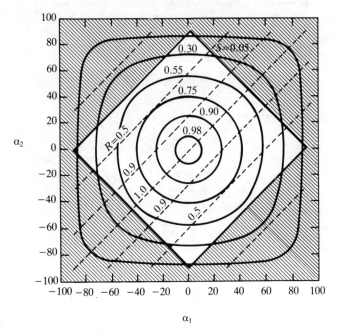

Figure 4.9. Contours of semangularity (———) and reangularity (– – –) as a function of α_1 and α_2. DPs have not been selected properly if the design is represented in the shaded region of the figure. (From Rinderle, 1982.)

between the uncoupled and coupled designs?

$$\begin{Bmatrix} FR_1 \\ FR_2 \end{Bmatrix} = \begin{bmatrix} 0.8 & -0.6 \\ 0.6 & 0.8 \end{bmatrix} \begin{Bmatrix} DP_1 \\ DP_2 \end{Bmatrix} \quad (a)$$

$$\begin{Bmatrix} FR_1 \\ FR_2 \end{Bmatrix} = \begin{bmatrix} 1 & 0 \\ 0 & 1 \end{bmatrix} \begin{Bmatrix} DP_1 \\ DP_2 \end{Bmatrix} \quad (b)$$

Solution

From the definition of reangularity given by Eq. 4.14, we can compute R for both Case a and Case b. The results show that R is unity in both cases. That is, the DPs are orthogonal to each other when plotted in function space. The semangularity can be computed by using Eq. 4.15. For the coupled case $S = 0.64$, whereas for the uncoupled case $S = 1$. It can therefore be seen that, for the uncoupled case, R and S are both unity. The coordinate axes of DPs for the coupled case are not parallel to FR-axes.

As this example illustrates, semangularity and reangularity are useful measures for determining the degree of coupling between FRs due to the particular set of DPs chosen as a result of the design process. Although the definitions of R and S given by Eqs. 4.14 and 4.15 are useful, they are not necessarily unique. One can conceive other measures which may be equally effective, since other descriptors may define the angular relationship among DPs and FRs equally well. Appendix 4C presents another set of measures which may also provide a useful metric coupling. However, these measures are not fully normalized, and are therefore less useful than those presented in this section.

4.6 The Effect of Scale of DPs and FRs on Measures of Coupling

Semangularity and reangularity, as defined by Eqs. 4.14 and 4.15, are insensitive to the scale (i.e., dimensional units) of DPs, since the columns of the design matrix are normalized in the expressions for R and S. These measures can also be made insensitive to the scale of FRs, in addition to the scale of DPs, by using dimensionless expressions for each FR. This may be done by dividing the equation for each FR_i by the value of the specified FR, FR_{i0}. In this way both R and S can be made independent of the scales used for the FR and DP variables.

R and S must be insensitive to the scale of DPs and FRs for two important reasons. First, changing the metric of one of the DPs (e.g., from kilograms to pounds) should not influence the measures S and R, since the design has not changed. Secondly, the coupling in design depends on the relative change in the individual FRs and on the relationship among FRs when a given DP is changed.

The measures of coupling are not sensitive to an absolute scale or tolerance of the FRs, only to the relative scale. Therefore, a proportionate change in scale of all the FRs does not affect the measures of coupling. When the design matrix is a diagonal matrix (i.e., the design is an uncoupled design), it is not necessary to be concerned about the scale of FRs or DPs, because the result of the analysis is so obvious. In this case the diagonal elements can be scaled to be unity, and the coupling measures are equal to unity.

Example 4.4: The degree of Coupling of Passive Filter Networks*

For the filter circuit design considered in Example 4.2, analyze the problem further to see how the measures of coupling in a design can be computed. Derive the functional requirements of the filter, cutoff frequency, and D.C. gain as nonlinear functions of the design parameters; that is, of capacitance and resistance. After suitable scales for the FRs have been adopted, compute the design matrix as a function of the DPs, by determining the rate of change of each FR with respect to each DP.

Solution

If we denote the filter cutoff frequency by ω_c, the d.c. beam light deflection by D, the capacitance by C, and the resistance by R, the elements of the

---
*Adapted from Rinderle (1982).

design matrix may be written as

$$A_{11} = \left(\frac{\partial \omega_c}{\partial c}\right) \omega_c \qquad (a)$$

$$A_{12} = \left(\frac{\partial \omega_c}{\partial R}\right) \bigg/ \omega_c \qquad (b)$$

$$A_{21} = \left(\frac{\partial D}{\partial C}\right) \bigg/ D \qquad (c)$$

$$A_{22} = \left(\frac{\partial D}{\partial R}\right) \bigg/ D \qquad (d)$$

The elements of the design matrix are given in Example 4.2; viz. Eq. e for Network a and Eq. f for Network b. These two equations can also be simplified, by a change in DP scale, as

For Network a

$$[\mathbf{DM}] = \begin{bmatrix} 1 & \dfrac{R_s}{R_g + R_2} \\ 0 & 1 \end{bmatrix} \qquad (e)$$

For Network b

$$[\mathbf{DM}] = \begin{bmatrix} 1 & -1 \\ 0 & 1 \end{bmatrix} \qquad (f)$$

Table a shows the computed values of the reangularity and semangularity using Eqs. 4.14 and 4.15. It also shows the component values. Note that S is equal to R because the design is a decoupled design. An examination of Table a enables the designer to choose the minimally coupled design, which in this case is Network a, with an LVDT and galvanometer. In many cases the designer will be able to identify the minimally coupled design without carrying out the detailed computations.

The values of S and R for a given design vary throughout the design space, and the minimally coupled network may therefore depend on the specific values of the components. In the case of Design a, the coupling of the passive filter depends only on FR_2, the beam deflection specification (see Eq.

TABLE a Measures of Coupling and Component Values for Passive Filter Studies

Network		LVDT and Galvanometer	Strain Gage and Galvanometer
Design a	S and R	1.00	0.9821
	$R_2\,(\Omega)$	222 k	527.2
	$C\,(\mu F)$	12.44	227.9
Design b	S and R	0.7071	0.7071
	$R_3\,(\Omega)$	0.823	22.31
	$C\,(\mu F)$	28.135	1,454

Note: Reangularity and semangularity have equal magnitudes because Designs a and b are decoupled designs.

The Independence Axiom and its Implications 121

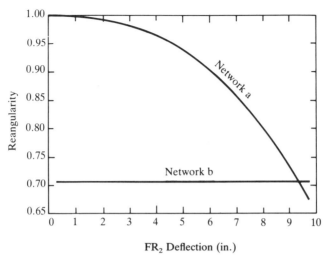

Figure a. Reangularity and semangularity for two network configurations matching the strain-gage displacement transducer and the type 348 galvanometer. (From Rinderle, 1982.)

e). The reangularity (and semangularity, since R and S are equal to each other) of Networks a and b are plotted as a function of the required beam deflection in Fig. a for the case of using a strain gage and a galvanometer. Network a is less coupled than Network b is over the allowable deflection range, but, when the deflection is large, the converse is true.

4.7 Functional Coupling and Tolerancing on FRs

Whether a given design is coupled or not depends on the tolerance imposed on the FRs. For example, the vertically hung refrigerator door (see Fig. 1.2) was considered to be coupled, since the cold air lost when the door is opened to remove food is assumed to be unacceptable, whereas the horizontally hung freezer door was deemed to be acceptable, since the small volume of cold air lost is tolerable. The concept of functional coupling is closely tied to the allowable tolerance associated with FRs. This is in turn related to the coupling measures S and R.

Let us again consider the following two-dimensional case

$$\begin{Bmatrix} FR_1 \\ FR_2 \end{Bmatrix} = \begin{bmatrix} A_{11} & A_{12} \\ A_{21} & A_{22} \end{bmatrix} \begin{Bmatrix} DP_1 \\ DP_2 \end{Bmatrix} \tag{4.16}$$

where A_{11}, A_{12}, A_{21}, and A_{22} are normalized elements of the design matrix. At a given design point, the change in FRs due to a change in DPs may be written as

$$\begin{Bmatrix} \Delta FR_1 \\ \Delta FR_2 \end{Bmatrix} = \begin{bmatrix} A_{11} & A_{12} \\ A_{21} & A_{22} \end{bmatrix} \begin{Bmatrix} \Delta DP_1 \\ \Delta DP_2 \end{Bmatrix} \tag{4.17}$$

That is, the system behavior is such that the changes in FR_1 and FR_2, when the changes in DPs are given, may be written as

$$\Delta FR_1 = \frac{\partial FR_1}{\partial DP_1} \Delta DP_1 + \frac{\partial FR_1}{\partial DP_2} \Delta DP_2 \qquad (4.18)$$

$$\Delta FR_2 = \frac{\partial FR_2}{\partial DP_1} \Delta DP_1 + \frac{\partial FR_2}{\partial DP_2} \Delta DP_2 \qquad (4.19)$$

From the designer's point of view, we must make sure that ΔFR_1 and ΔFR_2 are within the designer-specified tolerance; that is

$$\begin{aligned} \Delta FR_1 &= (\Delta FR_1)_0 \pm \delta FR_1 \\ \Delta FR_2 &= (\Delta FR_2)_0 \pm \delta FR_2 \end{aligned} \qquad (4.20)$$

where $(\Delta FR_i)_0$ is the desired change in FR_i due to the change in DPs, and δFR_i is the tolerance specified by the designer, respectively. In an uncoupled process, $(\Delta FR_1)_0$ must be equal to the first term of the right-hand side of Eq. 4.18. The deviation from this ideal condition is given by the second term of the right-hand side of Eq. 4.18. To be an uncoupled system, the designer-specified tolerance δFR_1 must be equal to or larger than the second term of the right-hand side of Eq. 4.18; that is,

$$\delta FR_1 \geq \frac{\partial FR_1}{\partial DP_2} \Delta DP_2 \qquad (4.21)$$

In the case of n design parameters, Eq. (4.20) may be generalized as

$$\delta FR_i \geq \sum_{\substack{j \neq i \\ j=1}}^{n} \frac{\partial FR_i}{\partial DP_j} \Delta DP_j \qquad (4.22)$$

Equation 4.22 specifies when the off-diagonal elements of the basic matrix may be neglected. Based on the foregoing discussion of designer-specified tolerance, Theorem 8 may be stated as follows:

> Theorem 8 (Independence and Tolerance)
> A design is an uncoupled design when the designer-specified tolerance is greater than
>
> $$\left(\sum_{\substack{j \neq i \\ j=1}}^{n} (\partial FR_i / \partial DP_j) \Delta DP_j \right)$$
>
> so that the nondiagonal elements of the design matrix can be neglected from design consideration.

Sometimes the tolerances for FRs are not specifically stated. In this case FRs are generally assumed to be satisfied when the actual values are within a certain multiple of the standard deviation of the specified FRs. The standard deviation σ is defined as

$$\sigma = \frac{1}{n-1}\left(\sum_{p=1}^{n}(\text{FR}_j^p - \overline{\text{FR}})^2\right)^{1/2} \qquad (4.23)$$

where n is the number of times FR_j is measured, $\overline{\text{FR}}$ is the average value of the n FR_j values measured, and FR_j^p is the pth measured value of FR_j. When $\overline{\text{FR}}$ is equal to the specified value $(\text{FR}_j)_0$, and if all the FRs that are within 3σ are acceptable, then for 99.7% of the time the design specification is satisfied by the design solution.

Since the concept of information content is closely related to probability, and since the probability is a function of tolerance, tolerance and information content must be interrelated. In Chapter 5 we show how the concept of tolerance is related to the information content of a design.

4.8 On Selection of DPs and Sensitivity Analysis

In Sec. 3.6 the use of the Independence Axiom in the design process is discussed. In searching for a new design solution, we must conceive a design that will yield a diagonal or triangular design matrix. Therefore, once the FRs are known, we have to seek a DP that can satisfy a given FR without affecting other FRs. In this section we expand on this search process, since we have defined the coupling measures, reangularity R and semangularity S, and discussed the relationship between tolerance and functional coupling in Secs. 4.5 and 4.6, respectively.

The search for an uncoupled design that has negligible off-diagonal elements, or a decoupled design with a triangular matrix, is analogous to performing a *sensitivity analysis*. This situation arises from the fact that the designer is seeking parameters that have minimal effects on FRs other than the specific ones that must be controlled (or satisfied). This can be done by analyzing the magnitude of each element of the design matrix as DPs are varied over an extreme range of permissible values. However, this process can be further generalized and systematized by using reangularity and semangularity.

For the purpose of generalization, let us assume that a set of FRs and constraints is given. Then the design problem reduces down to finding specific values of DPs that will satisfy FRs and constraints, maximize R and S, and at the same time have the least effect on R and S. In terms of mathematical expressions, the search process for specific DPs may be expressed as:

Given the design equation relating FRs to DPs:

$$\{\mathbf{FR}\} = [\mathbf{A}]\{\mathbf{DP}\}$$

find specific values of DPs subject to the following conditions:

(a) $\dfrac{\partial^2 R}{\partial DP_j^2} < 0 \quad \text{for } j = 1, \ldots, n$

(b) $\dfrac{\partial^2 S}{\partial DP_j^2} < 0 \quad \text{for } j = 1, \ldots, n$

(c) $\dfrac{\partial R}{\partial DP_j} \approx 0 \quad \text{for } j = 1, \ldots, n$ \hfill (4.24)

(d) $\dfrac{\partial S}{\partial DP_j} \approx 0 \quad \text{for } j = 1, \ldots, n$

(e) other design constraints are met

(In the classical treatment of optimization techniques, these relationships are predicated on a convex structure for R and S. Otherwise, difficulties may creep in; for example, points of local as well as global optimality will satisfy the equation.)

If the design is linear and satisfies Axiom 1, then these conditions are satisfied automatically. However, in the case of coupled designs there may be no design solutions that satisfy the above conditions, because there may be no local or global maxima for R and S. In this case one should seek a new design solution or compromise the design criteria specified in terms of the FRs, their tolerances, and constraints.

4.9 Experimental Determination of Principal DPs and the Role of Statistical Analysis

There are two situations where experimental techniques combined with statistical analysis must be used to analyze the design using the design axioms. First, unlike the cases treated in the two preceding sections, reliable analytical relationships between FRs and DPs may not be available, although FRs and DPs are known, and therefore the elements of the design matrix can only be determined experimentally. The second situation arises when the result of the design process in the form of a complex hardware or a complex manufacturing process exists for which the principal DPs have not been identified and the relationships among FRs and DPs are not known. In this case the principal DPs that control the FRs must be determined empirically from a large set of potential DPs.

In these situations the design axioms can be used to establish the theoretical framework for various specific methodologies that can be used to analyze the design. Some approaches are described in this section.

Experimental Determination of the Design Matrix and the Principal DPs through a Statistical Approach

In some cases the principal DPs that control a given set of FRs may not be known. In this case they have to be selected from a large number of potential DPs. In order to select the principal DPs that control the FRs, the variance σ^2 of a given FR_i can be determined by varying a DP. If we let DP_i assume a set of n different values around a set $(DP_i)_0$, the resultant FR_i will also consist of a set of n FR_i values. The variance of these n FR values is given by

$$\sigma^2 = \frac{1}{n-1} \sum_{p=1}^{n} (FR_i^p - \overline{FR_i})^2 \qquad (4.23)$$

where FR_i^p is the pth individual value of FR_i corresponding to the pth value of DP_i.

The mean value of the n FR_i^ps is denoted by $\overline{FR_i}$, which may be represented as

$$\overline{FR_i} = m = \frac{1}{n} \sum_{p=1}^{n} FR_i^p \qquad (4.25)$$

Taguchi (1987) defines the signal-to-noise SN, ratio, in terms of the variance and the mean of FR_i as (Taguchi and Phadke, 1984)

$$SN = 10 \log_{10} (m^2/\sigma^2) \qquad (4.26)$$

This definition of the SN ratio is adopted so that the unit of the measure is in decibels (dB). The DP that gives the largest value of the SN ratio has the largest effect on the FR, and hence is the principal DP. Through the evaluation of the SN ratio values corresponding to *all* potential DPs, the principal DPs corresponding to the given set of FRs can be determined. By Theorem 4, the number of principal DPs must be equal to the number of the FRs. (The Taguchi method is discussed in detail in Sec. 5.8).

Once the principal DPs are chosen, the magnitude of the diagonal elements of the design matrix (e.g., $A_{11}, A_{22}, \ldots, A_{nn}$) can be determined from the principal values that were generated to determine SN ratio values statistically. Similarly, the magnitude of the nondiagonal elements $(A_{12}, A_{13}, A_{21}, \ldots, A_{pq})$ can also be determined from the experimental values.

Experimental Determination of the Design Matrix through the Planned Experimental Evaluation of the Design Elements

In many situations the principal DPs corresponding to a set of FRs are known when an equal number of DPs are chosen to satisfy a given set of FRs per Theorem 4. In this case the magnitude of the diagonal elements of the design matrix can readily be found, since the DPs are chosen to satisfy the specific FRs. However, the determination of nondiagonal elements may not be straightforward when the physics of the causality is not clearly known. These off-diagonal elements can be determined experimentally by performing a set of critical experiments to clarify the underlying physics of the FR–DP relationship. Such an approach is illustrated in Sec. 6.6.

Establishment of the Optimum Operating Range of a Given Design

In nonlinear design the "design window" in which the principal DPs can effectively control the FRs may exist. However, outside of the "design window," the SN ratio may be small. Therefore, it is important to measure the functional independence by determining the reangularity R and semangularity S. This can be done by substituting the experimentally determined values of the elements of the design matrix, as discussed earlier.

In order to search for the design window, we have to evaluate the slopes and curvatures of S and R by systematically varying the values of the principal DPs. The criteria given by Eq. 4.24 can then be used to determine the subsequent values of DPs that can lead toward the optimum design point.

Elimination of the Bias b

The behavior of the product or the manufacturing process may be such that the mean of the FRs obtained by varying the principal DPs can be different from the desired value of $(FR_i)_0$. The difference between the mean m and the desired value is called the bias, which may be written as

$$b = (FR_i)_0 - \overline{FR} = (FR_i)_0 - m \tag{4.27}$$

In order to satisfy the FRs, we must reduce the bias b to zero. This can be done by changing either the system range so as to reduce the information content. This is discussed further in Sec. 5.7.

We can develop many specific problem-solving methodologies for various applications based on the two design axioms. All of these methodologies are acceptable as long as they do not violate the Independence Axiom and the Information Axiom. In Sec. 5.7 one such method, the Taguchi method, is illustrated by solving the passive filter network problem discussed in Examples 4.2–4.4.

4.10 Design for Manufacturability

In Example 2.1 the design of a new material structure called microcellular plastic is discussed; the structure was created to reduce materials consumption without adversely affecting the mechanical properties of the plastic. Once the material was designed, we had to design the manufacturing process that could make the material. The problem of designing a product that can be manufactured is a very important industrial concern. In this section the necessary conditions for *design for manufacturability* (or producibility) are developed, based on the Independence Axiom (Axiom 1) (Suh, 1988).

The issue of design for producibility is important both in mass-production and in batch-production. In mass-production the cost of manufacturing and the reliability of the product is amplified by the decisions made at the design stage, because of the large volume of identical goods produced in manufacturing. In small-lot batch-production the cost of design cannot be amortized over a large number of products, so the room for error is therefore small and subsequent corrective measures cannot easily be implemented, especially when the unit cost is high. It is generally believed that the decisions made at the stage of product design determine 70–80% of manufacturing productivity in all production environments, a fact that underscores the importance of design for producibility.

The central question is: "However do we design the product and its components so that they can be manufactured in the most efficient manner, regardless of the specific nature of the product to be made?" There are obviously many facets to this question. We need to understand them, explore rational approaches to dealing with them, and establish the basic foundation on which to generalize and build a scientific framework for design and manufacturing.

Some simple designs of discrete mechanical parts that can be readily manufactured, and others that cannot, are shown in Fig. 4.10 (Milacic, 1988). Given the limitations of existing machine tools, the parts shown at the top of the figure are difficult to make, whereas those at the bottom can

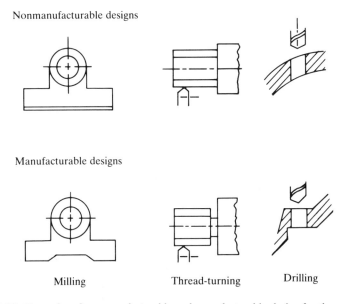

Figure 4.10. Examples of nonmanufacturable and manufacturable design for three machining operations. (From Milacic, 1988.)

be readily manufactured. Therefore, knowing what can be made using existing machine tools, one can come up with a better design. Obviously, one way of dealing with design for manufacturability is to develop a handbook of rules for various situations. These handbooks may be very useful for specific situations, but a more basic approach would be to establish principles for design for producibility which are applicable in all design/manufacturing tasks. The "handbook approach" is a specialized approach, in contrast to the general axiomatic approach discussed in this book.

A schematic representation of the relationship among the FRs of the product, DPs of the product, and the process variables (PVs) of the manufacturing operations is illustrated in Fig. 4.11. It shows that there is a relationship between the FRs of the functional domain and the DPs of the physical domain, between the DPs of the physical domain and PVs of the process domain, and thus between the FRs of the functional domain and PVs of the process domain. Therefore, if we understand the criteria for good design that these relationships must satisfy, then we will be able to develop designs that are producible and thereby enhance productivity.

In making a product, there are two design steps involved. First we have to design the product to satisfy the FRs by developing an acceptable set of DPs. These DPs of the product become the FRs of the manufacturing process. Therefore, the second design step it to satisfy the DPs of the product (the FRs of the manufacturing process) by developing PVs that satisfy the DPs and the Independence Axiom.

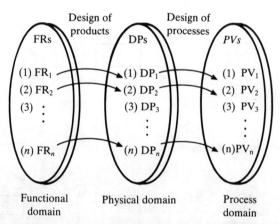

Figure 4.11. In the design/manufacturing world there are three domains: functional domain, physical domain, and process domain. Each of these domains is defined by multivariables or multiparameters. During the product design phase the FRs specified in the functional domain must be satisfied by choosing a proper set of DPs, whereas during the process design phase, the DPs in the physical domain must be satisfied by selecting an optimum set of PVs. Design for producibility requires the optimization of the relationships among the functional, physical, and process domains.

The design equation for the product design may be written as

$$\{FR\} = [A]\{DP\}$$

or

$$FR_i = \sum_j A_{ij} DP_j \qquad (3.1)$$

In view of Theorem 4, [A] is assumed to be a square matrix. The design equation for the process design is similarly given as

$$\{DP\} = [B]\{PV\}$$

or

$$DP_j = \sum_k B_{jk} PV_k \qquad (4.28)$$

The two matrix equations

$$[FR] = [A][DP] \qquad (3.1)$$
$$[DP] = [B][PV] \qquad (4.28)$$

can be combined into a single relationship linking the product requirements to PVs. The result is

$$[FR] = [C][PV]$$

where

$$[C] = [A][B] \qquad (4.29)$$

and each element of the resultant matrix is given by

$$C_{ik} = \sum_j A_{ij} B_{jk} \qquad (4.30)$$

To satisfy the Independence Axiom, the [C] matrix must also be diagonal or triangular.

The preceding equations yield the following relationship for each FR of the product in terms of PVs of the manufacturing system:

$$FR_i = \sum_k C_{ik} PV_k$$
$$= \sum_j \sum_k A_{ij} B_{jk} PV_k \qquad (4.31)$$

When [C] in Eq. 4.30 is neither a diagonal nor a triangular matrix, either [A] or [B] must be modified by changing the design of the product or the design of the process. For the purpose of discussion, let us consider the two-dimensional case represented by the following design matrices for the product and the process.

The *Design Matrix for the Product*

$$[A] = \begin{bmatrix} A_{11} & A_{12} \\ A_{21} & A_{22} \end{bmatrix} \qquad (4.32)$$

The *Design Matrix for the Process*

$$[\mathbf{B}] = \begin{bmatrix} B_{11} & B_{12} \\ B_{21} & B_{22} \end{bmatrix} \quad (4.33)$$

Then C_{ik} is given by

$$[\mathbf{C}] = \begin{bmatrix} A_{11}B_{11} + A_{12}B_{21} & A_{11}B_{12} + A_{12}B_{22} \\ A_{21}B_{11} + A_{22}B_{21} & A_{21}B_{12} + A_{22}B_{22} \end{bmatrix} \quad (4.34)$$

The design for producibility requires that the design/manufacturing matrix [C] satisfy one of the following six conditions:

$$\begin{aligned}
&\text{Condition 1: } A_{12} = B_{12} = 0 \\
&\text{Condition 2: } A_{21} = B_{21} = 0 \\
&\text{Condition 3: } A_{12} = B_{12} = A_{21} = B_{21} = 0 \\
&\text{Condition 4: } A_{11}B_{12} = -A_{12}B_{22} \quad\quad\quad (4.35) \\
&\text{Condition 5: } A_{21}B_{11} = -A_{22}B_{21} \\
&\text{Condition 6: } A_{11}B_{12} = -A_{12}B_{22} \\
&\quad\quad\quad\quad\quad A_{21}B_{11} = -A_{22}B_{21}
\end{aligned}$$

Conditions 1 and 2 require that both the product and process designs be decoupled (i.e., that the design matrices be triangular). Then the resulting design/manufacturing system also becomes a decoupled design. Therefore, the design of the product is producible if a certain sequence of manufacturing operations is followed. Condition 3 requires both the product and the process designs to be uncoupled designs, in which the design/manufacturing system is also an uncoupled design. Therefore, the design of products and processes that satisfy Condition 3 can always be produced. Conditions 4 and 5 require special relationships between some of the elements of the product and the process design matrices, in order to obtain a decoupled design/manufacturing design system. These conditions may be difficult to satisfy in practice. Similarly, Condition 6 specifies the relationship for an uncoupled design/manufacturing design system. Except for special circumstances (e.g., the mass of a paperweight being determined solely by the production process that changes only its height), it is not clear how Condition 6 can be satisfied in practice.

According to Eq. 4.34, the design/manufacturing situation represents a coupled design even when the FR–DP relationship is an uncoupled design, if the DP–PV relationship is coupled. Similarly, the FR–DP relationship must not be a coupled one even when the DP–PV relationship is an uncoupled one; that is, each component of the FR–DP–PV relationship must be noncoupled if the overall relationship is to fulfill the Independence Axiom.

The derivation leading to Eq. 4.35 and the discussions given in this section serve as the basis for Theorem 9, the *Design for Manufacturability Theorem*, which may be stated as follows.

Theorem 9 (Design for Manufacturability)
For a product to be manufacturable, the design matrix for the product, [**A**] (which relates the **FR** vector for the product to the **DP** vector of the product) times the design matrix for the manufacturing process, [**B**] (which relates the **DP** vector to the **PV** vector of the manufacturing process) must yield either a diagonal or triangular matrix. Consequently, when any one of these design matrices, that is, either [**A**] or [**B**], represents a coupled design, the product cannot be manufactured.

4.11 Summary

In Chapter 4 the first axiom and the concept of functional coupling are discussed in great detail. The uncoupled and decoupled designs are shown to satisfy the Independence Axiom, whereas the coupled design is shown to violate it. The physical implications of the Independence Axiom are explained by using a graphical representation of the two-dimensional case.

In order to evaluate the extent of coupling for multidimensional cases, two coupling measures—reangularity and semangularity—are defined; these measure, respectively, the orthogonality among design parameters, and the relationship between DPs and FRs. Finally, the tolerance on FRs is shown to be a critical factor in determining whether a design satisfies or violates the Independence Axiom. Theorem 7 specifies when a design is an uncoupled design in terms of the designer-specified tolerance, which is stated as

Theorem 8 (Independence and Tolerance)
A design is an uncoupled design when the designer-specified tolerance is greater than

$$\left(\sum_{\substack{j \neq i \\ j=1}}^{n} (\partial FR_i / \partial DP_j)\, \Delta DP_j \right)$$

so that the nondiagonal elements of the design matrix can be neglected from design consideration.

This chapter also discusses how the design matrix can be determined through experiments and through the use of statistical analysis of the system (or product or process) performance. The notion of the SN ratio is discussed, as well as the mean and the variance of the system performance.

The *design for producibility* is analyzed in terms of Axiom 1. It is shown that when the design of either the product or the process is coupled, the part cannot be manufactured. This is stated in the form of a theorem as follows:

> Theorem 9 (Design for Manufacturability)
> For a product to be manufacturable, the design matrix for the product, [A] (which relates the **FR** vector for the product to the **DP** vector of the product) times the design matrix for the manufacturing process, [B] (which relates the **DP** vector to the **PV** vector of the manufacturing process) must yield either a diagonal or triangular matrix. Consequently, when any one of these design matrices, that is, either [A] or [B], represents a coupled design, the product cannot be manufactured.

Two additional theorems and Corollary 8 are presented in Appendix 4D, "Theorems on Measures of Functional Independence," and are summarized here.

> Theorem 10 (Modularity of Independence Measures)
> Suppose that a design matrix [DM] can be partitioned into square submatrices that are nonzero only along the main diagonal. Then the reangularity and semangularity for [DM] are equal to the products of the corresponding measures for each of the submatrices.
>
> Theorem 11 (Invariance)
> Reangularity and semangularity for a design matrix are invariant under alternative orderings of the FR and DP variables, as long as the orderings preserve the association of each FR with its corresponding DP.
>
> Corollary 8 (Effective Reangularity for a Scalar)
> The effective reangularity R for a scalar design "matrix" or element is unity.

References

Hildebrand, F.B., *Methods of Applied Mathematics*. Prentice-Hall, Englewood Cliffs, NJ, 1952.

Milacic, V.R., "Fundamental Principles for Design and Producibility," Unpublished Report, University of Beograd, 1988.

Rinderle, J.R., "Measures of Functional Coupling in Design," Ph.D. Thesis, MIT, 1982.

Rinderle, J.R., and Suh, N.P., "Measures of Functional Coupling in Design," *Transactions of A.S.M.E./Journal of Engineering for Industry* **104**(4):383–388, 1982.

Suh, N.P., "Basic Concepts in Design for Producibility," *CIRP Annals* **37**(2):559–568, 1988.

Suh, N.P., and Rinderle, J.R., "Qualitative and Quantitative Use of Design and Manufacturing Axioms," *CIRP Annals* **31**(1):333–338, 1982.

Taguchi, G., *System of Experimental Design: Engineering Methods to Optimize Quality and Minimize Cost.* American Supply Institute, 1987.

Taguchi, G., and Phadke, M.S., "Quality Engineering Through Design Optimization," *Proceedings of I.E.E.E. GLOBECOM-84 Conference,* Atlanta, GA, 1106–1113, 1984.

Problems

4.1. Plot Fig. a of Example 4.4.

4.2. Consider the network configuration shown in the figure. Determine its reangularity and semangularity. Is this design better or worse than Network a, given in Example 4.2? All of the characteristics given for the displacement transducer and galvanometer apply to this problem.

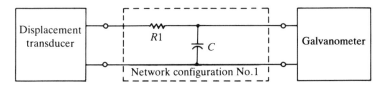

4.3. For the commercial RIM machine shown in Fig. 3.5, specify the tolerance on the metering ratio that will make the design an uncoupled design.

4.4. Determine the reangularity and semangularity of the gear pump-assisted extruder discussed in Chapter 3 (see Appendix 3B and Example 3.2).

4.5. Specify the tolerance on leakage that will always make the gear pump-assisted extruder an uncoupled or decoupled design.

4.6. Specify the conditions for *design for manufacturability* when there are three FRs.

4.7. The principal DP that controls a given FR varies randomly by $\pm \delta(DP)$ about its set point. Determine its effect on the FR.

4.8. Liquid metals are quenched on a rapidly rotating cold metal cylinder to make thin films with nonequilibrium structures. A typical arrangement is shown in the figure.

The FRs of the design are

FR_1 = Cooling rate for maximum production.

FR_2 = Nonsticking condition.

FR_3 = Wettability.

Wettability of the liquid metal is deemed to be important for effective heat transfer until the liquid metal solidifies. Once the metal is solidified, the film should separate from the wheel, due to the centrifugal force. However, if the

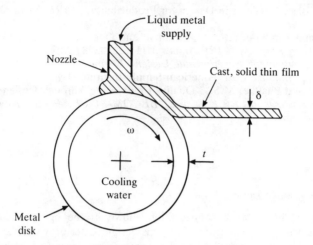

surface temperature of the wheel is higher than a critical temperature, the film sticks to the surface, creating an unacceptable situation.

Develop a design (through optimization of the design shown or by conceptualizing a new design) and justify your design through an appropriate analysis.

4.9. Derive Eq. 4.5.

4.10. In examples 4.2 and 4.4 we deal with only two FRs; that is, the cutoff frequency and the d.c. gain. With the decoupled circuit given by Network a of Example 4.2, it was found that the system was underclamped. Design a network that will provide 64% critical damping of the galvanometer, in addition to satisfying the original two FRs. Determine the design matrix and values of R and S.

4.11. The following design equation describes a coupled design:

$$\begin{Bmatrix} FR_1 \\ FR_2 \\ FR_3 \end{Bmatrix} = \begin{bmatrix} A_{11} & A_{12} & A_{13} \\ A_{21} & A_{22} & A_{33} \\ A_{31} & A_{32} & A_{33} \end{bmatrix} \begin{Bmatrix} DP_1 \\ DP_2 \\ DP_3 \end{Bmatrix}$$

where

$FR_1 = 10$ lb; $FR_2 = 36$ in.; $FR_3 = 45°F$
$DP_1 = 2$ in.; $DP_2 = 3$ cal; $DP_3 = 1$ in/s
$A_{11} = 2$ lb/in.; $A_{12} = 1.5$ lb/cal; $A_{13} = 1.5$ lb s/in.
$A_{21} = 4$; $A_{22} = 6$ cal^{-1}; $A_{23} = 10$ s
$A_{31} = 5°F/$in.; $A_{32} = 5°F/$Cal; $A_{33} = 20°F$ s/in.

(a) Normalize the design equation. (b) Determine R and S.

4A: DERIVATION OF THE DESIGN MATRIX FOR THE PASSIVE FILTERS

In Example 4.2 the design of a passive filter, that can provide a high-fidelity reproduction of the input signal created by a transducer such

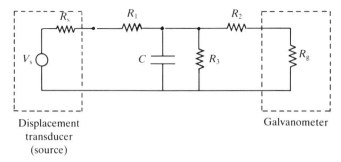

Displacement transducer (source)   Galvanometer

as a strain gage or LVDT, is considered. Two different kinds of passive filters are considered (shown in Example 4.2, Figs a and b). In this Appendix the elements of the design matrix, Eqs. e and f of Example 4.2, are derived with algebraic help from computer software MACSYMA (Rinderle, 1987).

Instead of analyzing the circuits shown in Figs a and b separately, a general displacement solution for the network shown above is obtained, from which the specific solutions for the circuit shown in Fig. a and Fig. b can be derived. Network a is obtained by letting $R_3 \to \infty$ and $R_1 = 0$. For Network b, $R_1 = R_2 = 0$. The voltage at the galvanometer can be obtained using equivalent resistances and the simple voltage-divider relation.

From the Kirchhoff rule the output voltage V_0 can be expressed as a function of the input voltage V_s and the components of electric circuit as

$$V_0 = V_s \frac{R_3 R_g}{(CsR^{III} + R^{II})} \quad (4A.1)$$

where
$$R^{III} = R_3 R_g R_s + R_2 R_3 R_s + R_1 R_3 R_g + R_1 R_2 R_3$$
$$R^{II} = R_g R_s + R_3 R_s + R_2 R_s + R_3 R_g + R_1 R_g + R_2 R_3 + R_1 R_3 + R_1 R_2$$
$s =$ Laplace variable

The galvanometer deflection of the network can be expressed as

$$D = \frac{V_0}{G_{sen}} = \frac{R_3 R_g V_{in}}{G_{sen}} \left(\frac{1}{CsR^{III} + R^{II}} \right) \quad (4A.1a)$$

We obtain the d.c. deflection (FR$_2$) by letting $s = 0$. The d.c. deflections for Network a and b are given by

Network a
$$D_a = \frac{R_g V_{in}}{G_{sen}(R_s + R_g + R_2)} \quad (4A.2)$$

Network b
$$D_b = \frac{R_3 R_g V_{in}}{G_{sen}[(R_g + R_3)R_s + R_3 R_g]} \quad (4A.3)$$

Eqs. 4A.2 and 4A.3 represent FR_2 for the networks. The elements of the design matrix are, in normalized form, $A_{ij} = (\partial FR_i/\partial DP_j)/FR_i$, so by differentiating Eqs. 4A.2 and 4A.3 with respect to R_2 and R_3, respectively, and then dividing by Eqs. 4A.2 and 4A.3, respectively, we obtain A_{22} for Networks a and b as

Network a

$$A_{22} = -\frac{1}{R_s + R_g + R_2} \qquad (4A.4)$$

Network b

$$A_{22} = \frac{R_g R_s}{(R_3 R_g + R_3^2)R_s + R_3^2 R_g} \qquad (4A.5)$$

The d.c. deflection does not depend on the capacitance C; therefore, $A_{21} = 0$ for both networks.

To obtain the first row of each design matrix, we first express the cutoff frequency in terms of the component values. The cutoff frequency ω_c for a first-order low-pass filter is the reciprocal of the time constant, τ; that is, $\tau = 2\pi/\omega_c$. The response of the first-order system is governed by an equation of the form

$$\tau \dot{x} + x = F_0 e^{j\omega_f t}/k. \qquad (4A.6)$$

where $F_0 \exp(j\omega_f t)$ is the external forcing function, ω_f is the frequency of the forcing function, and k is a spring constant. The response of such a system is

$$\frac{V_0}{V_{in}} = \frac{1}{\tau s + 1} \qquad (4A.6a)$$

The cutoff frequency for the general network, which is a first-order system, is

$$\omega_c = \frac{2\pi[(R_g + R_3 + R_2)R_s + (R_3 + R_1)R_g + (R_2 + R_1)R_3 + R_1 R_2]}{C[(R_3 R_g + R_2 R_3)R_s + R_1 R_3 R_g + R_1 R_2 R_3]} \qquad (4A.7)$$

The cutoff frequencies for Networks a and b are, respectively,

$$(\omega_c)_a = \frac{2\pi(R_s + R_g + R_2)}{C(R_g + R_2)R_s} \qquad (4A.8)$$

$$(\omega_c)_b = \frac{2\pi[(R_g + R_3)R_s + R_3 R_g]}{C R_3 R_g R_s} \qquad (4A.9)$$

By differentiating Eqs. 4A.8 and 4A.9 with respect to C and normalizing by dividing by $(\omega_c)_a$ and $(\omega_c)_b$, respectively, we obtain the A_{11} element of the design matrix, which is $-(1/C)$ for both networks.

Similarly, the A_{12} elements are obtained as

Network a

$$A_{12} = \frac{\partial(\omega_c)_a}{\partial R_2} \frac{1}{(\omega_c)_a} = \frac{R_s}{(R_g + R_2)R_s + R_g^2 + 2R_2R_g + R_2^2} \quad (4A.10)$$

Network b

$$A_{12} = \frac{\partial(\omega_c)_b}{\partial R_2} \frac{1}{(\omega_c)_b} = \frac{-R_gR_s}{(R_3R_g + R_3^2)R_s + R_3^2R_g} \quad (4A.11)$$

References

Rinderle, J.R., Personal Communication, 1987.

4B. SPECTRAL ANALYSIS OF THE DEMODULATED OUTPUTS AND FILTER POLE PLACEMENT

The relative magnitudes of original and carrier can be determined through a spectral analysis of the demodulator output when the displacement to be measured is a simple sinusoid (Rinderle, 1982). If the displacement, D, is (see Fig. c of Example 4.2)

$$D = A_s \sin(\omega_s t) \quad (4B.1)$$

where the subscript s denotes signal, then the transducer output, X, is the amplitude-modulated sine wave (see Fig. d of Example 4.2)

$$X = [A_s \sin(\omega_s t)][A_c \sin(\omega_c t)] \quad (4B.2)$$

where the subscript c denotes carrier. The demodulated output, Y, is (see Fig. e of Example 4.2)

$$Y = [A_s \sin(\omega_s t)] |A_c \sin(\omega_c t)| \quad (4B.3)$$

Y can be expanded into Fourier components by expanding the portion within the absolute value, carrying out the multiplication and then using a trigonometric identity to replace the (sin)(cos) terms, as

$$Y = \frac{2A_cA_s}{\pi}[\sin(\omega_s t) - \tfrac{1}{3}\sin(2\omega_c + \omega_s)t + \tfrac{1}{3}\sin(2\omega_c - \omega_s)t$$
$$- \tfrac{1}{15}\sin(4\omega_c + \omega_s)t + \tfrac{1}{15}\sin(4\omega_c - \omega_s)t + \ldots] \quad (4B.4)$$

If the low-pass filter can adequately attenuate the frequency component above $3\omega_c$, we may truncate the series given in Eq. 4B.4 to the first three terms. The second and third terms are at slightly different frequencies near $2\omega_c$, so they will superpose at the beat frequencies. When the two sidebands are 180 degrees out of phase, the carrier magnitude is 2/3 and the signal magnitude is unity. Therefore, in the worst case, the signal to carrier or SN ratio is 1.5.

The pole location of low-pass filters determines the degree to which a carrier frequency or high-frequency noise is suppressed, and the fidelity with which the signal is passed. The filter pole selection is a tradeoff between pass-band fidelity and reject-band suppression. The filter pole position can be selected in a rational manner by examining the errors introduced because of nonideal filtering.

If we elect to use a first-order low-pass filter, the single filter parameter is the cutoff frequency, ω_c. The magnitude response of a first-order low-pass filter is

$$\frac{|V_0|}{|V_{in}|} = \frac{1}{\sqrt{1 + (\omega/\omega_c)^2}} \qquad (4B.5)$$

where ω is the signal frequency and ω_c the filter cutoff frequency. To reject the carrier component (120 ± 2 Hz), we should choose ω_c to be small. However, as ω_c becomes smaller, the attenuation of the signal (0–2 Hz) increases. The relative amplitudes of the signal and carrier are given by Eq. 4B.4.

We can calculate the total error by considering both the signal attenuation and the carrier rejection, using Eq. 4B.5. The maximum error possible is

$$E_{max} = E_{pb} + 2A_n R_r \qquad (4B.6)$$

where E_{pb} is the maximum pass-band error in the 0–2 Hz frequency, A_n is the noise amplitude due to the carrier frequency in the 0–2 Hz range, and R_r is the rejection ratio of the noise frequency. If we divide by the signal magnitude to obtain the error ratio, E_{rat}, we obtain

$$E_{rat} = (1 - G_{bp}) + (2R_r/R_{sn}) \qquad (4B.7)$$

where G_{bp} is the minimum pass-band gain and R_{sn} is the signal-to-noise ratio. The SN ratio can be determined from the relative magnitudes of the signal and the carrier given in Eq. 4B.4. The pass-band gain and rejection ratio are then expressed in terms of the filter cutoff frequency, using Eq. 4B.5, as

$$G_{bp} = 1/[1 + (2 \text{ Hz}/\omega_c)^2]^{1/2}$$
$$R_r = 1/[1 + (120 \text{ Hz}/\omega_c)^2]^{1/2} \qquad (4B.8)$$

Substituting Eqs. 4B.8 into Eq. 4B.7, E_{rat} can be expressed in terms of the cutoff frequency:

$$E_{rat} = f(\omega_c) \qquad (4B.9)$$

By differentiating Eq. 4B.9 with respect to ω, or by numerical computation, we can determine that the minimum error ratio occurs when the curoff frequency is 6.84 Hz at an R_{sn} of 1.5. At this frequency the minimum error ratio is approximately 12%. A lower error could be obtained by using a higher-order filter.

References

Rinderle, J.R., "Measures of Functional Coupling in Design," Ph.D. Thesis, MIT, 1982.

4C: ANOTHER SET OF MEASURES FOR FUNCTIONAL INDEPENDENCE

The reangularity and semangularity are the most convenient set of metrics for coupling yet developed. However, there can be other measures for functional independence. One of those proposed earlier consists of orthogonality and alignment indices (Rinderle et al., 1981). They are defined as follows

4C.1 Orthogonality Index π_o

Each component of the FR vector may be expressed as

$$FR_1 = \mathbf{C}_1 \cdot \mathbf{DP}^T$$
$$FR_2 = \mathbf{C}_2 \cdot \mathbf{DP}^T$$
$$\vdots$$
$$FR_n = \mathbf{C}_n \cdot \mathbf{DP}^T$$

(4C.1)

where $\mathbf{C}_i$s and $\mathbf{DP}^T$ are vectors given by

$$\mathbf{C}_i = (A_{i1}, A_{i2}, \ldots, A)_{im}$$
$$\mathbf{DP}^T = (DP_1, DP_2, \ldots, DP_m)$$

the superscript T denotes vector transposition.

In terms of these vectors, the orthogonality index π_o, which measures the independence of FRs and the degree of coupling, can be defined as

$$\pi_o = 1 - \sum_{j=1}^{n} \sum_{k=1}^{j-1} \text{ABS}\left(\frac{\mathbf{C}_j \cdot \mathbf{C}_k}{\|\mathbf{C}_j\| \|\mathbf{C}_k\|}\right) \Big/ {}_nD_2$$

(4C.2)

where

ABS( ) $\equiv$ absolute value of ( ).
$\mathbf{C}_j \cdot \mathbf{C}_k \equiv$ the scalar product of vectors $\mathbf{C}_i$ and $\mathbf{C}_j$
$n \equiv$ number of FRs
$$\|\mathbf{C}_j\| = \left(\sum_{i=1}^{n} A_{ij}^2\right)^{1/2}$$

$${}_nD_2 = \frac{n!}{(n-2)!\,2!} \equiv \text{number of possible combinations of the FRs}$$

In the case of $n = 2$

$$\pi_o = 1 - \text{ABS}(\cos \theta)$$

(4C.3)

In the case of $n = 3$

$$\pi_o = 1 - \tfrac{1}{3}[\text{ABS}(\cos\theta_{12}) + \text{ABS}(\cos\theta_{23}) + \text{ABS}(\cos\theta_{13})] \quad (4\text{C}.4)$$

where θ_{ij} is the angle between $\mathbf{C}_i$ and $\mathbf{C}_j$. π_o is equal to unity when all of the C_is are mutually orthogonal.

4C.2 Alignment Index π_a

In order to measure the alignment of DPs to FRs, the alignment index π_a is defined as

$$\pi_a = \text{ABS}\left(\prod_{i=1}^{n} \frac{A_{ji}^2}{\|C_i\|}\right)^{1/2}$$

where A_{ij} are the elements of the design matrix. π_a is equal to the unity for the uncoupled system.

These measures are not used in this book, since they are not fully normalized and thus vary in the function space.

References

Suh, N.P., and Rinderle, J.R., "Qualitative and Quantitative Use of Design and Manufacturing Axioms," *CIRP Annals* **31**(1):333–338, 1982.

4D: THEOREMS ON MEASURES OF FUNCTIONAL INDEPENDENCE*

4D.1 Modularity of Independence Measures

For a design matrix [**DM**]† that can be partitioned into square submatrices which are nonzero only on the main diagonal, we can derive simple formulas for the reangularity and semangularity of [**DM**] in terms of those for the submatrices (Kim, 1985).

> Theorem 10 (Modularity of Independence Measures)
> Suppose that a design matrix [**DM**] can be partitioned into square submatrices that are nonzero only along the main diagonal. Then the reangularity and semangularity for [**DM**] are equal to the products of their corresponding measures for each of the nonzero submatrices.

---

* This appendix is from Appendix E of Professor Steve Kim's doctoral thesis (Kim, 1985).
† "Design matrix" and "coupling matrix" are used interchangeably throughout the book.

In other words, suppose that the $N \times N$ matrix C can be written in the form of a matrix as

$$C = \begin{array}{|c|c|c|c|} \hline C_1 & 0 & \ldots & 0 \\ \hline 0 & C_2 & \ldots & 0 \\ \hline \ldots & \ldots & \ldots & 0 \\ \hline 0 & 0 & 0 & C_k \\ \hline \end{array}$$

Here C_i is a square submatrix of order $N_i \times N_i$, and the 0s denote square submatrices composed of 0-values. Obviously, $N_1 + N_2 + \ldots + N_k = N$. For such a matrix C the reangularity R and the semangularity S are

$$R = R_1 * R_2 * \ldots * R_k$$
$$S = S_1 * S_2 * \ldots * S_k$$

where the R_i and S_i factors are the respective measures of reangularity and semangularity for submatrix C_i. The proof of the theorem follows.

Proof

Let $c(i)$ denote the ith column vector of the C matrix, and $c(i, j)$ the element of C corresponding to the ith row and the jth column. The reangularity R is defined as (Rinderle 1982; Rinderle and Suh 1982; Suh and Rinderle 1982)

$$R = \prod_{ij} [1 - A(i, j)]^{1/2} \tag{4D.1}$$

where the product is taken over the indices $i = 1, 2, \ldots, N-1$ and $j = i+1, \ldots, N$. Each $A(i, j)$ term denotes the square of the cosine of the angle formed by column vectors $c(i)$ and $c(j)$; that is,

$$A(i, j) = [c(i) * c(j) / \|c(i)\| * \|c(j)\|]^2 \tag{4D.2}$$

Here $c(i) * c(j)$ denotes the inner product of the ith and jth column vectors

$$c(i) * c(j) = c(1, i) * c(1, j) + c(2, i) * c(2j) + \ldots + c(N, i) * c(N, j) \tag{4D.3}$$

while $\|c(i)\|$ denotes the euclidean norm of the ith column

$$\|C(i)\| = [c(1, i)^2 + c(2, i)^2 + \ldots + c(N, i)^2]^{1/2} \tag{4D.4}$$

The singularity S is defined as

$$S = \prod_i |c(i, i)| / \|c(i)\| \tag{4D.5}$$

where the index $i = 1, \ldots, N$.

To calculate the reangularity for the coupling matrix C, we must calculate the inner products $c(i) * c(j)$ as the numerators of the $A(i, j)$ factors. If $c(i)$ and $c(j)$ correspond to the same submatrix, say, C_1, then the inner product is

$$c(1, i) * c(1, j) + \ldots + c(N_1, i) * c(N_1, j)$$

since the elements in rows $N_1 + 1$ and lower are all zero. However, this is the same expression as the inner product $\mathbf{c}(i) * \mathbf{c}(j)$ for calculating R_1 corresponding to submatrix $\mathbf{C}_1$.

Now suppose that $\mathbf{c}(i)$ and $\mathbf{c}(j)$ correspond to different submatrices, say $\mathbf{C}_1$ and $\mathbf{C}_2$. The zero elements in $\mathbf{c}(i)$ and $\mathbf{c}(j)$ are positioned in such a way that each term of the inner product is zero; that is,

$$\mathbf{c}(1, i) * \mathbf{c}(1, j) = \mathbf{c}(2, i) * \mathbf{c}(2, j) = \ldots = \mathbf{c}(N_1, i) * \mathbf{c}(N_2, j) = 0$$

Since $\mathbf{c}(i) * \mathbf{c}(j) = 0$, the corresponding angular measure is $A(i, j) = 0$.

From Eq. 4D.1 we see that this factor (within the brackets) equals unity, and therefore has no effect on the reangularity R. In other words, the inner products of columns vectors corresponding to different submatrices have no impact on R. Hence, the reangularity R for the whole matrix $\mathbf{C}$ can be written as the product of reangularities for the nonzero submatrices

$$R = R_1 * R_2 * \ldots * R_k$$

We can proceed analogously for the semangularity S. Consider the semangularity corresponding to any submatrix, say, S_1 for $\mathbf{C}_1$. S_1 is given by the product

$$S_1 = \mathbf{u}(1) * \mathbf{u}(2) * \ldots * \mathbf{u}(N_1)$$

where each $\mathbf{u}(i)$ is defined as

$$\mathbf{u}(i) = |\mathbf{c}(i, i)| / [\mathbf{c}(1, i)^2 + \ldots + \mathbf{c}(N_1, i)^2]^{1/2}$$

Since the elements $\mathbf{c}(N_1 + 1, i)$ and lower are all zero, the expression $\mathbf{u}(i)$ corresponding to the ith column of $\mathbf{C}_1$ is equal to the $\mathbf{u}(i)$ for the ith column of the full matrix $\mathbf{C}$.

A similar statement may be made for any column corresponding to any submatrix $\mathbf{C}_j$. Hence, the semangularity S for the whole matrix $\mathbf{C}$ equals the product of the semangularities for each of the submatrices:

$$S = S_1 * S_2 * \ldots * S_k$$

This completes the proof of the Modularity Theorem.

4D.2 Effective Reangularity for a Scalar Coupling Element

The indices in Eq. 4D.1 imply that at least two column vectors are required for the calculation of reangularity. The following corollary, however, shows that the effective reangularity may be defined in a simple way (Kim, 1985).

> Corollary 8 (Effective Reangularity for a Scalar)
> The effective reangularity R for a scalar coupling "matrix" or element is unity.

Proof

Consider the following design matrix **C**, where **D** is an $M \times M$ submatrix with $M > 1$, and d is a scalar "submatrix":

$$\mathbf{C} = \begin{vmatrix} \mathbf{D} & 0 \\ \hline 0 & d \end{vmatrix}$$

From the preceding Modularity Theorem, the rangularity for **C** is given by $R = R_D * R_d$, where R_D and R_d denote the reangularity measures for **D** and d, respectively.

Now consider the angular measure $A(i, M+1)$, defined in connection with Eq. 4D.1. Since the first M components of the last column are all zero, the value of $A(i, M+1)$ is zero. Hence, only the $A(i, j)$ corresponding to the first M factors contribute to R. In other words, $R = R_D$. By relating the last two equations for R, we obtain $R_d = 1$. Hence, the effective reangularity for a scalar coupling component is unity.

4D.3 Invariance Under Reordering

The following theorem states that the FRs and their associated DPs may be interchanged in the **FR** and **DP** vectors without changing the reangularity and semangularity of the design matrix (Kim, 1985). Note that the proof is for the general $N \times N$ design matrix; it is not restricted to, say, triangular matrices.

> Theorem 11 (Invariance)
> Reangularity and semangularity for a design matrix [DM] are invariant under alternative orderings of the FR and DP variables, as long as the orderings preserve the association of each FR with its corresponding DP.

To illustrate this, take the simple 2×2 case given by

$$x = \mathbf{a} * \mathbf{p} + \mathbf{b} * \mathbf{q}$$
$$y = \mathbf{c} * \mathbf{p} + \mathbf{d} * \mathbf{q}$$

where the DP p is most closely associated with the FR x, and the DP q with FR y. Then, the coupling matrix is given by

$$\mathbf{C} = \begin{vmatrix} a & b \\ c & d \end{vmatrix}$$

Suppose we now interchange the order of the equations while still ensuring that the FRs and DPs correspond as before. Then

$$y = \mathbf{d} * \mathbf{q} + \mathbf{c} * \mathbf{p}$$
$$x = \mathbf{b} * \mathbf{q} + \mathbf{a} * \mathbf{p}$$

The corresponding design matrix is

$$C' = \begin{vmatrix} d & c \\ b & a \end{vmatrix}$$

The invariance theorem says that the values of the reangularity and semangularity for **C'** are the same as those for **C**.

Proof

The proof is given in three steps.

1. The reordering of the FRs and DPs corresponds to one row and one column interchange of the elements of the coupling matrix. Moreover, it makes no difference whether the rows are interchanged first, then the columns, or vice versa.
2. Reangularity R is unaffected by any reordering.
3. Semangularity S is unaffected by any reordering.

Step 1. Suppose that a functional requirement FR* is the ith element of the **FR** vector, while the corresponding primary design parameter DP* is the ith element of the **DP** vector. When we move FR* to the jth position in the **FR** vector, we must also move DP* to the jth position in the DP vector in order to maintain correspondence between these variables (otherwise the whole concept of semangularity, or correspondence between any FR_k and DP_k, is meaningless). However, relocating FR* corresponds to an interchange of rows i and j in the $N \times N$ design matrix **C**, and relocating DP* corresponds to interchanging columns i and j.

From linear algebra, we recall that such interchange operations can be represented by pre- and post-multiplication by elementary matrices. Let $E(i, j)$ denote the elementary matrix obtained from the $N \times N$ identity matrix by interchanging rows i and j. Then the product $\mathbf{E}(i, j) * \mathbf{C}$ results in an interchange of rows i and j in **C**, whereas $\mathbf{C} * \mathbf{E}(i, j)$ results in an interchange of columns i and j. The compound operation of interchanging both rows and columns is effected by $\mathbf{E}(i, j) * \mathbf{C} * \mathbf{E}(i, j) = \mathbf{C}'$.

By the associative property of matrix multiplication,

$$[\mathbf{E}(i, j) * \mathbf{C}] * \mathbf{E}(i, j) = \mathbf{E}(i, j) * [\mathbf{C} * \mathbf{E}(i, j)]$$

Hence, it is immaterial whether the rows are interchanged first, then the columns, or vice versa.

Step 2. Let $\mathbf{c}(i)$ and $\mathbf{c}(j)$ denote the ith and jth column vectors of **C**. What are the effects of the row and column interchanges on these vectors?

Effect 1. Interchanging the column results in a swapping of the $\mathbf{c}(i)$ and $\mathbf{c}(j)$ vectors. Other columns $\mathbf{c}(k)$, for $k \neq i$ or j, are left unchanged.

Effect 2. Interchanging the rows results in a reordering of the ith and jth elements in all N column vectors.

The Independence Axiom and its Implications

Let $\mathbf{c}(m)$ and $\mathbf{c}(n)$ denote arbitrary columns of the design matrix. From Eq. (4D.2), the angular measure between any two columns is defined as

$$A(m, n) = [\mathbf{c}(m) * \mathbf{c}(n)/\|\mathbf{c}(m)\| * \|\mathbf{c}(n)\|]^2$$

The indices in Eq. (4D.2) imply that $m < n$. Without loss of generality, then, we can assume that $i < j$ in the following discussion.

Three cases may be identified, depending on the relationships between i, j, m, and n.

1. *Neither m nor n equal to i or j.* From Effects 1 and 2, the inner products as well as the euclidean norms of $\mathbf{c}(m)$ and $\mathbf{c}(n)$ remain unchanged under the transformation of the coupling matrix from $\mathbf{C}$ to $\mathbf{C}'$. Hence, $A(m, n)'$, the value of $A(M, n)$ for the transformed design matrix $\mathbf{C}'$, remains the same as $A(m, n)$ for the original matrix $\mathbf{C}$.
2. *m equal to i and n equal to j.* The positions of the column vectors are reversed, but their inner product and norms remain the same. Hence, $A(m, n)' = A(i, j)' = A(j, i)$.
3. *Exactly one of m or n equals i or j.* Without loss of generality, suppose that $m = i$, and $n \neq j$. From Effects 1 and 2, the new ith column contains the same elements as the old jth column, and all of the elements $(c(k, i)', c(k, n)')$ between columns i and n are paired as before. Since $A(i, n)' = A(j, n)$, every occurrence of $A(i, n)'$ may be replaced by $A(j, n)$, and $A(j, n)'$ by $A(i, n)$.

In short, there corresponds to every $A(m, n)'$ for $\mathbf{C}'$ exactly one $A(k, l)$ in $\mathbf{C}$, and vice versa. As a result, the value of the product in Eq. (4D.2) remains the same. We therefore see that reangularity is invariant under alternative orderings of the (FR_k, DP_k) pairs.

Step 3. From Eq. (4D.5), we see that the reangularity S is defined as the product of factors of the form $u(k) = |c(k, k)|/\|\mathbf{c}(k)\|$. We may identify two cases.

1. *k not unequal to i or j.* From Effect 2 in Step 2, we see that $c(k, k)$ will remain as the kth element of $\mathbf{c}(k)$. Since the new $\mathbf{c}(k)$ contains the same set of elements as before, $\|\mathbf{c}(k)\|$ is unaffected. Hence, $\mathbf{u}(k)$ remains the same.
2. *k equal to i or j.* Without loss of generality, we assume that $k = 1$. From Effects 1 and 2, we see that the new $c(k, k)$ $[= c(i, i)]$ will equal the old $c(j, j)$. Also, the new $\|\mathbf{c}(k)\|$ will equal the old $\|\mathbf{c}(j)\|$, since they contain the same set of elements. Therefore the new $\mathbf{u}(k)$ will equal the old $\mathbf{u}(j)$.

In summary, the only effect of the transformation from $\mathbf{C}$ to $\mathbf{C}'$ is that the values of $\mathbf{u}(i)$ and $\mathbf{u}(j)$ are swapped. Hence, semangularity is invariant under alternative orderings of the (FR_k, DP_k) pairs. This ends the proof of the Invariance Theorem.

References

Kim, S., "Mathematical Foundation of Manufacturing Science: Theory and Implications," Ph.D. Thesis, MIT, 1985.

Rinderle, J.R., "Measures of Functional Coupling in Design," Ph.D. Thesis, MIT, 1982.

Rinderle, J.R. and Suh, N.P., "Measures of Functional Coupling in Design," *Transactions of A.S.M.E./Journal of Engineering for Industry,* **104**(4):383–388, 1982.

Suh, N.P., and Rinderle, J.R., "Qualitative and Quantitative Use of Design and Manufacturing Axiom," *CIRP Annals* **31**(1):333–338, 1982.

5
THE INFORMATION AXIOM AND ITS IMPLICATION

5.1 Introduction

The second axiom refers to the minimization of some parameter called information. More specifically, it states that among all designs that satisfy functional independence (Axiom 1), the one that possesses the least information is best. Some immediate questions come to mind:

>What is the *Nature* of information?
>What is meant by information *content*?
>How do you *measure* information?

To answer these questions, this chapter explores the nature of information and defines the term *information content*. The stage is then set to develop a metric for information based on the concepts of probability, which is also examined in the context of information theory and entropy.

The notion of information is both easy and difficult to understand. The word "information" is a commonly used English word which typically means different things to different people, even under identical situations. It refers to the knowledge transmitted between two parties, or the messages transmitted. Information is also related to the notion of "complexity": the more complex a phenomenon or device is, the more information required to describe it.

In design and manufacturing, the word "information" has been used to mean all of these things. It is therefore necessary in this book to define the phrase "information content" more precisely, so that there can be no misunderstanding in the context of the axiomatic approach to design.

Information in engineering is associated with a multitude of different attributes. Sometimes it is simply a message transmitted over the wire. In the design/manufacturing world, there are many different kinds of attributes: length, hardness, surface finish, impurity level, cost, etc. Furthermore, design/manufacturing must be optimized with respect to several of these attributes at the same time. In order to deal with these different attributes, various different media have been employed to transmit information: visual, audio, tactile, olfactory, and thermal. These attributes must be treated objectively, using a common metric that is an effective descriptor of information.

There are two kinds of information: useful information and superfluous information. In order to understand the difference between these two kinds of information, consider the following question: "What is the information content in a wristwatch?" One can conceivably supply information on the accuracy of time, the geometrical shape of individual parts in the watch, the assembly of parts, material properties and their variations, etc. In fact, there can be an unbounded amount of information in the watch. However, much of the information is likely to be useless to a person who is only interested in the accuracy of the timepiece. The useful information relates solely to the satisfaction of a particular task. This task is, of course, specified in terms of the FRs and constraints.

In preceding chapters *design* was defined as the mapping process between a set of FRs in the functional domain and a specific physical embodiments in the physical domain. Since there is an infinite number of physical embodiments that satisfy the FRs, there is no unique process except when the specific physical embodiment is considered. Although Axiom 1 was concerned only with the independence of FRs in the functional domain, Axiom 2 must deal with the information contents in both the functional and the physical domains, if they are related to the specific FRs to be fulfilled (Suh, 1984).

The information content in design/manufacturing is more complex than that encountered in communications and many other fields. This situation arises from two major factors. The first is that if the designer specifies, for example, a very tight tolerance that cannot be produced by the manufacturing system, it would take an infinite amount of information. In the past, when the overlap between the designer's specification and the manufacturing machine's capability was small, skilled operators were employed to supply the additional information required to produce an acceptable part. Therefore, in design/manufacturing the information content is closely related to the designer's specification and the capability of the manufacturing system.

The second reason for the complexity of the design/manufacturing case is that the possible number of variables is infinite and they are, for the most part, unconstrained. For example, the number of chemical species that one must consider in analyzing the nonequilibrium combustion problem is finite and constrained by the laws of nature, but in design the designer starts out with an infinite number of possibilities and must gradually reduce the number of variables by introducing more constraints as the design progresses. Certainly, intellectual challenges abound in the design/manufacturing field!

In design/manufacturing, information is stored in both hardware and software. Prior to the era of mechanization, all human knowledge was stored in the software possessed by skilled artisans. This era was followed by a long period of human history in which most of the human effort was made to store the information that used to be possessed by the skilled artisans in hardware such as mechanical cams, gears, timing belts, and machine tools. In recent decades, with the advent of digital computers, more information than ever before is being stored as software in computers. One of the goals of the design field is to partition *the total information* properly into hardware and software so that the total information content is minimized per the Information Axiom.

5.2 Definition and Metric for the Information Content

In Chapter 3 information content is defined as the measure of knowledge required to satisfy a given FR at a given level of the FR hierarchy. The knowledge required to achieve a task depends on the *probability* of success. If the task is so configured that it can always be satisfied without any prior knowledge or additional knowledge, then the probability of success is unity while the requisite information is zero. In actual design/manufacturing problems we always try to transmit the *sufficient* amount of knowledge so that the probability of achieving the task is as high as possible, although usually less than unity. Since the probability of success depends on the complexity of the task, information is related to complexity.

Preliminary Model of Information

We can acquire better physical insight into the meaning of information content, and thereby develop a proper metric for information content, by considering a simple example. Suppose we measure a rod to a length $L \pm (\Delta L/2)$, as discussed in Chapter 3. One way of satisfying the length requirement is to measure the rod with a precision gage block of length ΔL. If we measure off n increments of the bar with the gage block, where n is an integer closest to $(L/\Delta L)$, then we can be certain that we will be within the specified tolerance of $\pm(\Delta L/2)$. However, the uncertainty in the measurement increases as n increases. When $L = \Delta L$, the uncertainty in counting n is less than unity, but when $L = 1,000 \, \Delta L$, the probability of making the error is 1,000 times greater. We may therefore define the information content I as

$$I = n = L/\Delta L \qquad (5.1)$$

Without further prior knowledge, we may assume the following regarding the measurement of the rod: the event that the length of the rod will be L is as likely to occur as the event that the device will be any fraction of L. Under this assumption of a uniform prior distribution of the probability, the probability that the measured length will lie within tolerance of the actual length is given by $p = \Delta L/L$. Then we see from Eq. 5.1 that the information is inversely proportional to the probability of success:

$$I = 1/p \qquad (5.2)$$

In other words, the probability of committing an error in counting is proportional to the total number of increments, n.

We can extend the preceding argument to a slightly more complicated case of cutting a rectangular plate to specifications $x = L_1 \pm (\Delta L_1/2)$ and $y = L_2 \pm (\Delta L_2/2)$. We assume that the relationship between x and y is orthogonal: that is, that the errors in the two axes are independent of each other. We also assume that a uniform probability distribution exists for both the x and the y measurements. Then the information content required to satisfy the FRs for each dimension of the plate may be written as

$$I_x = 1/p_x = L_1/\Delta L_1$$
$$I_y = 1/p_y = L_2/\Delta L_2 \tag{5.3}$$

The probability that both of these events occur is given by the product of these two probabilities; that is, the total information content according to Eq. 5.2 is

$$I = I_x I_y = 1/p_x p_y \tag{5.3a}$$

The information content defined by Eq. 5.2 requires that the information contents of separate events be multiplied rather than added. This is counter to our intuitive understanding of complexity. For example, when we take measurements along the two coordinate axes, the complexity should be the sum of the complexity of the two events, rather than the product of the two.

In the foregoing discussion on the measurement of length, we assumed that the uncertainty in measurement exists over a *range* of length of the rod. In the specific example of measuring it with a gage block of length ΔL, the probability distribution was assumed to be uniform over the entire

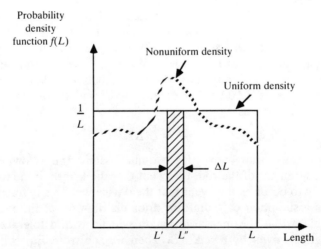

Figure 5.1. Probability distribution f in making the length measurement of a rod of length L. The probability of the measurement being within the shaded area is $p = \int_{L'}^{L'+\Delta L} f(L)\,dL$. Note that $\int_0^L f(L)\,dL = 1$

length of the rod, L, as illustrated by the solid line in Fig. 5.1. The measurement error was assumed to be in increments of ΔL over the entire length L. However, depending on the specific measurement technique and implements used, both the *uncertainty* and the *range* over which the uncertainty exists can be quite different from that assumed in Fig. 5.1. In the general case, the probability of success is given by the integral of the probability density function f within the tolerance band:

$$p = \int_{L'}^{L''} F(L)\,dL \qquad (5.4)$$

Information as Logarithmic Probability

The previous section discusses one shortcoming of defining information as inverse probability: the joint probability of two or more independent events is multiplicative, but our intuitive notion requires an additive function. A better approach is to define the information content as the logarithm of the inverse of probability as (Shannon, 1948)

$$I = \log_2(1/p) \qquad (5.5)$$

According to this logarithmic definition, the information content is zero when the probability is equal to unity. The base of the logarithm is taken to be 2 so that the information content has the unit of *bits*. Based on this definition, the information content of measuring the rectangular plate to the specified tolerances is

$$I = \log_2(p_x p_y) = \log p_x + \log p_y \qquad (5.5a)$$

In this model, the total information content is the sum of individual events.

If the measurement in the y-dimension is contingent upon the prior accurate measurement of the x-dimension, then the probability of the second measurement (i.e., the y measurement) being correct must be expressed as a conditional probability. The conditional probability is designated as $p(y\,|\,x)$. The probability of a compound event x and y occurring is equal to the absolute probability of one event times the conditional probability of the other event under the condition that the first event has occurred.

Let $p(x, y)$ denote the joint probability that events x and y both occur. On the other hand $p(y\,|\,x)$ refers to the conditional probability that event y occurs, after event x is known to have transpired; similarly $p(x\,|\,y)$ is the conditional probability of x given y. These quantities are related by

$$p(x, y) = p(x)p(y\,|\,x) = p(y)p(x\,|\,y) \qquad (5.5b)$$

The conditional probability $p(x\,|\,y)$ reduces to $p(x)$ if the events do not affect each other in any way; and similarly for $p(y)$ as a special case of $p(y\,|\,x)$.

Throughout this book, the information is defined as the logarithm of the inverse probability defined by Eq. 5.5. For convenience, the base 2 will be eliminated but assumed: hereafter, log means $\log_2$.

Information may also be defined in terms of the natural logarithm, ln, as discussed in Chapter 3. When the natural logarithm is used, the unit for information is *nats*. These units are equivalent, the only difference being that 1 bit = 1.443 nats. Both of these units of measurement are used throughout the book. In other words

$$I = K \ln(1/p) = \log(1/p) \tag{5.5c}$$

where $K = 1.443$.

What is the information content of a message? Suppose that the probability of success prior to receiving the message is p_1, and the probability after receipt is p_2. Then the informational value of the message is the difference in the prior and posterior information required for success:

$$I = I_1 - I_2 = \log p_2 - \log p_1 = \log(p_2/p_1) \tag{5.5d}$$

The definition of information given by Eq. 5.5 may also be expressed in terms of the number of equiprobable outcomes. Let N be the number of events, each of which is equally likely. The prior probability that any particular event will occur is $p = 1/N$. The corresponding information value is

$$I = -K \ln p = K \ln N \tag{5.6}$$

What is the information content of a compound event? Suppose that a successful event is defined by Q out of N events. The probability of success is $p = Q/N$. If a message or device increases the probability from (Q_1/N_1) to (Q_2/N_2), then the corresponding information value is

$$I = I_1 - I_2 = K \ln(Q_2/Q_1)(N_1/N_2) \tag{5.6a}$$

For example, suppose that a card player has to guess the identity of a card selected at random from a normal deck of 52. Suppose that the successful event E is defined by guessing the suit of the selected card. Then $Q_1 = 13$ and $N_1 = 52$, and $I_1 = \log(N_1/Q_1) = 2$ bits. If the player is given a clue that the suit is red, then $Q_2 = 13$, $N_2 = 26$, and $I_2 = \log(N_2/Q_2) = 1$ bit. Therefore the information content of the clue is

$$I = I_1 - I_2 = \log(N_1/Q_1) - \log(N_2/Q_2)$$
$$= \log(Q_2/Q_1)(N_1/N_2)$$
$$= 1 \text{ bit}$$

a result that illustrates the use of Eq. 5.6a).

The preceding formulation of information is consistent with the following viewpoints:

1. All values of information content, whether they are associated with the same or different attributes (e.g. length, hardness, or cost), are comparable as long as the underlying probabilities are the same.

The Information Axiom and its Implications   153

Hence, they can be added or subtracted without regard to the original units that define the probabilities of success, since all of these probabilities are directly related to the success of achieving a given design task.

2. The information content is equal to the complexity of the task (i.e., the FRs) involved. As the complexity of a task increases, so the probability of success decreases.
3. The information content of a message is the minimum information required to satisfy an FR within its specified tolerances.
4. The reduction in uncertainty due to a message is related to the ratio of prior and posterior probabilities due to the message.

The reasoning that led to the definition of information content in Eq. 5.5 indicates that the information content is simply the logarithm of the *range* of that variable divided by its *tolerance*. Assuming that the ratio range/tolerance is uniform throughout the range, the information content defined by Eq. 5.5 may be written as (Wilson, 1980)

$$I = \log(\text{range/tolerance}) \quad (5.7)$$

The concept embodied in Eq. 5.7 is used extensively throughout this book to measure the information content. However, before we discuss the computation of the information content for practical problems, more-philosophical aspects of information are examined in the next section.

Based on the logarithmic definition adopted for the information content, important theorems that form the basis for subsequent computation of information content can be stated.

> **Theorem 12** (Sum of Information)
> The sum of the information for a set of events is also information, provided that proper conditional probabilities are used when the events are not statistically independent.

Proof
Assume that I is the information for a pair of events, given by $I_1 = \ln(N_{11}/N_{12})$ and $I_2 = \ln(N_{21}/N_{22})$, where:

> N_{11} = a priori number of possible alternatives of event 1
> N_{12} = a posteriori number of possible alternatives of event 1
> N_{21} = a priori number of possible alternatives of event 2
> N_{22} = a posteriori number of possible alternatives of event 2
> N_{12}/N_{11} = probability of event 1 occurring
> N_{22}/N_{21} = probability of event 2 occurring

Then $I_1 + I_2 = \ln(N_{11}N_{21}/N_{12}N_{22}) = I$. The product $N_{11}N_{21}$ is the a priori number of alternatives and the product $N_{12}N_{22}$ is the a posteriori number of alternatives of the compound event. According to the definition of information, I is also information.

Theorem 13 (Information Content of the Total System)
If each FR is probabilistically independent of other FRs, the information content of the total product is the sum of the information of all individual events associated with the set of FRs that must be satisfied.

Therefore, the information of a system is defined as the sum of the information of each FR of the system in the case of product design. The information of the system represents the probability of meeting the design specifications through proper control of the system.

5.3 The Information Axiom and the Principle of Maximum Entropy

The concept of information minimization has similarities to information theory, entropy (i.e., the second law of thermodynamics), and the principle of maximum entropy. This is an important observation, in that some of the mathematical tools developed in information theory and the principle of maximum entropy may be adopted to deal with information in the design and manufacturing context. Information theory was advanced by Shannon (1948), and the principle of maximum entropy developed by Jaynes (1957), based on Shannon's work. Information theory was also used to explain classical thermodynamics by Tribus (1961). Although the second axiom—the Information Axiom for design—was advanced independently of these works, it appears that there is a great deal of similarity.

When there is a number of discrete events occurring in an ensemble with probabilities $p_1, p_2, \ldots, p_i$, the definition adopted for the information content in Eq. 5.5 can be extended to compute the *average* information content of the discrete events as (Brillouin, 1962)

$$I = -\sum_i p_i \log p_i \qquad (5.8)$$

The right-hand side of Eq. 5.8 is also proportional to the Gibbs entropy. Thus, it can be stated that the information content of a system is equal to the entropy of that ensemble (Shannon, 1948; Gallager, 1968). Equation 5.8 represents any phenomena where *uncertainty* is involved.

The principle of maximum entropy of Jaynes (1957) states that natural phenomena occur such that the entropy function given by Eq. 5.8 is the maximum. This argument enables the probabilities to be found for various components of an ensemble by solving Eq. 5.8 subject to such constrains as

$$\sum_i p_i = 1 \qquad (5.9)$$

$$\sum_i C_i p_i = C \qquad (5.10)$$

where C may be the total energy and the C_i are the energy associated with individual components. There can be many other constraints, depending on the problem. The greater the number of constraints is, the smaller the entropy.

The central question here is: "Does the principle of maximum entropy state the same thing as the Information Axiom?" The Information Axiom states that, among all proposed designs which satisfy the functional independence principle, the best design is the one with minimum information content. Since there is an infinite number of designs that satisfy Axiom 1, we can choose the best design on a relative basis by comparing only the designs proposed. That is, information content is a relative quantity, just as entropy is.

In design the information content may be minimized in several different ways: (1) by choosing designs and tolerances which yield larger p_is; (2) by minimizing the number of variables i, when other things are nearly equal; (3) by imposing the maximum number of constraints to the proposed design. The minimization of information content in design is an open-ended problem, because of the inherent nature of design, whereas the maximization of entropy of natural phenomena has an optimum solution.

The difference between the principle of maximum entropy and the minimum information axiom is that the former looks for a probability distribution for the $\{p_i\}$ which will maximize the function I of Eq. 5.8 subject to constraints, whereas the Information Axiom requires that we choose a set of p_is that will minimize the function I subject to constraints. The Information Axiom may be stated more formally as: minimize the information content I given by Eq. 5.8. However, unlike the case of the maximum entropy, the relationship

$$\sum_i p_i = 1$$

does not hold in the realm of design, because all probable events occur independently of others. In addition, cost is a major factor in design. The cost is normally higher when the probability is low. Therefore, a limiting constraint on cost may be expressed as

$$\sum_i C_i f_i(p_i) = C \tag{5.11}$$

where C_i is the cost of the ith event when the probability of that event occurring is unity, and $f_i(p_i)$ is a function that depends on probability, p_i. When $p_i = 1$, $f_i(p_i = 1) = 1$. $f_i(p_i)$ may be of the form $1/p_i$. Equation 5.11 states that, if the cost associated with a given event being within the design specification is C_i, then the total cost C is the sum of the cost of each event multiplied by a probability function for that event. C cannot exceed the total cost.

Further research is required to put the ideas presented in this section on a firm ground for rigorous mathematical treatment of the Information Axiom. In the following sections of this chapter, a more "practical" approach to the use of Axiom 2 is presented. It may lack mathematical rigor, but it is nevertheless a powerful way of comparing proposed designs quantitatively, based on the Information Axiom. Chapter 8 also presents numerous case studies.

5.4 Measurement of Information in a Systems Context

As stated in the preceding section, most design/manufacturing applications require the designer to specify geometry, material hardness, etc., which must be satisfied by manufacturing operations. When the designer's specifications can be satisfied by the manufacturing system 100% of the time, the information required is zero. When the specifications cannot be satisfied at all by the manufacturing system, even an infinite amount of information supplied to the system will not yield a satisfactory product. Therefore, the requisite information is related to the designer's specifications and the capability of the manufacturing system.

In the case of the problem of cutting a bar to a specified tolerance, the machine must cut the bar to the designer's specification of $L \pm (\Delta L/2)$. However, some machines may not be sufficiently accurate to cut the bar within the tolerance of $\pm(\Delta L/2)$. When there is uncertainty in meeting the designer's specification because of the inaccuracy of the machine itself, additional information will be required to meet the specification. In the past, human operators provided the additional information required to produce the part within the specification.

Therefore, in design/manufacturing the information content of a design, as specified by the designer, is different from the information content of the total system, because of the interaction among subelements (e.g., designer and machine) of the system. In this section, the general approach used in computing the information content of simple systems is presented (Nakazawa and Suh, 1984; Nakazawa, 1987).

Information content is a measure of the *probability of success* of achieving the specified FRs (in the case of *product* design) or the specified DPs (in the case of process design). The probability of success of a given design endeavor is obtained by considering all the FRs to be satisfied in the case of FR/DP mapping or all the DPs to be satisfied in the case of DP/PV mapping. Then, the total information content is obtained by summing up individual Is corresponding to a set of FRs or DPs to be satisfied. The probability of success of individual FRs is governed by how well the designer specified FR is satisfied by the product, whereas in the case of process it is determined by how well the designer specified DP is satisfied by the process. This can be graphically illustrated using the process design (i.e. the DP/PV mapping case) as an example.

The relationship between the designer's specification and the capability of a manufacturing system is illustrated in Fig. 5.2. This is a plot of the probability distribution of a given system parameter versus the absolute

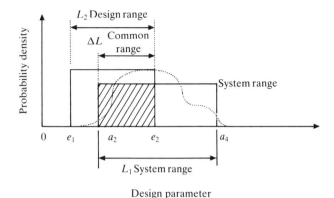

Figure 5.2. Probability distribution of a system parameter. The dotted line is for the case of a nonuniform probability distribution. The shaded area is the common range where the designer-specified tolerance and the system tolerance overlap.

value of the design parameter. The figure defines the *design range*, which is the tolerance associated with the DP specified by the designer, and the *system range*, which is the capability of the manufacturing system given in terms of tolerances. (For convenience, a uniform probability density function, as indicated by the solid lines in Fig. 5.2, is assumed. The dotted curve would represent a more complex probability density.)

When these two ranges overlap (the shaded area), the outcome is determined by the amount of the overlap (the *common range* shown in Fig. 5.2). Then, the definition for the information content given by Eq. 5.7 can be rewritten as

$$I = \log\left(\frac{\text{system range}}{\text{common range}}\right) \qquad (5.12)$$

The information content defined by Eq. 5.12 represents the probability that the system can produce the part as specified by the designer, using the particular manufacturing system chosen for making the part. If we wish to be certain that the designer's specification is always satisfied, then the amount of information specified by Eq. 5.12 must be supplied by using an operator or a sensing–control device.

For example, if the design range covers the entire system range, then $L_1 = \Delta L$ and $I = 0$; that is, the system can produce the part within the specified production parameter every time and all of the time without expending any additional effort. On the other hand, if the design range does not overlap the system range, then $\Delta L = 0$ and $I = \infty$. This means that the system requires an infinite amount of information in order to meet the designer's specification. Therefore, the probability that an acceptable part can be produced by the system is zero.

The foregoing examples are extreme cases. In general, according to Eq. 5.12, there are two ways of reducing the information content: either by reducing the system range or by increasing the common range. One way of increasing the common range is to let the system range be inside the design range by removing the bias between the two ranges. Once the bias is made zero, the system range (which is approximately twice the variance) should be reduced.

The information measure used in axiomatic design is independent of the specific nature of DPs (i.e., surface finish, cost, etc.). Therefore, all of the information of a system can be added together, even when events are statistically dependent. According to Axiom 2, the design that has the least total information content is the best one among those proposed that do not violate Axiom 1. It then follows that, if an uncoupled design requires more information than a coupled design, one should look for another uncoupled design (Corollary 7). If Axioms 1 and 2 are valid, there should always be an uncoupled design that requires less information than a coupled design.

Since we have to evaluate the information content associated with many different kinds of DPs, some typical cases are considered here.

Information Associated with Geometrical Precision

The most commonly encountered design (or production) parameter is the geometrical dimension. Consider the case of cutting a rod to length L. The tolerance specified by the designer is from $(L - e_1)$ to $(L + e_2)$. The system tolerance range of the machine that is used to cut the rod is from a_3 to a_4. These ranges are illustrated in Fig. 5.3.

For the case shown, the common range is ΔL_1, which represents the tolerance band overlap within which the manufacturing system must operate to produce a satisfactory part. In this case ΔL_1 happens to be equal to $(e_1 + e_2)$. Therefore, the information required to produce the part within the designer's specification is

$$I = \ln(L_1/\Delta L_1) \tag{5.13}$$

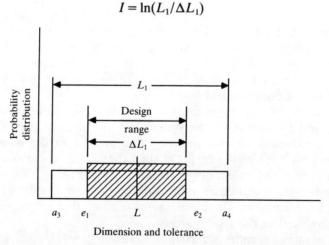

Figure 5.3. Probability distribution of a dimension. L_1 is the system range.

Equation 5.13 assumes that the other end of the rod can be precisely located on the machine and, therefore, no information is required for positioning. If this is not the case, then the information content must be increased to produce the rod to within the designer's specifications.

Information for Surface Quality

Surface roughness can be chosen as a design (or production) parameter as shown in Fig. 5.4. The design tolerance range is $[0, a_1]$ and the system range L_1 corresponds to the range of surface roughness that can be produced by the manufacturing system, consisting of a machine tool and various cutters. The common range is ΔL. Information I for surface roughness can be obtained using Eq. 5.12.

Information for Heat Treatment

Another design (or production) parameter may be hardness. The design tolerance for hardness is specified by the designer. The system range can be determined by evaluating the hardness obtained using a given heat treatment process and equipment. For example, the design range for Brinell hardness may be $200 \pm 10 \text{ kg/mm}^2$, and the normal system range of the heat treatment process may be $190 \pm 20 \text{ kg/mm}^2$. The common range is then 20 kg/mm^2. Equation 5.12 can again be used to determine quantitatively the information associated with heat treatment.

Information for Part Flow

Queueing day may be chosen as a production parameter. In this case the design range is defined in terms of the delivery time of the part, whereas the system range can be defined in terms of the queueing days of a machine tool. Using these quantities, information can be determined by using Eq. 5.12.

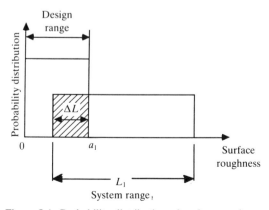

Figure 5.4. Probability distribution of surface roughness.

Information for Production Cost

Machining cost can also be a DP. The system range of Eq. 5.12 is determined, establishing the range of the operating cost and the overhead of a specific machine. The common range is the overlap in cost between the cost specified by the designer (or management) and the manufacturing cost given by the system range. Again, the information can be determined by using Eq. 5.12. When it is difficult to establish the design range for the manufacturing cost, the mean value of the maximum and the minimum costs of all the alternative systems (machines) may be used as the upper limit of the design range for the cost.

The foregoing discussions and examples were primarily confined to the case of process design. In the case of product design, the probability distribution is given for individual FRs and the common range is determined by determining the overlap between the designer specific range for each FR and the actual FR range of the probability distribution the product is able to deliver. An example of this type of information content measure is illustrated in Section 8.4, which deals with the selection of the best sub-compact car from the consumer's point of view.

5.5 Process Planning and the Information Axiom

The goal of manufacturing is to produce products according to the designer's specifications. An important step in accomplishing this goal is *process planning,* which involves choosing the best process plan from among many possible processes. In this section, a process planning method based on the Information Axiom (Nakazawa and Suh, 1984) is presented as an example of the use of information content in making design decisions.

Process planning is typically done either intuitively by experienced planners or by using computer-based algorithms. Notwithstanding its importance in manufacturing engineering and the research done in the past on the subject, it has not yet evolved into a science, as the thorough review by Weill et al. (1982) clearly indicates. The current state of knowledge is best summarized by them: "In repetitive, relatively simple derivations, like determination of machining conditions for elementary operations or in time calculations, the computer is certainly very helpful. On the contrary, when more intelligent decisions are to be made, like analyzing a drawing, the man still has the advantage over the computer."

There are three major problems in making process planning a science rather than ad hoc technology for a specific firm or plant. First of all, a rational decision-making principle or rule has not been available in choosing the best process plan, especially when the process plan has to be optimized with respect to more than one criterion. Therefore, many process plans are ad hoc and plant specific. Secondly, the information that has to be incorporated in process planning comes from diverse sources and involves different types of quantities such as geometry, material properties, delivery dates, etc. Therefore, a process plan must be optimized with respect to many DPs. Finally, the information cannot be treated in a deterministic manner, since the designer's specifications and the capability of manufacturing processes and systems interact with each other, involving tolerances. Therefore, process planning must be treated as a probabilistic problem.

There are many algorithms written for computers as listed in the review paper by Weill et al. (1982). However, many of the existing algorithms—AMP (IBM) AUTOPROS (Nissen, 1969), and TEPS (Christoffersen et al.)—can only choose the best process when the process is optimized with respect to a single criterion such as the manufacturing cost or the delivery time. However, when there is more than one criterion for which the process has to be optimized (e.g., when manufacturing cost *and* delivery time have equal priority), the conventional algorithms do not, in general, yield an optimum plan. This problem stems from the fact that there was no appropriate theoretical criterion for determining optimum process plans. Furthermore, the use of deterministic algorithms and input data may sometimes lead to erroneous conclusions, since there may be differences between the estimated data and the real system values. Real system values often have a broad band of tolerances associated with each variable. Therefore, it is difficult to assign a single value as the input data, since the resultant process plan depends on the specific values chosen for each parameter.

A critical prerequisite in developing an intelligent, generative process planning methodology is the rational decision-making principle that can generally be applied to a large family of process planning tasks. The Information Axiom provides such a principle. Assuming that the first axiom on the independence of FRs is not violated by the proposed process plans, the second axiom, which deals with information, can be used as the basis for choosing the most optimal process plan. The information accounting system presented in this section treats the interaction between the designer's specifications and the capability of a manufacturing system as a probability problem for all production parameters.

Process Planning Procedure

As a means of illustrating the basic concept involved, the procedure for using this process planning technique for a machined part will be presented here. The steps involved in developing a process plan are as follows (Nakazawa and Suh, 1984):

1. Listing of all the design (or production) parameters to be evaluated.
2. Categorization of the surfaces to be produced into *surface groups,* each group consisting of surfaces that can be machined by a single machine.
3. Listing candidate machines for each surface group.

4. Calculating the information for each production parameter and machine.
5. Obtaining the total information content and selecting the best machine combination based on the Information Axiom.

As the first step in this process planning procedure, all of the production (or design) parameters that constitute the manufacturing system must be listed. (Note that the DPs in this manufacturing context are equivalent to FRs with respect to PVs). Selection of a proper set of production parameters is important, since the inclusion of insignificant items requires extensive calculations without providing useful insight, and the exclusion of an important parameter may produce erroneous results.

Once the production parameters are chosen, surface groups must be defined. Transformation of a material from its initial to the final state by machining involves the generation of many surfaces. These surfaces may be grouped in terms of the surfaces that can be produced by the same machining operation (e.g., flat surfaces that can be machined by a milling machine). There may be several kinds of groupings in a typical manufacturing operation. If there is more than one group, calculations must be done for each group.

The third step is to list as many candidate machines as are available for the machining of each surface group.

The fourth step consists of the calculation of the information for each parameter, using the method outlined in this chapter.

The last step in the procedure is to sum up the information for each candidate machine of each surface group, followed by selection of the candidate machines with minimum total information. This combination of machines is optimum according to Axiom 2.

Through the selection process for an appropriate set of machines, Axiom 1 (functional independence) must not be violated. For example, heat treatment should not be done after the final grinding operation, since heat treatment may change the geometrical shape. If heat treatment is done before grinding it is a decoupled (or a quasi-coupled) system. In this case Axiom 1 is not violated if the specific sequence of manufacturing processes is followed strictly, in order to maintain the independence of FRs.

When calculating the information associated with part flow, it should be done for a combination of machines. However, it may take extensive calculation to evaluate all possible combinations of the machines for all surface groups, since we have not yet developed a generalized theory that can choose the best combination of machines a priori without performing detailed calculations.

In order to apply this methodology to practical problems, a demonstration program which can be used in a minicomputer was developed. The program is named PPINC (Process Planning by INformation Concept) and is written in BASIC and can be processed by a microcomputer. The user can input data interactively. Using PPINC, the best combination of machines can be determined automatically. Only three production parameters (i.e., dimensional accuracy, surface quality, and production cost) have so far been taken into consideration. In the next section examples of process planning by PPINC are given. The quantity of the part to be manufactured is assumed to be one in these case studies, although PPINC can deal with the lot size as one of the variables.

Example 5.1: Case Studies

Hypothetical Factory and System Ranges: This hypothetical factory has seven machine tools, the specification of which are shown in Table a. Seven items (i.e., machine identification no., kind of machine, maximum precision, best surface finish, power, efficiency, labor, and depreciation rate) are specified to determine the system range. The upper limit for the dimensional accuracy and the surface roughness is assumed to be 10 times the lower limit (i.e., the best value). In these examples the design range for cost is set as the mean of the maximum and minimum costs calculated for all the candidate machines for each surface group. It is assumed that either the events (i.e., accuracy, surface roughness, and cost) are statistically independent or the proper conditional probability data are used when the events are not statistically independent.

Case 1. The part to be manufactured is a rectangular parallelopiped block, shown in Fig. a. This part is produced from a bar. Information was calculated for two possible sets of surface groups that can generate the desired final shape as illustrated in Fig. b. Minimum and maximum handling times for all candidate machine tools and surface groups are assumed to be 8 and 40

TABLE a Machine Tool Data

Machine ID No.	Machine kind	Max. Precision (μm)	Best Surface Roughness (μm)	Power (kW)	Efficiency	Labor & Depreciation Rate ($/minute)
#1	Shaper	75	10	4	0.58	1
#2	Vertical milling machine	75	8	3.7	0.50	2
#3	Horizontal surface grinding machine	2.5	0.5	1	0.65	3
#4	Lathe	20	4	4	0.70	2
#5	High-precision lathe	10	2	4	0.75	2.5
#6	Horizontal milling machine	75	10	3.7	0.50	1.5
#7	Cylindrical grinding machine	2.5	0.5	1	0.65	3

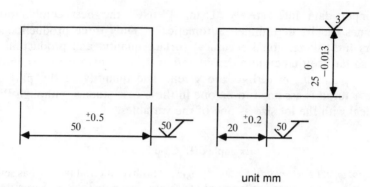

unit mm

Figure a. Block (material: alloy steel/aluminum alloy).

minutes, respectively. Design tolerance ranges are shown in Table b. The system range of cost for each machine tool is calculated, based on the machining time, which is calculated from the specific cutting energy, power, efficiency, volume to be removed, and labor and depreciation rate (Boothroyd, 1975). (The details of the calculation are illustrated in Sec. 8.2)

The calculated total information of each candidate machine for precision, surface quality, and production cost is shown in Table c. The result shows

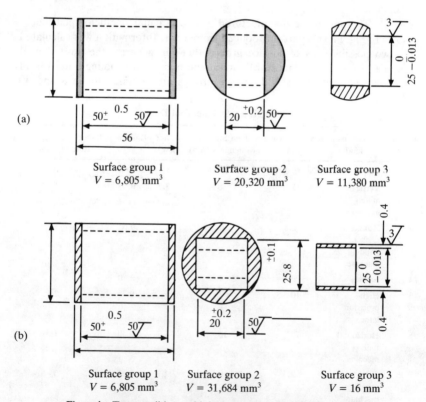

Figure b. Two possible machining sequences for surface groups

TABLE b Design Ranges for Each Surface Group of a Block

Surface Group	Tolerance (μm)		Surface Roughness (μm)	
	Lower	Upper	Lower	Upper
1	−500	500	0	50
2	−200	200	0	50
3	−13	0	0	3

that the production of the alloy steel part machined in the sequence shown by the set of surface groups indicated in part (a) of Fig. b can most effectively be done using the shaper (#1) for surface group 1 and the horizontal surface grinding machine (#3) for surface groups 2 and 3. If the best selection cannot be implemented for whatever reason, then the next best machine selection defined by the next-larger value of total information should be adopted. It is interesting to note, that, whereas most process designers may intuitively select the vertical milling machine for surface groups 1 and 2, and the horizontal surface grinding machine for surface group 3, this combination is not optimal in this case.

The machining sequence shown in part (b) of Fig. b differs from that shown in part (a), in that most of the material is removed during the second manufacturing step (i.e., surface group 2) and only the last 0.2-mm thick layer is removed during the final machining stage (i.e., surface group 3). In this case it is best to use the shaper (#1) for surface groups 1 and 2, and the horizontal surface grinding machine (#3) for surface group 3 in manufacturing the alloy steel part. Also the total information of this sequence is reduced from 7.470 to 7.008; that is, this sequence is better than that in part (a) of the figure.

TABLE c Total Information for Manufacturing System of Blocks

Surface Groups	Candidate Machines	Total Information			
		Sequence a		Sequence b	
		Alloy Steel	Al Alloy	Alloy Steel	Al Alloy
1	#1	1.274*	1.274*	1.274*	1.274*
	#2	1.575	1.530	1.575	1.530
	#3	1.748	1.278	1.748	1.278
2	#1	2.497	2.497	2.497*	2.497
	#2	2.522	2.398	2.601	2.414
	#3	2.005*	1.006*	2.613	1.157*
3	#1	∞	∞	∞	∞
	#2	∞	∞	∞	∞
	#3	4.192*	3.436*	3.237*	3.236*
Best combination of machines		7.470	5.715	7.008	5.666

* Minimum information value among the candidate machines in each category.

The change of workpiece material from steel to aluminum does not alter the best combination of machine tools between the first and the second sequence. However, the information content for the aluminum alloy was slightly less, being 5.72 and 5.67 for the first and the second sequence of operations, respectively.

The foregoing example suggests that this method of process planning based on information measurement can also be applied at the design stage to evaluate the manufacturability of various proposed designs. The interaction between the designer and the computer, based on Axiom 2, provides a powerful tool for discriminating poor designs from alternate sets of solutions. This method offers a possibility of choosing the best design from all of the proposed designs that satisfy Axiom 1, without actually having to make the part.

This example is an extremely simple case, in that we are making only one piece of the block. If the required quantity of the block is in the thousands, then we may use an entirely different manufacturing process. For example, forging of the bar to a near-net shape, possibly followed by machining operations, may be better than the best machining sequence given in Table c. Even if the machine tools specified in this example are the only ones available for the production purpose, we must consider the production time as one of the DPs, in addition to the parameters listed in Table b, so as to maximize the overall productivity. Nevertheless, the example given here illustrates how a generic process plan can be developed based on Axiom 2.

Case 2. This example is also a simple one: the manufacture of the bolt shown in Fig. c. The surface groups are shown in Fig. d. Design ranges are given in Table d.

The calculated values of information for the candidate machines are shown in Table e. These results show that the high-precision lathe (#5) should be used for machining of the surface groups 1 and 2, the cylindrical grinding

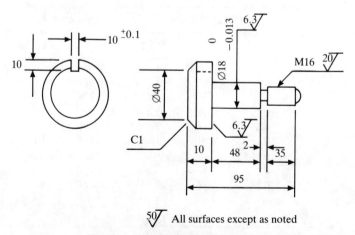

Figure c. Bolt (material: alloy steel).

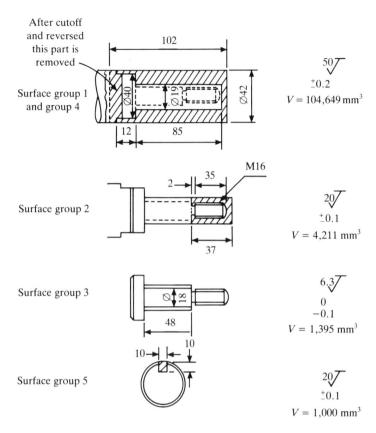

Figure d. Surface groups (i.e., machining sequence) and the volumes to be removed.

machine (#7) for the surface group 3, the lathe (#4) for the surface group 4, and the horizontal surface grinding machine (#3) for the surface group 5.

It is very interesting to note that the high-precision lathe is preferable to the conventional precision lathe, and the horizontal surface grinding machine to the milling machine, for the surface group 5, the high labor and depreciation rate notwithstanding.

TABLE d Design Ranges for Each Surface Group of a Bolt

Surface Group	Tolerance (μm)		Surface Roughness (μm)	
	Lower	Upper	Lower	Upper
1	−200	200	0	50
2	−100	100	0	20
3	−13	0	0	6.3
4	−200	200	0	50
5	−100	100	0	20

TABLE e Candidate Machines and Total Information

Surface Group	Candidate Machines	Total Information
1	#4	0*
	#5	0*
	#7	1.981
2	#4	2.144
	#5	0.855*
3	#4	∞
	#5	6.173
	#7	2.504*
4	#4	0.524*
	#5	0.853
5	#1	5.493
	#2	5.848
	#3	1.569*
	#6	5.840

* The minimum information among the candidate machines.

The foregoing example shows that process plans can be developed through the evaluation of the information content. The specific method developed considers all of the production requirements, such as design specifications and manufacturing cost, in developing the process plan. It also appears that the process planning by information measurement is reliable and reasonable, since all production requirements are specified in probabilistic terms.

5.6 Functional Coupling and Information Content

Chapter 4 demonstrated the relationship between functional independence and the allowable tolerance of the FRs specified by the designer (see Sec. 4.7). Specifically, it was stated that the change in the FRs of the system, ΔFR_i, must be within the designer-specified value of FR_i, $(\Delta FR_i)_0$, and the tolerance δFR_i as

$$\Delta FR_i = (\Delta FR_i)_0 + \delta FR_i \quad (5.14)$$

Theorem 8 states that when the deviation in ΔFR_i from the specified value $(\Delta FR_i)_0$ is greater than the designer-specified tolerance, the design is not an uncoupled system; that is, for an uncoupled system,

$$\delta FR_i \geq \sum_{\substack{j \neq i \\ j=1}}^{n} \frac{\partial FR_i}{\partial DP_j} \Delta DP_j \quad (4.22)$$

The Information Axiom and its Implications   169

Furthermore, it is shown in Sec. 4.2 that the change in the states of FRs is *path-dependent* for decoupled and coupled designs, whereas for an uncoupled design it is *path-independent*. Based on these observations, Theorem 14 may be stated as follows:

> Theorem 14 (Information Content of Coupled versus Uncoupled Designs)
> When the state of FRs is changed from one state to another in the functional domain, the information required for the change is greater for a coupled process than for an uncoupled process; that is,
>
> $$(\Delta I_{A \to B})_{\text{coupled}} > (\Delta I_{A \to B})_{\text{uncoupled}} \qquad (5.15)$$

Proof
For a coupled process the condition specified by Eq. 4.22 is violated; that is, for a coupled process,

$$\sum_{\substack{j=1 \\ j \neq i}}^{n} \frac{\partial FR_i}{\partial DP_j} \Delta DP_j > \delta FR_i$$

Therefore, the probability of FR_i being outside the designer-specified tolerance is greater (i.e. the probability of meeting the designer's specification is smaller). Therefore, it requires more information to satisfy the FRs in the case of a coupled system than an uncoupled system when the state of one FR is changed from one state to another.

5.7 Reduction of the Information Content of a Product, Process, or System

The preceding sections of this chapter describe the specific methodologies for quantitative measurement of the information content, and the use of the information content in selecting the best manufacturing process as an example. In this section, methods of reducing information content during the design stage or during the actual testing of the designed products are presented.

In Sec. 4.9 the SN ratio, variance σ^2, and the mean m are discussed as a means of evaluating the performance of the design product, based on a statistical approach. It is shown in this section that the mean m and the variance σ^2 are related to the information content. The information content can be reduced by proper selection of the DPs and manufacturing systems.

Figure 5.5 illustrates the system range and the design range of the FR of a product or DP (which is equivalent to the FR of a process). The design range is assumed to have a uniform distribution, whereas the system range of the product (or the manufacturing process or a system) is assumed to

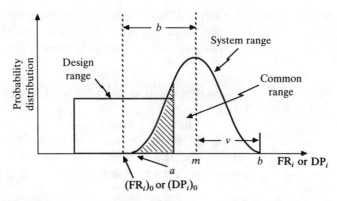

Figure 5.5. The relationship among the design range, the common range, the system range, the mean, variance, and bias of a design. $(FR_i)_0$ is the desired FR_i; m is the mean; v is proportional to the standard deviation, and thus to the variance; $(DP_i)_0$ is the DP_i that yields $(FR_i)_0$; and b is the bias between the design range and the system range.

have a nonuniform probability distribution. The mean and variance of the system range are indicated by m and v, respectively, and the bias b is shown to be the difference between the mean m and the desired FR, denoted by $(FR_i)_0$.

The information content is defined in Sec. 5.4 as

$$I = \log\left(\frac{\text{system range}}{\text{common range}}\right) \quad (5.12)$$

For the case shown in Fig. 5.6, Eq. 5.12 may be written as

$$I = \log\left[\left(\int_a^b f(\text{FR})\, d(\text{FR}) \Big/ \int_a^m f(\text{FR})\, d(\text{FR})\right)\right]$$

$$= \log\left(1 \Big/ \int_a^m f(\text{FR})\, d(\text{FR})\right) \quad (5.15)$$

The function $f(\text{FR})$ describes the probability distribution of the system range associated with the FR in question.

In order to reduce the information content, we must increase the common range and reduce the system range, according to Eq. 5.12. In order to achieve the goal of reducing the information content, we must first reduce the bias b to zero by moving either the design range to the right or the system range to the left. Certain redundant designs will allow the design range to change as illustrated through the wheelcover case study given in Chapter 7. Once the bias is made zero, then the variance should also be reduced. The variance does not have to be equal to zero, as long as the system range is inside the design range, although the smaller the variance is, the greater will be the reliability of the design in being able to deal with unforeseen events better.

Taguchi (1987) introduced the concept of "quality loss function," which is shown in Fig. 5.6. He assumes that the cost to society (or a company) of

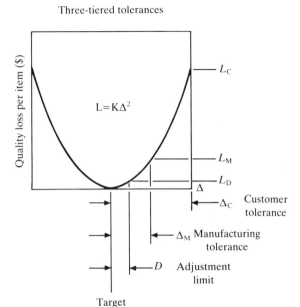

Figure 5.6. Taguchi's quality loss function in terms of the deviation from the target value.

making products that deviate from the target value [i.e., $(FR_i)_0$ or m] increases as the square of the deviation. The cost to society is expressed as loss L (i.e., units of $) which may be expressed as

$$L = K(y - m)^2 = K\Delta^2 \qquad (5.16)$$

where K is a proportionality constant, value y is the actual (dimension, etc.) obtained, and Δ is the deviation. Equation 5.16 states that, as the deviation from the designed FR increases, so the loss will increase. Therefore, the variance (or the standard deviation) should be made as small as possible. Furthermore, since the cost of monitoring a tight tolerance is much less than that of repairing in the field, the manufacturing operation must be done under a much tighter tolerance than the customer would be willing to tolerate.

5.8 The Design Axioms and the Taguchi Method

The design axioms are basic *principles* that are true in all design situations. Based on such a set of basic principles, we can develop many specific methodologies for analysis and problem solving. In Chapter 4 a functional analysis method of determining the functional independence is developed, using the concept of design matrix [**DM**], reangularity (R) and semangularity (S). Such a functional analysis method provides a specific quantitative technique for Axiom 1. In this chapter quantitative measures for information content are defined in order to deal with Axiom 2.

Another specific methodology that satisfies the design axioms was developed by Taguchi (1987) during the past three decades. It is based on a statistical approach discussed in Sec. 4.9. The Taguchi method prescribes that the SN ratio be maximized, the bias (b) be minimized, and that the standard deviation (σ) be minimized. The Taguchi method does satisfy the design axioms. This method is useful in analyzing and improving the performance of an existing design such as manufacturing operations.

In this section the Taguchi method is described, followed by an example. The example problem is the design of the passive filter network discussed in Examples 4.2 and 4.3. The same example was chosen so as to compare these two methodologies (Filippone, 1988). Both methodologies yield the same DPs, although they use different terminologies. The functional analysis described in Chapter 4 is more useful than the Taguchi method when new designs are being developed. It also provides an explicit quantitative measure for functional independence.

Introduction to the Taguchi Method

The Taguchi method characterizes the design in terms of four factors and an output response, as shown in Fig. 5.7 (Filippone, 1988). The *response* is the output of the design, which is characterized by FRs. *Signal factors* (M) are parameters that can be set to achieve a specific *response* (i.e., FRs). *Control factors* (z) are DPs with which the designer can control the output response (i.e., FRs). *Scaling factors* (R) are special cases of control factors that can be altered to achieve a desired relationship between a signal factor and output response. *Noise factors* (x) are those variables in a design that are uncontrolable or unpredictable, and represent an uncertainty that the desired output response will be achieved. The essence of this design method is to reduce the effect of this unpredictable component of the output response (Taguchi and Phadke, 1984).

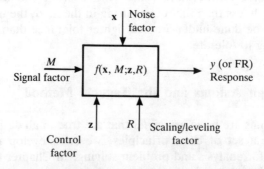

Figure 5.7. Taguchi design representation: a design representation showing the output response y of a design being the function of noise, signal, control, and scaling/leveling factors.

A metric developed by Taguchi (1987) in order to optimize a design is the ratio of the mean in output response resulting from control factors to variation resulting from unpredictable or noise factors.

The output response may be represented by a function $y = g(M; z, R) + e(x, M; z, R)$, where g is predictable and e is unpredictable. To optimize y, the z and R can be chosen such that the predictability ratio $\Delta g/\Delta e$ is maximized. The SN ratio η is expressed as the $\log_{10}$ of this predictability ratio, multiplied by 20 as defined for Eq. 4.26, which may be written as

$$\eta(x, R) = 20 \log_{10}[\Delta g(z, R)/\Delta e(z, R)] \quad (5.17)$$

The SN ratio is based on the analysis of variance technique that was introduced by Fisher (1925) and others. Measurements with several kinds of interderpendent effects working simultaneously are analyzed by the variance technique in order to determine the relative importance of the various effects. Scheffé (1959) provides a good, detailed description of the analysis of variance technique. The SN ratio for each FR (i.e., output response), where a nominal value is the preferred output response, is expressed as a ratio of the mean value μ to the standard deviation σ, which may be written as

$$\eta = 20 \log_{10} E(\mu/\sigma) = 10 \log_{10} E(\mu^2/\sigma^2) \quad (5.18)$$

In Eq. 5.18, unlike in Eq. 4.26, the estimated value E of the actual mean squared μ^2 and the standard deviation σ^2 of the output response are used to deal with the limited sample size of the experiments, in place of their precise values. Orthogonal arrays are used to assure a representative sample, as explained later in this section.

Since the actual mean μ is not known precisely, an estimate based on the sample of output responses $y_1, y_2, \ldots, y_n$ can be made to approximate μ as

$$E(\mu^2) = (nm^2 - V)/n \quad (5.19)$$

where m is the sample mean $[=(\Sigma_n y)/n]$, V is the total variation of y over the entire range of n samples, and n is the number of samples.

The error variation V is used to represent the expected value of the standard deviation squared. This is an estimate of the unpredictable part of the output response given in Eq. 5.17. This variation is calculated from the sample responses as

$$E(\sigma^2) = V = \left(\sum y^2 - S_m\right) \Big/ (n-1) \quad (5.20)$$

where $S_m = (\Sigma y)^2/n$. From Eqs. 5.18–5.20, the SN ratio as a function of the output response samples $y_1, y_2, \ldots, y_n$ can be expressed as

$$\eta = 10 \log_{10} E(\mu^2/\sigma^2) = 10 \log_{10}[(S_m - V)/nV] \quad (5.21)$$

The accuracy of this statistical technique is dependent on the sampled output response. A sample of responses must be sufficiently large to represent fully the broad range of possible outcomes. However, it should not be so large that the experimentation becomes overly time-consuming and impractical. Taguchi (1987) suggests the use of orthogonal arrays for this purpose.

Orthogonal Arrays

The use of orthogonal arrays for the design of experiments was first studied during the Second World War, although the origin of orthogonal arrays goes back to Euler's Greco-Latin squares.

Table 5.1 shows an orthogonal array with nine rows of experiments, denoted by L_9. A, B, C, and D in the table represent the control factors (i.e., DPs).

The L_9 orthogonal array is constructed from the Latin squares L_1 and L_2 shown in Table 5.1. For each of the four control factors A, B, C, and D of L_9 there are three levels that are represented in the columns. Each of the nine rows represents one combination of control factor levels (or one experiment).

Any two columns of L_9 have nine combinations of levels, (11), (12), (13), (21), (22), (23), (31), (32), and (33), for which each combination appears with the same frequency. Any two such columns are said to be balanced or orthogonal.

Orthogonal arrays are used in design of experiments in order to find the average effect of a given control factor when all other factors are varied. For example, consider factor A, which has three levels (i.e., A_1, A_2, and A_3). When the effect of A_1, A_2 and A_3 on the outcome is large, this influence will dominate, even when the conditions of the other factors are varied.

The main effect of control factors can be estimated by summing the squares of the effects at various levels and averaging over the number of degrees of freedom at each level. The number of degrees of freedom is

TABLE 5.1 An L_9 Orthogonal Array, and Two 3×3 Latin Squares, Denoted by L_1 and L_2, that were used to Construct the L_9 Array

Exp.	L_9				L_1			L_2		
	A	B	C	D						
1	1	1	1	1	1	2	3	1	3	2
2	1	2	2	2	2	3	1	2	1	3
3	1	3	3	3	3	1	2	3	2	1
4	2	1	2	3						
5	2	2	3	1						
6	2	3	1	2						
7	3	1	3	2						
8	3	2	1	3						
9	3	3	2	1						

defined as the total number of experiments (or observations) minus one. We illustrate the procedure by determining the main effect of control factor A on the variation in output response $y_1, y_2, \ldots, y_9$ shown in the L_9 orthogonal array of Table 5.1.

The effect of A on the total variation in the output y is the sum of the following three variations: the total variation in the output while A is at level 1 [i.e., $(y_1 - m) + (y_2 - m) + (y_3 - m)$], the output variation while A is at level 2, and the output variation while A is at level 3. These output variations are estimated as the sum of the difference between the output $y_1, y_2, \ldots, y_9$ from the sample mean m for the outputs corresponding to the same level of A. Each of these variations is squared to represent the absolute effect due to A on the variation in output response at the respective levels of A. These squared sums are divided by their respective degrees of freedom and summed, giving a measure for the total effect of A on the variation in output response y.

This *sum-of-the-squares* calculation measuring the effect of A for the nine outputs $y_1, y_2, \ldots, y_9$ corresponding to the nine experiments shown in Table 5.1 is

$$\text{Effect of A} = [(y_1 - m) + (y_2 - m) + (y_3 - m)]^2/2$$
$$+ [(y_4 - m) + (y_5 - m) + (y_6 - m)]^2/2$$
$$+ [(y_7 - m) + (y_8 - m) + (y_9 - m)]^2/2 \quad (5.22)$$

where $m = (\Sigma_n y)/n$, which is the sample mean at a given level.

In the following circuit example, this measurement of the effect of the various control factors described here will be calculated for both the SN ratio and sample mean m of the output response. These measures will be used to optimize the design by minimizing coupling effects among the control factors and by minimizing the effects of noise on the desired output response.

Example 5.2: Analysis of the Passive Filter Network

Statement of the Problem: A passive filter network is to be designed to measure the displacement signal generated by a strain-gage transducer. As discussed in Examples 4.2–4.4, the network provides the interface between the strain-gage transducer/demodulator and the recording instrument with a galvanometer/light-beam deflection indicator. The network conditions the signal generated by a strain-gage transducer with demodulated output and measures the original displacement signal by filtering out the carrier frequency. This arrangement was modeled in Example 4.2, which is reproduced in Fig. a. The characteristic values are given in Table a.
The FRs of the network are :

FR_1 = Minimize output distortion by placing the filter pole at 6.84 Hz.
FR_2 = Achieve a full-scale beam deflection of ± 3 in. by adjusting the filter gain.

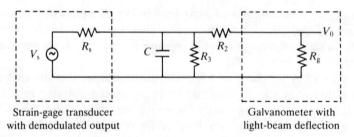

Strain-gage transducer with demodulated output

Galvanometer with light-beam deflection

Figure a. A single passive filter network for the measurement of displacement with a strain-gage transducer, demodulator, and galvanometer.

The design problem to be solved here is the determination of a correct set of DPs (i.e., control factors) that can satisfy the FRs. The Taguchi method must be used in solving this problem.

Solution

The solution to this problem was presented by Filippone (1988) as part of his master's thesis at MIT. In Example 4.2 the transfer function (V_0/V_s) was obtained using the Kirchhoff current law as

$$\frac{V_0}{V_s} = \frac{R_g R_3}{(R_2 + R_g)(R_s + R_3) + R_3 R_s + (R_2 + R_g) R_3 R_s C s} \quad (a)$$

where s is the Laplace variable. From this transfer function, the filter cutoff frequency ω_c and the galvanometer full-scale deflection D are derived as

$$\omega_c = \frac{(R_2 + R_g)(R_s + R_3) + R_3 R_s}{2\pi (R_2 + R_g) R_3 R_s C} \quad (b)$$

$$D = \frac{|V_0|}{G_{sen}} = \frac{|V_s| R_g R_s}{G_{sen}[(R_2 + R_g)(R_s + R_3) + R_s R_3]} \quad (c)$$

We must now determine which are the important DPs among the possible DPs (i.e., R_2, R_3, and C), and the values of them that will satisfy the Independence Axiom and the Information Axiom. The first step is to choose a control factor (i.e., DP) to adjust each desired output to its desired FR, using the Taguchi method.

An analysis of parameters will be done, using orthogonal arrays to measure the relative influence of each control factor on the output response. Control factors that are both insensitive to noise factors (i.e., have large η)

TABLE a Characteristic Values and Tolerances for the Transducer and Galvanometer Shown in Fig. a

Characteristic	Nominal Value	Tolerance
R_s (Ω)	120	±0.15%
V (mV)	15	±0.15%
R_g (Ω)	98	±0.15%
G_{sen} (μV/in.)	657.58	±0.15%

over a broad range of values and have a linear relationship to the output response make desirable adjustment factors. Two control factors (DPs) are required, per Theorem 4: one to adjust the frequency response ω_c and one to adjust the maximum light-beam deflection D. In choosing the control factors, the first axiom is applied to the results provided by the Taguchi analysis: the control factors (DPs) that maintain the independence of the FRs are chosen.

Other control factors that are not used as adjustment factors will be used to maximize the SN ratio. By experimenting with different levels of control factors, the levels that minimize the desired output response sensitivity to noise can be determined. Minimizing the output sensitivity to noise (i.e., maximizing the SN ratio) is equivalent to tightening the system range. It is therefore related to the second axiom of minimizing the information content.

Noise sensitivity and linear dependence over the entire range of possible control factor values must be checked in order to complete the analysis. Orthogonal arrays limit the various combinations of control factor values to a finite and reasonable number, while choosing a sample that is representative of the entire range of possible combinations. For the three control factors of the filter, low, medium, and high levels of values are chosen as shown in Table b.

To choose the levels of control factors given in Table b, the broadest range of values that are capable of fulfilling the desired design objectives is selected. However, we must avoid choosing levels for which the preferred output response becomes impossible and may yield impractical results. Conversely, we should not choose too narrow a range, which may not yield an optimal design because the optimum lies outside of the range of levels chosen. We must choose a range of control factor levels that is representative of the range of practical possibilities.

If we had to test all possible combinations of levels for the three factors shown in Table b, it would take $3^3 = 27$ experiments. An L_9 orthogonal array can be employed in this example to reduce the number of experiments from 27 to only nine experiments, as shown in Table 5.1. Since there are only three factors in this example, the first column of the L_9 orthogonal array shown in Table 5.1 can be omitted. The resultant array of combinations for the three control factors, each having three levels, is shown in Table c.

For each of the nine combinations of control factor levels shown in Table c, separate calculations must be made for each desired output response (i.e., ω_c and D).

A second orthogonal array is defined for each row in the L_9 orthogonal array of Table b as the outer array, in order to measure the variation in output response due to variation in the tolerance of the control factors. Each of these experiments yields a mean value m and an SN value η for each

TABLE b Levels for Control Factors

	Level		
	1	2	3
$R_3\ (\Omega)$	20	50,000	100,000
$R_2\ (\Omega)$	0.01	265	525
$C\ (\mu F)$	1,400	815	231

TABLE c Combinations of Control Factor Levels in an L_9 Orthogonal Array

Experiment	$R_3\,(\Omega)$	$R_2\,(\Omega)$	$C\,(\mu F)$
1	20	0.01	1,400
2	50,000	265	815
3	100,000	525	231
4	20	265	231
5	50,000	525	1,400
6	100,000	0.01	815
7	20	525	815
8	50,000	0.01	231
9	100,000	265	1,400

output response as calculated in Eqs. 5.19 and 5.20 respectively. The goal is to determine how the output response varies as a given combination of levels (i.e., 20 Ω, 0.01 Ω, and 1400 μF) are changed over their tolerance range.

The variation in output due to noise factors is determined using the SN metric described in Eq. 5.21, which gives an indication of how much the desired output response changes as the tolerances for the control factors are varied over their respective ranges. There are seven noise factors in the filter network as shown in Table d, and three levels that are chosen for each factor at their high-, medium-, and low-tolerance limits.

An exhaustive test of every combination possible would require $3^7 = 2,187$ experiments, which may be an impossible task. The use of orthogonal arrays can help to minimize the number of combinations necessary while maintaining a representative sample. Since there are seven control factors with three levels each, the L_9 orthogonal array described earlier will not be sufficient. An L_{27} orthogonal array can accommodate as many as 13 control factors of three levels each, which will be more than sufficient for this example.

An L_{27} orthogonal array may be constructed from the L_9 array of Table 5.1 as shown in Table e. As with the L_9 array, the nine combinations of levels between any two columns of the L_{27}, (11), (12), (13), (21), (22), (23), (31), (32), and (33) appear with the same frequency. Therefore any two such columns are said to be balanced or orthogonal.

The L_{27} orthogonal array is used in this example to form an outer array for each experiment (i.e., combination of control factors) shown in the L_9 array

TABLE d Levels for Noise Factors

Factor	Level		
	1	2	3
$R_3\,(\Omega)$	$R_3 - 5\%$	R_3	$R_3 + 5\%$
$R_2\,(\Omega)$	$R_2 - 5\%$	R_2	$R_2 + 5\%$
$C\,(\mu F)$	$C - 5\%$	C	$C + 5\%$
$R_s\,(\Omega)$	119.82	120	120.18
$R_g\,(\Omega)$	97.853	98	98.147
$G_{sen}\,(\mu V/in.)$	656.594	657.58	658.566
$V_d\,(V)$	0.014,978	0.015	0.015,023

TABLE e Construction of an L_{27} Orthogonal Array. An L_{27} Orthogonal Array Can be Used to Experiment With up to 13 Factors (A, B, ..., M), Each Having Three Levels. The Bold Type Shows How the L_9 Orthogonal Array of Table 5.6 was Used in Constructing the L_{27} Depicted

Exp.	A	B	C	D	E	F	G	H	I	J	K	L	M
1	1	1	1	1	1	1	1	1	1	1	1	1	1
2	1	1	1	1	2	2	2	2	2	2	2	2	2
3	1	1	1	1	3	3	3	3	3	3	3	3	3
4	1	2	2	2	1	1	1	2	2	2	3	3	3
5	1	2	2	2	2	2	2	3	3	3	1	1	1
6	1	2	2	2	3	3	3	1	1	1	2	2	2
7	1	3	3	3	1	1	1	3	3	3	2	2	2
8	1	3	3	3	2	2	2	1	1	1	3	3	3
9	1	3	3	3	3	3	3	2	2	2	1	1	1
10	2	1	2	3	1	2	3	1	2	3	1	2	3
11	2	1	2	3	2	3	1	2	3	1	2	3	1
12	2	1	2	3	3	1	2	3	1	2	3	1	2
13	2	2	3	1	1	2	3	2	3	1	3	1	2
14	2	2	3	1	2	3	1	3	1	2	1	2	3
15	2	2	3	1	3	1	2	1	2	3	2	3	1
16	2	3	1	2	1	2	3	3	1	2	2	3	1
17	2	3	1	2	2	3	1	1	2	3	3	1	2
18	2	3	1	2	3	1	2	2	3	1	1	2	3
19	3	**1**	3	2	**1**	3	2	**1**	3	2	**1**	3	2
20	3	**1**	3	2	**2**	1	3	**2**	1	3	**2**	1	3
21	3	**1**	3	2	**3**	2	1	**3**	2	1	**3**	2	1
22	3	**2**	1	3	**1**	3	2	**2**	1	3	**3**	2	1
23	3	**2**	1	3	**2**	1	3	**3**	2	1	**1**	3	2
24	3	**2**	1	3	**3**	2	1	**1**	3	2	**2**	1	3
25	3	**3**	2	1	**1**	3	2	**3**	2	1	**2**	1	3
26	3	**3**	2	1	**2**	1	3	**1**	3	2	**3**	2	1
27	3	**3**	2	1	**3**	2	1	**2**	1	3	**1**	3	2

of Table c. Table f shows the L_{27} outer array for the first experiment of Table c, from which the output responses ω_c and D are calculated for each of the 27 combinations of noise levels, using Eqs. a and b, respectively. Because there are only seven noise factors in this example and the L_{27} in Table c can accommodate 13 factors, only the last seven columns are utilized in Table f.

From the output data in Table f, the SN ratio value for each output response (i.e., ω_c and D) can be calculated from Eq. 5.22. The sample mean value for each output response is also calculated over the various combinations of noises, by summing the outputs and dividing by the number of points. This procedure is repeated for each of the nine experiments indicated in Table c. In Table g the SN ratio η and median output response m is listed for each output response.

The SN ratios in Table g represent a measure of output response sensitivity to noise for nine combinations of control factor levels. The sample mean m is an indication of the variation in the output response over the nine different combinations of control factor levels. From the results in Table g, an

TABLE f 27 Combinations of Noise Factors Tested for Experiment 1 Values from Table c with the Output Responses Calculated in the Last Two Columns

Datum no.	Noise Factor Combinations							Outputs	
	R_3 (Ω)	R_2 (Ω)	C (μF)	R_g (Ω)	R_s (Ω)	G_{sen} (μV/in.)	V_s (mV)	D (in.)	ω_c (Hz)
1	21	0.0105	1,470	84.42	120.18	566.4582	15.075	3.266	7.339
2	20	0.01	1,400	84	120	563.64	15	3.157	7.985
3	19	0.0095	1,330	83.58	119.82	560.8218	14.925	3.045	8.728
4	21	0.01	1,400	84	119.82	560.8218	14.925	3.272	7.715
5	20	0.0095	1,330	83.58	120.18	566.4582	15.075	3.150	8.411
6	19	0.0105	1,470	84.42	120	563.64	15	3.046	7.883
7	21	0.0095	1,330	83.58	120	563.64	15	3.265	8.127
8	20	0.0105	1,470	84.42	119.82	560.8218	14.925	3.164	7.599
9	19	0.01	1,400	84	120.18	566.4582	15.075	3.039	8.282
10	19	0.0105	1,400	83.58	120.18	563.64	14.925	3.021	8.289
11	21	0.01	1,330	84.42	120	560.8218	15.075	3.304	8.113
12	20	0.0095	1,470	84	119.82	566.4582	15	3.146	7.606
13	19	0.01	1,330	84.42	119.82	566.4582	15	3.034	8.714
14	21	0.0095	1,470	84	120.18	563.64	14.925	3.247	7.345
15	20	0.0105	1,400	83.58	120	560.8218	15.075	3.186	7.991
16	19	0.0095	1,470	84	120	560.8218	15.075	3.074	7.889
17	21	0.0105	1,400	83.58	119.82	566.4582	15	3.253	7.722
18	20	0.01	1,330	84.42	120.18	563.64	14.925	3.140	8.396
19	20	0.0105	1,330	84	120.18	560.8218	15	3.169	8.403
20	19	0.01	1,470	83.58	120	566.4582	14.925	3.010	7.896
21	21	0.0095	1,400	84.42	119.82	563.64	15.075	3.291	7.709
22	20	0.01	1,470	83.58	119.82	563.64	15.075	3.174	7.612
23	19	0.0095	1,400	84.42	120.18	560.8218	15	3.057	8.276
24	21	0.0105	1,330	84	120	566.4582	14.925	3.235	8.120
25	20	0.0095	1,400	84.42	120	566.4582	14.925	3.128	7.978
26	19	0.0105	1,330	84	119.82	563.64	15.075	3.062	8.721
27	21	0.01	1,470	83.58	120.18	560.8218	15	3.277	7.352

TABLE g Signal-to-Noise and Mean Output Response for Each Control Factor Combination Given in Table c

Experiment	ω_c		D	
	SN (dB)	m (Hz)	Sn (db)	m (in.)
1	25.821	8.008	30.388	3.156
2	27.443	2.195	32.218	4.761
3	27.517	6.893	30.234	3.066
4	25.378	42.295	26.756	0.873
5	27.517	1.138	30.233	3.063
6	27.591	3.960	43.857	10.952
7	25.309	11.748	26.069	0.511
8	27.591	13.980	43.858	10.947
9	27.443	1.277	32.220	4.765

indication of the relative influence that each of the control factors had on both the SN ratio and the mean value for each output response can be estimated. For example, the effect of control factor R_3 on the output sensitivity to noise (i.e., SN ratio) can be calculated by taking the average SN ratio for R_3 ($\pm \Delta R_3$) at each level (i.e., 20 Ω, 50 kΩ and 100 kΩ) to obtain three data points. These data can also be calculated for the other control factors, and compared for relative magnitudes and variations over the range of levels.

Similarly, the relative influence of each control factor on the mean value can be calculated by looking at an average value for the sample mean m at each level of the control factor. Control factors showing a constant SN ratio response over the entire range of levels while showing a linear relationship to the mean output response are desirable adjustment factors.

A measure of the influence of each control factor on each output response can be measured using the summation of squares method described in Eq. 5.4. For example, to determine the summation of squares SR_3 for the SN ratio of the factor R_3 and response ω_c, the results in the first column of Table g (indicated by $SN_1, SN_2, \ldots, SN_9$) will be used as

$$SR_3 = \tfrac{1}{2}[(SN_1 - m_{sn}) + (SN_4 - m_{sn}) + (SN_7 - m_{sn})]^2$$
$$+ \tfrac{1}{2}[(SN_2 - m_{sn}) + (SN_5 - m_{sn}) + (SN_8 - m_{sn})]^2$$
$$+ \tfrac{1}{2}[(SN_3 - m_{sn}) + (SN_6 - m_{sn}) + (SN_9 - m_{sn})]^2 - CF \quad (5.23)$$

where m_{sn} is the sample mean for the calculated SN values in Column 1 of Table g. The correction factor CF is defined as $[\Sigma_9 (SN_n - m_{sn})]^2/9$. These calculations are repeated for each control factor with both the SN ratio and the mean response of each output, and the results are reported in Tables h(a), h(b), h(c), and h(d).

Table c shows that level 1 of R_3 was used during experiments 1, 4, and 7, which yielded SN ratios for the cutoff frequency output ω_c of 25.821, 25.378 and 25.309 dB, respectively, as shown in Table g. The mean of these three SN ratios is taken to yield an average output sensitivity for R_3 at level 1 of 25.503 Ω as indicated in Table h(a). Similarly, the mean values for ω_c obtained while R_3 is at level 1 are calculated by averaging the respective data points in Table c, which yielded an average of 20.684 Hz for ω_c, level 1 as indicated in Table h(b). This is repeated for all three control factors and for both output responses, and the results are presented in Tables h(a)–h(d).

The independence axiom can be used to select the principal control factors (i.e., DPs). Control factors must be insensitive to noise factors (i.e., have large and constant η) over a broad range of values and must have a linear relationship to the output response. In addition, the independence of these outputs (i.e., FRs) relative to their respective control factors (DPs) must be maintained.

Based on the Independence Axiom, the control factor with the most significant contribution to the mean response of ω_c, while contributing a minimum to the mean response of D, should be chosen as a control factor for ω_c. The control factor that best fits this description is clearly C, as indicated by its minimal contribution to the sum of squares in Table h(d) and its significant contribution to the sum of squares in Table h(b). Thus, the sum-of-squares metric can be used as an approximation of the coupling of

TABLE h (a) Analysis of Variance for the SN Ratio of ω_c

Factor	Level Means			Sum of Squares
	1	2	3	
R_3	25.503	27.517	27.517	12.170
R_2	27.001	26.755	26.781	0.164
C	26.927	26.781	26.829	0.050

(b) Analysis of Variance for the Mean of ω_c

Factor	Level Means			Sum of Squares
	1	2	3	
R_3	20.684	5.771	4.043	753.402
R_2	8.649	15.256	6.593	184.358
C	3.474	5.968	21.056	814.464

(c) Analysis of Variance for the SN Ratio of D

Factor	Level Means			Sum of Squares
	1	2	3	
R_3	27.738	35.436	35.437	177.818
R_2	39.368	30.398	28.846	290.378
C	30.947	34.048	33.616	25.392

(d) Analysis of Variance for the Mean of D

Factor	Level Means			Sum of Squares
	1	2	3	
R_3	1.513	6.257	6.261	67.569
R_2	8.352	3.466	2.213	94.684
C	3.661	5.408	4.962	7.411

FRs by the various DPs or control factors, although it is not as explicit as determining the elements of the design matrix reangularity and semangularity.

The results shown in Tables h(a)–h(d) are plotted in Figs. b–d to gain some two-dimensional insight as to how the control factors are related over the entire range of values to the mean value and SN ratio of the output response.

The graphs in Fig. d indicate that C is a good choice for an adjustment factor for ω_c, since the SN ratio is constant over the broad range of its level values and since ω_c varies as a sensitive function of C. The mean output response for ω_c does not vary exactly linearly with respect to C, but does have a monotonic and almost linear relationship. For these reasons C is the

Noise Sensitivity of response for three levels of R_3

(1)

Linearity of response for three levels of R_3

(2)

Figure b. (1) The variation in output response sensitivity (i.e., SN ratio) as R_3 is varied over three values from 20 to 100 kΩ. (2) The mean output response measured as R_3 is varied over three levels.

most appropriate choice of control factor for the frequency output response ω_c.

The results in Table h(b) indicate that the contribution to the variation of the cutoff frequency response ω_c is significantly greater for R_3 than it is for R_2. The contribution of R_2 to the variation in the mean response of D is greater than the contribution of R_3, as indicated in Table h(d). This indicates that R_2 is coupled by a lesser degree to the output ω_c than is R_3, while also contributing more toward the variation in the deflection output D. Therefore, based on the Independence Axiom, R_2 would make a better control factor than R_3 for the deflection output response D of this filter network.

These analytical results based on the Taguchi method have been used to aid in the selection of DPs (i.e., control factors) that form a design with the least coupling. This is the same as choosing DPs that maintain the independence of FRs. The sum-of-squares column of Table h(b) shows that DP R_2 affects ω_c, and thus the design is partially coupled. With R_2 and C chosen as DPs, the optimization and the consequent selection of parameter

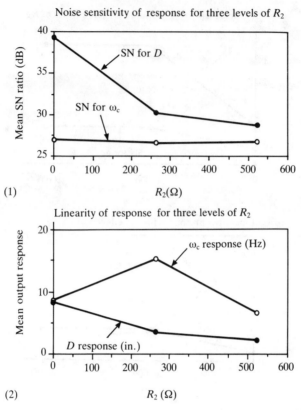

Figure c. (1) The variation in output response sensitivity (i.e., SN ratio) as R_2 is varied over three values from 0.01 to 525 Ω. (2) The mean output response measured as R_2 is varied over three levels.

values can be done using the SN ratio metric and applying the Information Axiom.

According to the Information Axiom, the information content of this design must be minimized. To minimize information, the bias must be minimized and the system range of Fig. 5.5 must be completely within the design range. This can be achieved partly by minimizing the standard deviation if the bias is small.

The desired response for ω_c was defined to be 6.84 Hz, which has the most value to the customer. A little less or a little more than this target of 6.84 Hz would represent a loss to the value of the design from the perspective of the customer. A tolerance for the output response is determined by deciding how much variation in ω_c would constitute an acceptable range to the majority of customers. The system range is usually not a step function, and is more likely to be approximated by a continuous function as illustrated in Fig. e.

As the component tolerances in the filter circuit are reduced, the system range is narrowed around the target value of 6.84 Hz and the information content is decreased, since the bias is zero for this case.

From the sum-of-squares column of Tables h(a) and h(b), the DP that

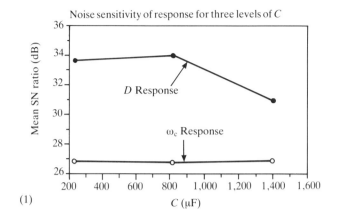

(1)

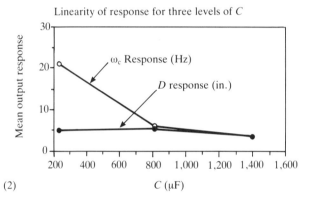

(2)

Figure d. (1) The variation in output response sensitivity (i.e., SN ratio) as C is varied over three values from 1400 to 231 μF. (2) The mean output response measured as C is varied over three levels.

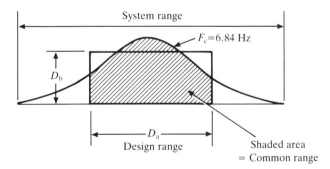

Figure e. The system range is for the frequency output response ω_c is approximated with a continuous function. As the system range is tightened around the target value of 6.84 Hz, the ratio of the system range to the common range is decreased.

contributes the most to variations in the SN ratio can be determined. Table h(a) indicates that DP R_3 has the most significant effect on the SN ratio. The tolerance of the DP R_3 must be reduced to increase the SN ratio.

Example 5.2 describes how the Taguchi method can yield the same result as the functional analysis described in Chapter 4 and the information content measurement described in this chapter. In that sense, the Taguchi method satisfies the design axioms. However, care must be exercised in using the Taguchi method, so that we do not simply decrease the variance without making the bias very small, as described in Sec. 4.9.

5.9 Summary

In this chapter the meaning of information content is studied further. Information is defined in terms of probability p as

$$I = \log_2(1/p)$$

The useful information that we must be concerned with is that related to achieving a given task as expressed in the form of FRs. When there are a number of discrete events occurring in an ensemble with probabilities p_i, the *average* information content of the discrete events is given by

$$I = -\sum_i p_i \log p_i$$

In a design situation, information can be broadly interpreted to be

$$I = \log\left(\frac{\text{range}}{\text{tolerance}}\right)$$

In design and manufacturing situations, the information content is determined by the tolerance specified by the designer and the tolerance of the manufacturing system. There must be an overlap in these tolerances in order to fabricate a part which satisfies the designer-specified FRs; otherwise the task cannot be achieved.

The concepts of *design range, system range,* and *common range* were introduced in this chapter, in order to deal with the overlap between the specified tolerance and the capability of the system in meeting the specification. The overlap is highly nonlinear with respect to changes in the design specification and the system capability. Therefore, the information content, which is a function of this overlap, can vary from a finite value to infinity when the design range and the common range do not overlap.

Information can be stored in both hardware and software. Therefore, the total information is the sum of the software and hardware information that must be available to achieve the task specified by the FRs. A proper partitioning of the information between software and hardware may be necessary in order to minimize the information content and satisfy the Information Axiom.

The minimum information criterion is a powerful tool in optimization and simulation of design/manufacturing processes when there are several variables with respect to which the solution must be optimized. Unlike many other optimization techniques, which typically deal with one variable, the information content measure can select the best solution among those proposed, regardless of the number of variables involved. In terms of the Information Axiom, the solution that satisfies Axiom 1 and possesses the minimum information is the best solution. Several case studies are presented in Chapter 8 to illustrate this point.

This chapter discusses a number of examples for the actual computation of information content, including the use of Axiom 2 in process planning. Finally, the information content of a coupled process is shown to be larger than that of an uncoupled process.

The theorems derived in this chapter are:

Theorem 12 (Sum of Information)
The sum of information for a set of events is also information, provided that proper conditional probabilities are used when the events are not statistically independent.

Theorem 13 (Information Content of the Total System)
If each DP is probabilistically independent of other DPs, the information content of the total system is the sum of the information of all individual events associated with the set of FRs that must be satisfied.

Theorem 14 (Information Content of Coupled vs. Uncoupled Designs)
When the state of FRs is changed from one state to another in the functional domain, the information required for the change is greater for a coupled process than for an uncoupled process.

Methods of reducing the information content were described in terms of both the system and the common and design range, and in terms of the mean, bias, and variance of the system response.

Finally, this chapter shows that the Taguchi method is another way of determining the principal DPs among many possible DPs, and that the method is consistent with the design axioms.

References

Boothroyd, G., *Fundamentals of Metal Machining and Machine Tools*. McGraw-Hill, NY, 1975.
Brillouin, L., *Science and Information Theory*. Academic Press, NY, 1962.
Christoffersen, J., Falser, P., Suonisvu, E., and Svardson, B., "TEPS—A Technical Economical Production Planning System," The Scandinavian Council for Applied Research.

Filippone, S., "The Principles of Design Applied to Engineering and Policy," SM Thesis, MIT, 1988.
Fisher, R.A., *Statistical Methods for Research Workers,* 1st edn. Oliver & Boyd, Edinburgh, 1925.
Gallager, R.G., *Information Theory and Reliable Communication.* John Wiley, NY, 1968.
IBM, "Automated Manufacturing Planning," IBM Technical Report GE20-0146-0.
Jaynes, E.T., "Information Theory and Statistical Mechanics," *Physical Review* **106**:620, 1957.
Nakazawa, H., *Information Integration Method.* Corona, Tokyo, Japan, 1987 (in Japanese).
Nakazawa, H., and Suh, N.P., "Process Planning Based on Information Concept," *Robotics and Computer Integrated Manufacturing* **1**(1):115–123, 1984.
Nissen, C.K., "AUTOPROS—Automated Process Planning System," *CIRP International Conference,* Vol. 8, 1969.
Scheffé, H., *The Analysis of Variance.* John Wiley, NY, 1959.
Shannon, C.E., "A Mathematical Theory of Communication," *The Bell System Technical Journal* **27**:379–623, 1948.
Suh, N.P., "Development of the Science Base for the Manufacturing Field Through the Axiomatic Approach," *Journal of Robotics and Computer Integrated Engineering* **1**(3/4):397–415, 1984.
Taguchi, G., *System of Experimental Design: Engineering Methods to Optimize Quality and Minimize Cost.* American Supply Institute, 1987.
Taguchi, G., and Phadke, M.S., "Quality Engineering through Design Optimization," *Proceedings of I.E.E.E. GLOBECOM-84 Conference, November, 1984, Atlanta, GA,* 1106–1113.
Tribus, M., "Information Theory as the Basis for Thermostatics and Thermodynamics," *Journal of Applied Mechanics* **28**:1–8, March, 1961.
Weill, R., Spur, G., and Eversheim, W., "Survey of Computer-aided Process Planning System," *CIRP Annals* **31**(2):539–551, 1982.
Wilson, D.R., "An Exploratory Study of Complexity in Axiomatic Design," Ph.D. Thesis, MIT, August, 1980.

Problems

5.1. Prove Theorem 12.

5.2. Joe Doe is looking for a house. He is considering two different towns, which are located near the town in which his company is located. His "design range" in making his decision is based on three factors: (1) land price, (2) commuting time, and (3) environment. He is willing to pay up to $5,000 per acre, to spend as much as 1.5 hours for commuting, and to live in an environment which meets 80% of his expectations. The two towns selected offer the following possibilities:

	Town A	Town B
Land price ($/acre)	2,000–4,500	4,800–6,000
Commuting time (hours)	1.0–1.7	0.8–1.0
Environment (%)	80–90	60—85

Which is a better town for Joe Doe?

The Information Axiom and its Implications

5.3. In Example 5.1, if the design range for surface roughness given in Table b is reduced by 40%, which is the best set of machines?

5.4. Consider three different kinds of beverage containers: aluminum cans, polyethylene terephthalate (PET) plastic bottles, and glass bottles. Which is the best container for you?

5.5. Prove Theorem 13.

5.6. Derive Eq. (5.8).

5.7. Derive a solution for the manufacturing process problem described in Example 5.1, using the Taguchi method described in Section 5.8.

6

MANUFACTURING-RELATED CASE STUDIES: DESIGN OF PRODUCTS, MANUFACTURING PROCESSES, AND INTELLIGENT MACHINES

6.1 Introduction

The fundamentals of the design axioms are discussed in the preceding chapters. Many examples, albeit simple ones, are given in those chapters to clarify the basic concept associated with axioms, theorems, and corollaries. In this chapter many cases related to design for manufacturability, manufacturing processes, and intelligent machines are presented in order to illustrate the use of the axioms, especially Axiom 1, the Independence Axiom. In Chapter 7 additional case studies are presented, primarily to illustrate the use of axioms in product design. The quantitative use of Axiom 2, the Information Axiom, is illustrated through case studies in Chapter 8. Many of these case studies involve real industrial problems that were solved based on the axioms and their corollaries. In some cases solutions provided before the advent of the axioms are analyzed using the axioms.

As stated earlier, the axioms and corollaries apply at all levels of the FR and the DP hierarchies. At the highest level the axioms deal with the overall concept or approach to problem solving. Many examples given in this chapter deal with issues at this level of hierarchy, since the decisions made early in the design process have the most important effect on the design and dictate the details of the overall solution.

It is shown in Chapter 3 that, in an ideal design, the design matrix must be a square matrix (see Theorem 4); that is, the number of FRs and DPs must be equal to avoid having a coupled or a redundant design. Furthermore, it is shown that DPs must be so chosen that the design matrix is a diagonal matrix, or a triangular matrix in order to satisfy Axiom 1.

The number of FRs must be kept at a minimum, according to Axiom 2. When the initial selection of FRs had not included an important FR, it can be added to the original list, but such an addition of a new FR requires that the entire design cycle be repeated. This iteration of the design cycle may be as simple as adding a new DP to satisfy the new FR when and only when the original design matrix was a diagonal matrix or when a sufficient number of FRs are independent so as to yield a triagular matrix (see Theorem 2 of Chapter 3).

The power of Axiom 1 is to enable quick evaluation of a proposed design and elimination of bad designs. This will make it possible to devote attention to creative design activities, and to prevent carrying unproductive ideas to prototyping and testing, which can be very costly. As shown in this chapter, many design decisions can be made based on a qualitative or approximate analysis of a proposed design. However, when the relationship between FRs and DPs is not obvious, a careful quantitative analysis of the causality between the DPs and FRs must be done.

In Chapter 4 design for producibility is discussed. It is shown that the FRs of the product, the DPs of the product, and the PVs of the manufacturing process are related as:

$$\{\mathbf{FR}\} = [\mathbf{A}]\{\mathbf{DP}\} \tag{3.1}$$

$$\{\mathbf{DP}\} = [\mathbf{B}]\{\mathbf{PV}\} \tag{4.28}$$

$$\{\mathbf{FR}\} = [\mathbf{A}][\mathbf{B}]\{\mathbf{PV}\} \tag{4.29}$$

Just as the FR–DP relationships must satisfy the Independence Axiom, so do the DP–PV relationships; that is, the DPs in the DP–PV relationship are counterparts to FRs in the FR–DP relationship. Therefore, when we choose PVs to satisfy DPs, the independence of DPs must be maintained by the PV selected. Based on the nature of Eq. 4.29, Theorem 9 (the design for producibility theorem) was derived. The theorem, in essence, states that neither the relationship between FRs and DPs nor that between DPs and PVs can violate Axiom 1 by being coupled designs. Therefore, both the design of the product (given by the FR–DP relationship) and the design of the process (given by the DP–PV relationship) must satisfy the Independence Axiom.

In this chapter case studies involving the design of the product and the design of manufacturing processes are presented. Although FRs may be directly related to PVs, for the sake of clarity FRs are related to DPs first and then DPs are related to PVs, wherever appropriate. The relationship between DPs and PVs is such that DPs represent the requirements that the PVs must satisfy in process design, just as FRs are the requirements that the DPs must satisfy in product design.

6.2 Design and Manufacture of a Multilense Plate

Statement of the Problem

A university professor invented an information storage device that can store 200 pages of an $8\frac{1}{2} \times 11$ in. manual by creating a unique microfiche reader. The surface area of the $8\frac{1}{2} \times 11$ in. document sheet is divided into 64 subsections, and the image of each subsection is reduced into a small dot on a negative film. This dot is then remagnified to construct the image of the entire sheet. On each subsection of the negative film 200 individual dots can be developed by moving the film to 200 different positions by a

Manufacturing-Related Case Studies

preset mechanical method. The goal of this case study is the design of a multilense plate that can magnify the 64 images and reconstruct them on a glass plate so that the entire page of the document can be read.

The FRs of the multilense plate are:

$(FR_1)_m$ = The image of each subsection must be magnified with good resolution.

$(FR_2)_n$ = The images from all subsections must be contiguous to reproduce the entire $8\frac{1}{2} \times 11$ in. sheet accurately.

$(FR_1)_m$ represents m number of FR_1s related to lense quality. Similarly, $(FR_2)_n$ indicates that there are n number of spacing between the lenses that must be controlled. Therefore, there are $(n + m)$ number of FRs that must be satisfied. Since there are 64 individual lenses, the number of FRs is very large.

In order to satisfy the FRs of the optical part of the readers, it was decided to produce a product as shown in Fig. 6.1. This design makes use of 64 lenses placed at the correct locations so as to magnify the images in a contiguous fashion. The DPs and the constraints of the multilense plate are:

$(DP_1)_m$ = Each lense must have a specific radius of curvature within a few wavelengths of visible light.

$(DP_2)_n$ = The distance between every pair and all pairs of lenses must be within ± 0.0005 in. (12.2 μm).

C_1 = The manufacturing cost of the multilense plate must be less than $10.00 each when the production rate is 100,000 per year.

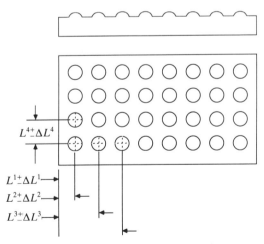

Figure 6.1. Multilense plate. The DPs of this multilense plate are: (1) the quality of each and every lense which must be within three wavelengths of the visible light, and (2) the positions of the lenses which must be within a tolerance of 0.0005 in. The total number of DPs is very large. One of the main constraints is that it must be made for less than $10.00.

This design is an uncoupled design that satisfies the FRs and Axiom 1. Having designed the product, we must now develop the manufacturing process.

Obviously, there are many different ways of making the multilense plate defined by $(DP_1)_m$ and $(DP_2)_n$. We can cast glass into rough shape and then start machining and polishing it into the final shape. This type of manufacturing operation satisfies Axiom 1, since FRs are not coupled by the processes. However, this solution may be unacceptable because it cannot satisfy the cost constraint; that is, the $10.00 limit. We therefore have to search for a new solution. Another method that does not violate Axiom 1 may be to make individual lenses, and then assemble them together by bonding them to a substrate at the precise location specified. This approach may again produce some difficulty in meeting the cost constraint, although Axiom 1 is satisfied.

In view of this cost constraint, it was decided by XYZ Corporation that the entire multilense plate should be made by injection molding a thermoplastic called polymethyl methacrylate (PMMA) in a carefully made metal mold to the final dimensions and finishes specified by the FRs. The engineers of the Corporation designed and tested a molding system shown in Fig. 6.2. In this approach, the plastic is injected into the mold through an edge gate. After the plastic solidifies in the mold, the part is taken out of the mold.

After 2 years of repeated trial and error and a great deal of expense, they still could not produce acceptable parts consistently and reliably. The problem was that it was difficult to satisfy the DPs (which are the FRs for the process design) related to the position of the lenses. When it was found that some of the lense positions were not within the specified tolerance, the mold was taken out and new lense cavities were machined at different locations, using the previous test results. Then it was found that the lenses that used to satisfy the position requirement were no longer at the specified positions. This process of modifying the mold was tried several times without much success.

In this example we are concerned with both product and process design. Although the design of the product appeared to be satisfactory, the product could not be manufactured using the injection molding process. It

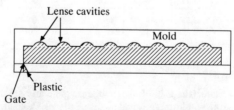

Figure 6.2. A mold design for a multilense plate. The shaded area shows the plastic that is injected into the mold through an edge gate from a reciprocating simple screw plasticator.

is therefore clear that either the product must be redesigned or a new process must be used, or both.

What is wrong with the design of the manufacturing process? Would you, as the supervisor of the engineers, have approved such an approach from the beginning? How would you satisfy the FRs, DPs and constraints through an alternate design?

Solution

The basic problem with their design is that the proposed manufacturing method based on the injection molding process violates the independence of the FRs of the product. The PVs couple the DPs and thus FRs, violating Theorem 9 (the design for manufacturability theorem).

The injection molding process typically works as follows. Plastic granules in the hopper are fed into a plasticator, which consists of a reciprocating single screw rotating in a barrel, very similar to the extruder shown in Fig. 3.1, except that the screw can also move axially. The plastic granules melt and mix in the plasticator, due to the rotational motion of the screw. The molten plastic is prevented from flowing out of the plasticator and flows into the mold when the valve at the exit nozzle is closed while the screw is rotating. Consequently, the molten plastic accumulates in front of the screw and pushes the screw backward. When a sufficient amount of plastic has accumulated in front of the screw, the valve is opened and the plastic is injected into a cold mold using a hydraulic piston that is attached to the screw. Figure 6.3 shows a typical injection molding machine.

When the hot, molten plastic is injected into the mold, the velocity distribution of the plastic front is as shown in Fig. 6.4. The radial velocity component of the fluid is the highest at the center of the gap, because of the wall friction. When the liquid stream near the center reaches the plastic–air interface (i.e., the plastic flow front), the fluid changes its flow

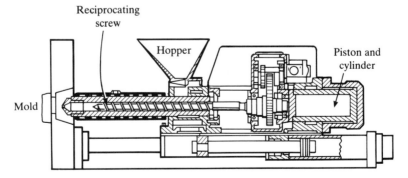

Figure 6.3. Schematic view of a conventional injection molding machine. (From Tadmor and Gogos, 1979).

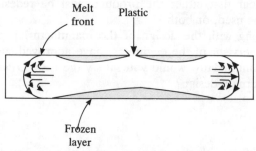

Figure 6.4. Velocity distribution of molten plastic in a mold cavity during injection molding. Since the plastic is injected into a cold mold, plastic begins to freeze as soon as it contacts the mold. The thickness of the frozen layer grows as a function of the contact time. The radial velocity is maximum at the center of the mold cavity. The plastic at the flow front changes direction toward the mold surface. This is known as the "fountain effect."

direction toward the mold surface. This flow phenomenon is sometimes referred to as the "fountain effect."

Furthermore, since the mold is maintained at a temperature below the glass transition temperature of the plastic, T_g, the plastic freezes along the wall. The thickness of the frozen layer increases with time. Because the solidified layer becomes thicker, the temperature of the plastic at the melt front decreases with the flow distance from the gate. If the mold is very large and thin, the plastic may freeze before the mold is completely filled. The pressure drop also changes as the plastic flows into the mold. Therefore, both the temperature and the pressure of the plastic at every point in the mold undergo a different thermodynamic history.

After the mold is filled with plastic, the molten plastic is subjected to a high packing pressure so as to put as much plastic as possible into the mold by compressing the plastic elastically. As the plastic freezes (normally the plastic at the gate is made to freeze first so that the plastic in the mold would not flow back out of the mold when the injection pressure is released), the volume of the plastic decreases due to thermal shrinkage or phase change.

This thermal shrinkage is partially compensated by the elastic compression that the plastic was subjected to under the high packing pressure. Since the temperature and pressure at every point in the plastic at the completion of the filling process are different, the amount of dimensional change is different from point to point. Furthermore, since all the lenses are interconnected, the dimensional change at one region affects the distance between the lenses elsewhere. As a consequence, the distance between the lenses is not the same as the distance between mold cavities for individual lenses.

The final dimensions between the lenses is affected by four factors: the spacing of lense cavities in the mold; the elastic compression of the molten plastic at the completion of the mold-filling process (which is controlled by the local temperature and the maximum pressure*); the shrinkage of the plastic during cooling; and, finally, the stress relaxation of the plastic. Since these four factors are different at every point in the mold, the final dimensions also differ from point to point. Since there are at most only five process variables (i.e., temperature of the mold, temperature of the plastic at the gate, injection pressure, packing pressure, and gate-freezing time), which is smaller than the number of FRs, this is a coupled design by Theorem 1.

The design equation for this proposed design can be written in terms of the DP vector, the PV vector and the design matrix as

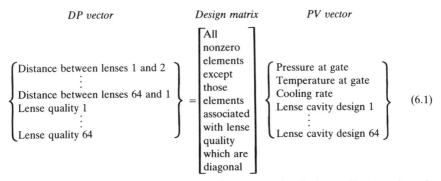

In this design equation it can be recognized immediately that it represents a coupled system, since the DP vector (and thus the FR vector) has more components than the PV vector (Theorem 1). The coupled system does not satisfy Axiom 1.

Knowing that the processing technique tried by XYZ Corporation is not a rational design since it violates Axiom 1, we have to search for or design a new solution. We now have a much better basis to come up with a good design, since we know what the problem is. The problem is caused by the coupling introduced by nonuniform shrinkage of plastics throughout the mold. For a *coupled* design to produce an acceptable part, one must be extremely lucky, or loosen up the tolerance of the FRs to uncouple the design (per Theorem 8) if the perceived needs can still be satisfied even with loose tolerances. However, since the latter cannot be done in this case, we have to develop a new design.

The new solution that we should seek is an uncoupled design that is based on Corollary 1. This can be done if we eliminate the variable shrinkage of the plastic that occurs in the mold. A possible solution that is consistent with Axiom 1 is shown in Fig. 6.5. It is a two-step process. The

---

* The equation of state for amorphous polymers is given by the Spencer–Gilmore equation as:

$$(p + \pi)(v + \omega) = RT \tag{a}$$

where p is the pressure, v the specific volume, T the temperature of the plastic, and π, ω, and R are material constants. As the pressure is increased isothermally, the specific volume of the plastic descreases, according to Eq. a.

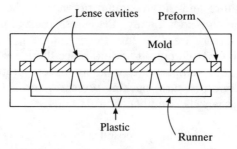

Figure 6.5. Example solution for multilense plate. It uses a preform with holes located approximately near the lense cavities. Since the quality and the position of each lense is controlled by the lense cavity and the distance between the lenses is no longer affected by the shrinkage of the plastic, it satisfies Axiom 1.

first step consists of making a preform by punching holes in a plastic or cardboard sheet in regions where the lenses must be placed. The second step is to insert the preform into a mold and inject plastic into separate lense cavities simultaneously. In this process the location of individual lenses is fixed by the position of the lense cavity in the mold, which is not affected by the molding process. Also, the quality of each lense is independently controlled by the quality of lense cavities. Therefore, Axiom 1 is satisfied. The design equation for this proposed solution may be written in terms of the DP vector, the PV vector, and the design matrix as

$$\begin{Bmatrix} \text{Location of lense 1} \\ \vdots \\ \text{Location of lense } N \\ \text{Quality of lense 1} \\ \vdots \\ \text{Quality of lense } N \end{Bmatrix} = \begin{bmatrix} \times & 0 & 0 & 0 & 0 & 0 \\ 0 & \times & 0 & 0 & 0 & 0 \\ 0 & 0 & \times & 0 & 0 & 0 \\ 0 & 0 & 0 & \times & 0 & 0 \\ 0 & 0 & 0 & 0 & \times & 0 \\ 0 & 0 & 0 & 0 & 0 & \times \end{bmatrix} \begin{Bmatrix} \text{Location of lense cavity 1} \\ \vdots \\ \text{Location of lense cavity } N \\ \text{Quality of cavity 1} \\ \vdots \\ \text{Quality of cavity } N \end{Bmatrix} \quad (6.2)$$

The design matrix of this design is a diagonal matrix, indicating that the manufacturing process is an uncoupled design. Since the design of the product was also an uncoupled design, this design satisfies the axiom in accordance with Theorem 9.

This solution is similar to another idea discussed earlier but discarded for cost reasons, which was to make high-quality individual lenses and bond them to a substrate individually at a specified location. However, in this case the information content is much smaller for the injection molding process, since the location of the lenses, as well as the quality of the lenses, is controlled by the mold, the bonding of the lense to the substrate is done as part of the injection molding process, and since the preform does not have to be accurate. This injection molding process with preforms appears to be the most inexpensive process so far considered.

Going from the coupled design shown in Fig. 6.2 to the uncoupled design shown in Fig. 6.5, the information content has decreased significantly. The former design required information in terms of an accurate description of by how much plastic shrinks, the exact mold temperature, the pressure and temperature of the plastics, etc., which is much more than we can realistically provide for the system. Even if we could supply all of the necessary information, there is no guarantee that the system would have produced good parts, since the system is too coupled. In the uncoupled design, however, the information required is associated only with the location and quality of each cavity in the mold, and the temperature and pressure history of each molding cycle.

The total information content of the uncoupled system may be written as

$$I = \sum_{i}^{2N} (I_l)_i + \sum_{i}^{N} (I_q)_i + I_p + I_t \tag{6.3}$$

where I_l is the information content associated with the location of each cavity in the mold, measured along both the x-axis and y-axis; I_q is the information content associated with the quality of each lense, I_p is the information content associated with the peak pressure exerted on the molten plastic in the mold by the injection molding machine; and I_t is the information content associated with the maximum temperature of the plastic enters the mold. The information content associated with the location is determined by what the designer specifies and what the machine can deliver in terms of position accuracy (in this case a precision milling machine may be used to locate the position and machine the cavity).

Figure 6.6 shows how the tolerance specified by the designer (i.e., $L_{xi} \pm \Delta L$) and the tolerance the machining system can provide overlap in locating the ith cavity along the x-axis. The probability of the manufacturing system meeting the designer-specified dimensional requirement is the ratio of the tolerance overlap ΔL_0 and the system tolerance ΔL_{xi} (i.e., $\Delta L_i / \Delta L_{xi}$). The figure also shows the overlap in the maximum pressure specified by the designer and what the injection molding machine can deliver. Similar arguments can be made for the FRs related to the quality of lense and the temperature of plastic. Then, each one of the information contents may by expressed as

$$\sum_{i}^{2N} (I)_i = \sum_{i}^{N} \log(\Delta L_{xi}/\Delta L_i) + \sum_{i}^{N} \log(\Delta L_{yi}/\Delta L_i)$$
$$\sum_{i}^{N} (I_q)_i = \sum_{i}^{N} \log(\Delta R_i/\Delta R_0)$$
$$I_p = \log(\Delta P_m/\Delta P_0)$$
$$I_t = \log(\Delta T_m/\Delta T_0) \tag{6.4}$$

where ΔL_{yi} is the system tolerance in locating the ith cavity along the y-axis, ΔL_0 is the tolerance overlap between the designer-specified and the system tolerances, ΔR_i is the tolerance on the radius of curvature of the cavity that the system provides, ΔR_0 is the overlap between the designer-

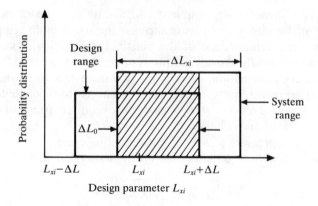

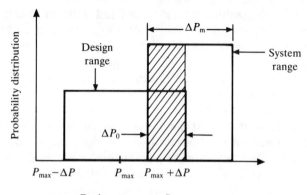

Figure 6.6. Probability distribution of L_{xi} and P_{max}. In each case it shows the tolerance the design range, and the system range that is the tolerance the machine can provide.

specified tolerance of the radius of curvature of the cavity and the system tolerance, ΔP_m is the system tolerance on the maximum pressure, ΔP_0 is the overlap between the designer-specified and the system pressure tolerance, ΔT_m is the system tolerance on the maximum temperature of the plastic, and ΔT_0 is the tolerance overlap between the designer-specified and the system temperature tolerance.

In expressing the information content associated with the quality of lense, it was assumed that the variation in lense quality due to thermal shrinkage of the plastic is a second-order effect. Equation 6.5 may be further simplified if ΔL_{xi}, ΔL_{yi}, ΔL_0, ΔR_i, and ΔR_0 are the same for all cavities.

It should be noted that the uncoupled design shown in Fig. 6.5 is just one of many possible uncoupled solutions. For example, another solution may be to heat the mold to the same temperature as the incoming plastic temperature for isothermal flow of plastic, and then cool the mold and plastic upon completion of the filling process. When a large number of uncoupled solutions are available, the optimal solution among the known solutions is the one with the least information content. We cannot come up with *the best* design, since we cannot conceive of *all* possible solutions.

Concluding Observations

In this case study we had to modify the design of the product in order to be able to make the product without affecting the FRs of the product. This was necessary because the proposed manufacturing process (i.e., PVs) coupled the FRs of the product. If the FRs of the product were such that they could not be altered, a different manufacturing process, which maintains the independence of the FRs of the product, should have been designed. This fact may be stated as a theorem as follows.

> Theorem 15 (Design–Manufacturing Interface)
> When the manufacturing system compromises the independence of the FRs of the product, either the design of the product must be modified or a new manufacturing process must be designed and/or used to maintain the independence of the FRs of the product.

This theorem is a consequence of Theorem 9 (design for manufacturability).

6.3 Vented Compression Molding

Statement of the Problem

The external tank of the Space Shuttle Program of NASA is shown in Fig. 6.7. This tank, which carries liquid fuels (i.e., oxygen and hydrogen), is made of aluminum shell. The exterior of the tank has to be insulated with a material of low thermal conductivity, in order to prevent the formation of ice and to limit excessive heat transfer to the liquid hydrogen and oxygen. The FRs of the design of the product are *thermal insulation* and *light weight*. Figure 6.8 shows the insulation layer placed on top of aluminum substrate. One of the materials used as an insulator is a lightweight composite consisting of silicone rubber and various fillers, such as hollow glass spheres. This is a very expensive material, costing about $50.00 per pound. The insulating material in an uncured, original state has the consistency of wet sand, and therefore does not flow easily in the mold.

The thermal conductivity and diffusivity are controlled by the thickness of the insulating layer and its density. The density must be less than that of the fully compacted material, so as to leave a right amount of voids. As a result, the density of the materials must be controlled accurately during the compaction process so that after curing the specific gravity is between 1.6

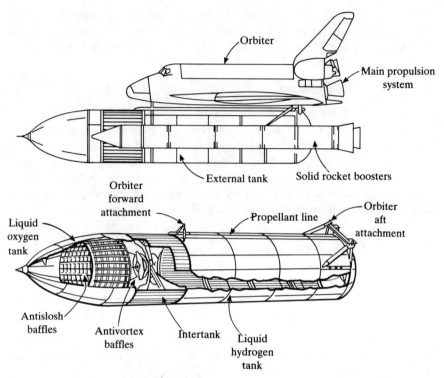

Figure 6.7. External tank of the Space Shuttle before separation from the orbiter.

and 1.7. The desired thickness is $\frac{1}{2}$ in. Other variables are not as critical. Therefore, the control of the density and the thickness of the foam are two DPs of the manufacturing process.

The original process that was used to mold the insulating material on the aluminum substrate was as follows:

1. Distribute and pack by hand the uncured insulating material over the aluminum shells.

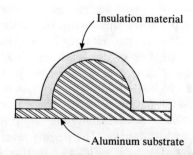

Figure 6.8. Thermal protection system (TPS) for the Space Shuttle's external tank. The FRs for the manufacturing processes are: thickness of the insulation layer, and the thermal conductivity and diffusionivity of the insulation layer, which are controlled by the density of the insulation material.

2. Cover the entire assembly with polyethylene sheet and draw a vacuum to apply uniform pressure over the insulating material and thus control the density. (Note: this process is sometimes known as vacuum bagging.)
3. Cure it in an oven.
4. Machine off the excess cured insulating material using numerical-control milling machines specifically equipped with plastic cutters so as to prevent possible damage.

The prime contractor for the external tank asked MIT to review this original manufacturing process and see if improvements could be made in the process, so as to lower the manufacturing cost and to minimize the material wastage.

An analysis of the entire manufacturing operation indicates that the original process does not violate Axiom 1. The process is a decoupled (i.e., quasi-coupled) process (i.e., the design matrix is a triangular matrix). However, the process involves many unnecessary and redundant steps. Therefore, the process uses excess information to satisfy the two FRs: the density and thickness of the cured insulating material. Since there are more PVs than DPs in the original process, it is not an ideal design per Theorem 4. Is there a better manufacturing process?

Solution

This problem was solved by McCree and Erwin (1984) as part of the MIT–Industry Polymer Processing Program. They noted that the DPs of the product (or the FRs of the process) are:

DP_1 = The thickness of the insulating material: 0.5 ± 0.05 in.
DP_2 = The specific gravity of the molded material: 1.65 ± 0.05. (Since the density is a monotonic function of the compaction pressure, DP may be stated as: control the pressure to 20 ± 0.05 p.s.i.)

Since these are two DPs, the ideal design requires that there be two PVs to render an uncoupled design. To minimize information content per Axiom 2, it is desirable to develop a process that is as simple as possible. Conventional compression molding process cannot be used, however, since the material is compressible, and the density can vary throughout the molded part, depending on the distribution of material.

Knowing what the FRs and DPs are, we have to search for PVs that will yield a diagonal matrix. That is, we have to look for a PV that can be used to change DP_1 and DP_1 only, and, similarly, a PV that changes only DP_2. The process of selecting these PVs is the most important element in the design of the manufacturing process.

A solution to the problem is the vented compression molding shown in Fig. 6.9 (McCree, 1981). It consists of a screen mold which is made of wire net with a predetermined opening. When the "net" mold is pushed

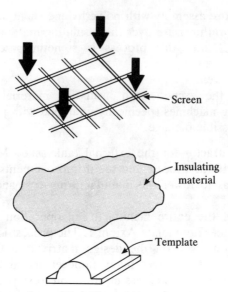

Figure 6.9. Vented compression molding. The mold is made of wire screen. Therefore, when the mold pushes through the loose, uncured insulating material, the excess material extrudes through the opening of the wire mesh. The pressure in the mold is controlled by the size of the opening of the wire mesh.

through the unpacked and uncured insulating material, a constant pressure is developed in the material below the moving net, yielding a constant packing density. Throughout the molding process, the pressure is constant since the pressure is simply a function of the size of the net opening. As the screen mold advances through the insulating material, the excess material extrudes through the opening of the net. When the vented screen mold reaches a preset distance away from the substrate, the excess material above the wire net is vacuumed away, leaving only the molded part. This molded part is then cured, yielding the final molded product which does not require any machining. This simple process replaced all of the complex manufacturing operations originally used.

The PVs for this simple process design are:

$$PV_1 = \text{Mold gap control.}$$
$$PV_2 = \text{Wire net opening.}$$

Then, the design equation may be expressed in terms of the DP vector and the PV vector for this process as follows:

$$\begin{Bmatrix} \text{Thickness } t \\ \text{Density } \rho \end{Bmatrix} = \begin{bmatrix} \times & 0 \\ 0 & \times \end{bmatrix} \begin{Bmatrix} \text{Mold gap control} \\ \text{Wire net opening} \end{Bmatrix} \qquad (6.5)$$

Without putting in detailed expressions for ×s (see McCree and Erwin, 1984, for a detailed investigation), it is obvious that this process is an uncoupled process, satisfying Axiom 1. Strictly, the nondiagonal elements are not absolute zeros, but they are assumed to be negligible, based on Theorem 8.

The information content of the vented compression molding technique is only associated with the pressure and the thickness control. If we denote the system ranges as $P_0 \pm \Delta P_s$ for pressure and $t_a \pm \Delta t_s$ for thickness, and similarly the design ranges as $P_0 \pm \Delta P_d$ for pressure and $t_0 \pm \Delta t_d$ for thickness, the tolerances may overlap as shown in Fig. 6.10. The information content for the process may then be written for a uniform probability distribution function as

$$I = I_p + I_t = \log(\Delta P_s/\Delta P_d) + \log(\Delta t_s/\Delta t_d) \qquad (6.6)$$

When $\Delta P_s < \Delta P_d$ (or $\Delta t_s < \Delta t_d$), the information required for the pressure (or thickness) is equal to zero, since the overlap in the tolerances becomes equal to the system capability. That is, every time the screen mold compresses the material, an acceptable product is produced without the intervention of human operators.

Concluding Observations

This case study illustrates several important points. First, when there are more DPs than FRs (or, similarly, more PVs than DPs), the design is redundant (per Theorem 3) or more complicated than is necessary. Therefore, one should look for a better design solution, although the original design satisfies the Independence Axiom. Secondly, the designer must deliberately search for DPs and PVs that yield a diagonal design matrix for both the FR–DP and the DP–PV relationships. Thirdly, many existing manufacturing operations may be much more complex than is necessary.

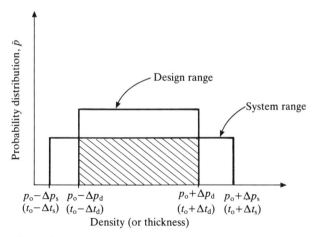

Figure 6.10. The probability distribution functions for the designer-specified pressure range (and thickness), and for the manufacturing system tolerance range. Subscript d denotes the designer-specified tolerances, and subscript s denotes the system tolerance.

204   The Principles of Design

It is worth noting there that 18 months after this project was undertaken by the MIT–Industry Polymer Processing Program, the part was commercially made and flew on the Space Shuttle. McCree and Erwin were cited for their contribution. Their process saved NASA tens of millions of dollars.

6.4 Robot Application in Assembly of Molded Parts

This case study is somewhat different from the two preceding case studies, in that the design of the product is considered separately from the design of the manufacturing operations. This case study is nevertheless related to the manufacturing of products.

Statement of the Problem

A major corporation in the United States has been manufacturing an instrument box by injection molding two halves of the box with impact-grade polystyrene, and by welding them together ultrasonically (see Fig. 6.11). The original process used by the firm involved the following manufacturing operations: the use of conveyor lines on which the injection molded parts were dropped upon demolding, orientation devices to line up and stack the parts properly in a queueing line (a buffer), and the ultrasonic device that welds the two halves together.

One of the engineers reasoned that the process was not very efficient, since the parts in the mold with known orientation and location were dropped on the conveyor belt, losing the information content associated with the orientation of the part, and thus requiring the addition of new information to reorient the parts properly before welding. As a means of overcoming this shortcoming, the engineer suggested to his management that the company purchase a robot to pick up the parts from the mold, assemble and weld them together (see Fig. 6.12). The engineer justified his recommendation as being more economical, since the use of the robot eliminated the need for the conveyor and the orientation device.

The engineer's recommendation appears to be reasonable at a first glance. If you were the chief engineer who had to make the decision,

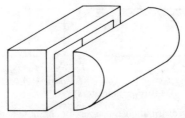

Figure 6.11. Two injection molded parts made of polystyrene. These two parts are welded together ultrasonically to manufacture an instrument box.

(a)

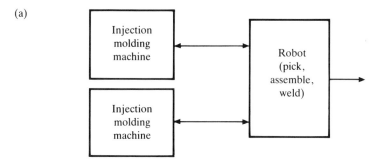

(b)

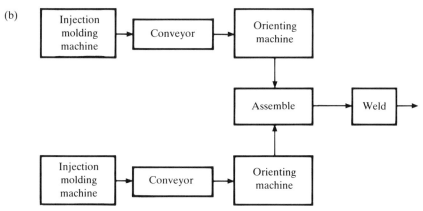

Figure 6.12. Assembly of two injection molded parts. (a) New proposed manufacturing system; (b) original manufacturing system.

would you approve the request? Indeed, the company went ahead and implemented the concept. Contrary to their expectation, the productivity went down when the robot was used! What was wrong with the idea?

Solution

The problem stems from the fact that the new system based on the robot is a quasi-coupled (or decoupled) system, as is shown later. The entire production line goes down when any one of the injection molded machines breaks down or makes imperfect parts. This problem was taken care of in the original system by using the conveyor belt as a buffer. The buffer gave sufficient tolerances to the production rate of the injection molding machines that, even when they broke down occasionally, the operation of the entire system was not affected. Also, a bad part could be eliminated from the conveyor belt before bonding it to the other part, increasing productivity at the same time.

The FR of the manufacturing system, at the highest level of the FR hierarchy is

FR = Make the instrument box with polystyrene.

The DP selected to satisfy this FR is

DP = Use the injection molding process to make the box.

Then, at the next level of the FR hierarchy, the original FR is decomposed into the following set of FRs for the new manufacturing process:

FR_1 = Make part A.
FR_2 = Make part B.
FR_3 = Assemble A and B.
FR_4 = Bond the parts A and B.

The DPs selected to satisfy the above set of FRs are

DP_1 = Injection mold part A.
DP_2 = Injection mold part B.
DP_3 = Pick and assemble the parts with a robot.
DP_4 = Weld the parts with the ultrasonic device.

The relationship between FRs and DPs is given by the design equation as

$$\begin{Bmatrix} FR_1 \\ FR_2 \\ FR_3 \\ FR_4 \end{Bmatrix} = \begin{bmatrix} A_{11} & 0 & 0 & 0 \\ 0 & A_{22} & 0 & 0 \\ A_{31} & A_{32} & A_{33} & 0 \\ A_{41} & A_{42} & A_{43} & A_{44} \end{bmatrix} \begin{Bmatrix} DP_1 \\ DP_2 \\ DP_3 \\ DP_4 \end{Bmatrix} \quad (6.7)$$

A_{31}, A_{32}, A_{41}, and A_{42} are not equal to zero, since without Part A and Part B, assembly and welding can not take place. Similarly, A_{43} is not equal to zero, since the parts must be assembled before they can be welded. Since this is a quasi-coupled system, the DPs must be controlled in the sequence given in order to be able to vary FRs independently of each other. Consequently, the system cannot operate when any one of its components fails. Furthermore, the production rate is controlled by the slowest machine.

The information contents involved for FR_3 and FR_4 are large, since conditional probabilities are involved. The information may be expressed as

$$I = -[\log p_{11} + \log p_{22} + \log p_{33} + \log(p_{33}/p_{32})$$
$$+ \log(p_{33}/p_{31}) + \log p_{44} + \log(p_{44}/p_{41})$$
$$+ \log(p_{44}/p_{42}) + \log(p_{44}/p_{43})] \quad (6.8)$$

where p_{ii} is the probability associated with satisfying FR_i by varying DP_i, whereas (p_{ii}/p_{ij}) is the conditional probability of satisfying FR_i by varying DP_j. Obviously, when p_{ij} is small because of the occasional malfunctioning

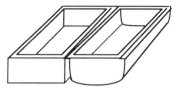

Figure 6.13. One-piece injection molded part that can be folded over and assembled.

of the injection molding machine, the information required becomes very large.

The original FR could also have been decomposed in a different manner. For example, we could have made Part A and Part B in one piece as shown in Fig. 6.13, folded it over, and welded. The FRs and DPs for this design decision may be stated as

FR_1 = Make a one-piece part.
FR_2 = Make into the box shape.
FR_3 = Join the upper and lower parts together.

The corresponding DPs are

DP_1 = Injection mold the integrated piece.
DP_2 = Fold along the hinge.
DP_3 = Weld ultrasonically.

In this example we have made use of a very important corollary—Corollary 3. This corollary states that parts should be *integrated* into a single part if the integration of the parts (in the physical domain) can be done without compromising the independence of FRs. By joining two halves of the part using a web, all of the information associated with part orientation, selection, and assembly were eliminated, thus minimizing the information content.

Concluding Observations

This case study illustrates the need to analyze design decisions, since some correct decisions are counterintuitive. A qualitative design analysis such as that presented in this section may not be sufficient when the relationship between the FRs and DPs is complex.

6.5 Use of Axioms in the Development of Intelligent Brakeforming Machines

In some manufacturing-related situations, the design of the product cannot be altered. In this case the machine must be so designed that it accommodates the differing requirements of the product. For this reason, various different machines have been designed. This case study involves the modification of one of these machines to make it more intelligent.

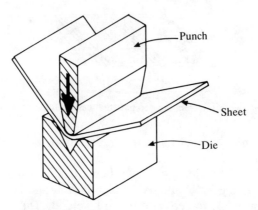

Figure 6.14. Typical brakeforming operation.

Statement of the Problem

It is clear that the factory of the future will make use of more-intelligent machines that are capable of making their own decisions. This will minimize the required data base and simplify the information flow between the central computer and individual machines. One of the basic requirements of the intelligent machine is that it must be an uncoupled system. Otherwise, the machine will be difficult to control and will require continuous fine tuning. A coupled machine cannot be reliable or dependable. The development of an intelligent brakeforming press is used to illustrate the basic approach to intelligent machines.

The brakeforming operation is one of the most commonly used sheet metal forming processes. For example, a large number of airplane parts are made of bent aluminum sheets. In a brakeforming operation a sheet metal is bent into a bend angle θ_f, using a die and punch (see Figs. 6.14 and 6.15). The punch pushes the sheet metal, which is supported by the die. It is a three-point bending operation. This is done by an experienced

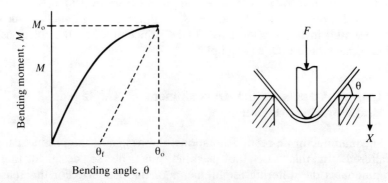

Figure 6.15. Moment–curvature relationship in sheet metal forming. The bending angle θ and the displacement X are defined by the schematic diagram on the right.

operator who chooses the correct die and punch, and then sets the punch for correct displacement through a lengthy trial-and-error process.

It is not easy to achieve the specified bend angle. The sheet must be overbent to an angle θ_0 so that, when the load is released, it springs back by an angle $(\theta_0 - \theta_f)$ as shown in Fig. 6.15. The unloading follows the dotted line, which shows that the unloading is elastic. The difficulty in achieving the correct bend angle is caused by the variation and anisotropy in the materials properties, especially the work-hardening characteristics, which results in a different amount of elastic springback.

The design problem to be solved is the development of an intelligent brakeforming machine that will always yield the correct bend angle without the intervention of a skilled operator, regardless of the variation in the material properties of the sheet metal. How should be design such a machine? Should we generate a large data bank on material properties and then preset the machine before an individual sheet is bent?

Solution

The FR at the highest level of the FR hierarchy is that the bend angle be $\theta_f \pm \Delta\theta_f$, regardless of how the material properties vary.* Based on the constitutive relationship between the bending moment, M, and the bending angle, θ (shown in Fig. 6.15), the FR, θ_f, may be decomposed into the following three FRs:

$$FR_1 = M_0$$
$$FR_2 = \theta_0 \quad (6.9)$$
$$FR_3 = \tan^{-1} EI$$

where M_0 is the maximum bending moment reached before unloading, θ_0 is the corresponding maximum bend angle, and $\tan^{-1} EI$ is the slope of the unloading curve. The problem then becomes one of choosing the correct DPs to yield a diagonal matrix.

It is easy to see that M_0 can be satisfied if the force exerted by the punch, F, can be measured for a fixed die shape; θ_0 can be controlled by regulating the displacement of the punch, X_a, and the elastic springback can be determined by monitoring the springback displacement of the punch, ΔX_a, when the load is released from any M and θ. This can be

---
* In this case study the product that we want to make has already been decided—bent sheet metal. Therefore, we have defined $\theta_f \pm \Delta\theta_f$ to be the FR. On the other hand, if the problem is not constrained a priori to a sheet-forming operation, the FR may be stated as: produce a bent metal part with a constant thickness. Then, an acceptable design solution may be a form of die casting or sheet metal bending or extrusion process. For whatever reason, if we choose the sheet metal bending operation, then the DP may be stated as: control the bend angle to be $\theta_f \pm \Delta\theta_f$. We will then choose appropriate PVs. We could therefore replace all FRs with DPs and all DPs with PVs without any effect on the solution. The DP–PV relationship must satisfy Axiom 1, just as teh FR–DP relationship must, per Theorem 9.

represented mathematically in the form of a design equation as

$$\begin{Bmatrix} M_0 \\ \theta_0 \\ \tan^{-1} EI \end{Bmatrix} = \begin{bmatrix} \times & 0 & 0 \\ 0 & \times & 0 \\ 0 & 0 & \times \end{bmatrix} \begin{Bmatrix} F \\ X_a \\ \Delta X_a \end{Bmatrix} \quad (6.10)$$

Such an intelligent brakeforming machine was built at the MIT Laboratory for Manufacturing and Productivity (Stellson, 1981).* The machine's operating sequence was to bring down the punch and then stop instantaneously and unload in the midst of the forming operation, to measure the springback. Once the springback is determined, the relationship between M_0 and θ_f can be used to calculate the final punch displacement X_a. The machine produced the desired bend angle within 1 degree, regardless of the variations in the work material, the first time and all of the time, which is better than what the experienced operator could produce.

This kind of intelligent machine does not require an extensive data bank on materials and on operating conditions of the machine. It also does not need a closed-loop feedback control involving the measurement of the actual final bend angle and the control of individual machine operations from a central computer. It needs sensors and a local computational capability that may be connected to a central computer for inventory control purpose only.

Concluding Observations

The same concept can be used to develop other kinds of intelligent machines, such as injection molding machines for plastics. The basic methodology of the axiomatic approach, to reiterate the key points, is to define FRs to be fulfilled by the machine and then choose DPs that yield an uncoupled machine; that is, a machine that can be characterized by a diagonal design matrix.

6.6 USM Foam Molding Process

Statement of the Problem

USM Corporation was interested in selling machines that can make a foamed plastic shoe sole, shown in Fig. 6.16. The core of the sole is to made of foamed polyvinylchloride (PVC) of uniform density ρ for light weight, comfort, and flexibility. The outer skin is to be a solid PVC layer

---

*A detailed analysis of the forming operation and the feedback control of the bending operation was done by Stellson (1981) as part of his doctroal thesis at MIT. However, the detailed analysis is not given here, since the approximate analysis presented here is sufficient for general conceptual design of the intelligent breakforming machine.

Manufacturing-Related Case Studies

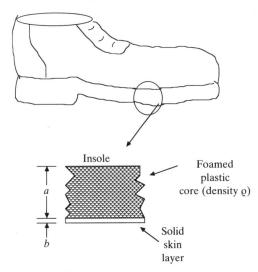

Figure 6.16. Construction of the USM foamed plastic shoe sole. The core is to be a foamed structure and the outer layer is to be an unfoamed, solid plastic layer.

of thickness b for good wear resistance. It was decided that the machine should be based on the injection molding process (see Fig. 6.3), since the firm was the major supplier of injection molding machines to the shoe-making industry. A young engineer was asked to design a manufacturing process that can manufacture the shoe sole inexpensively.

Solution

In response to the problem statement, a new process was developed (Suh, 1961), which is known as the USM high-pressure foam molding process. The process is now extensively used to make furniture panels, automobile components, and television cabinets. The solution is a straightforward application of the axioms, although at the time of invention these principles were intuitively used, since no formal axiomatic statement was available.

Based on the perceived needs of the marketplace that the Corporation was trying to satisfy, the FRs of the product may be written as

FR_1 = Flexibility.
FR_2 = Light weight.
FR_3 = Wear resistance.

The product that was designed to satisfy these FRs is shown in Fig. 6.16. The DPs of the product are

DP_1 = Thickness of the foamed core, a.
DP_2 = Foamed plastic core of uniform density ρ.
DP_3 = Solid plastic layer of uniform thickness, b.

The design matrix relating these DPs to FRs is a diagonal matrix, satisfying Axiom 1.

Knowing what the DPs are, we must now search for PVs that can yield a diagonal design matrix. Since the decision was made that the manufacturing process be the injection molding process, the PVs that can satisfy DPs and FRs per Theorem 9 and Eq. 4.27 must now be determined. In order to obtain a uniform density ρ throughout the core of the shoe sole, where the thickness varies significantly, especially the heel section, it is clear that the material must be uniformly distributed in proportion to the thickness of each section of the sole. However, a compressible material such as foamed plastic cannot be uniformly distributed by injection molding unless uniform pressure can be applied to the compressible material, which is not possible in the conventional injection molding process since the pressure is different throughout the mold. On the other hand, the incompressible material such as solid plastic can be distributed to fill a given space, even when the pressure distribution is not uniform. This reasoning leads to the solution.

However, before explaining the solution any further, the process of injection molding foamed plastic is described here briefly for those readers who are not familiar with polymer processing. Plastic granules are fed into the plasticator (see Fig. 6.3) with blowing agents. The blowing agents can be in the form of a chemical compound (typically a fine powder) dusted on the pellets, that decomposes into gases at the processing temperature in the plasticator, or a physical compound that vaporizes at the processing temperature. In processing PVC, the chemical blowing agent in the form of fine powder is used. This blowing agent is thoroughly mixed in the plasticator with PVC, but the decomposed blowing agent cannot expand in the plasticator, due to the high pressure imposed on the plastic. When the plastic is extruded or injected into a low-pressure region, it expands, forming a cellular structure because of the expansion of the gas. However, if this plastic with foaming agent is injected into the mold, the density of the part will vary a great deal, due to the variation in pressure and high compressibility of the foamed plastic.

Now, returning to the specific solution sought for the problem at hand, it is clear that the molten plastic with foaming agent must be prevented from foaming until the incompressible material is uniformly distributed throughout the mold. When the incompressible, unfoamed plastic is uniformly distributed, we can let the plastic foam under uniform pressure to create foamed plastic with a desired density. All of these conditions can be satisfied by using an expandable mold. The idea is as follows:

1. Make the mold such that its cavity depth can be varied by moving one side of the mold as shown in Fig. 6.17.
2. Narrow at the gap of the mold before injecting the plastic, and then inject the plastic at high velocity. The narrow gap and the high injection rate will increase the pressure distribution throughout the molten plastic flowing into the mold, except at the plastic flow front, which is at

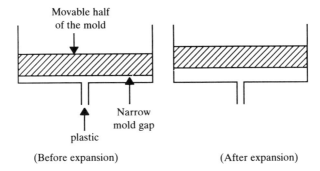

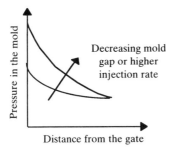

Figure 6.17. Schematic illustration of the proposed solution. Initially the mold gap is narrowed to prevent foaming due to high pressure, and to fill the mold with unexpanded plastic. Then the mold is expanded. The thickness of the solid skin layer is determined by the temperature of the movable half of the mold and the time elapsed before expansion.

the ambient pressure. This is illustrated schematically in Fig. 6.17. Because of the high pressure, foaming will not take place until the mold is completely filled with incompressible plastic.

3. Knowing how incompressible material is distributed in the mold, the mold volume can be expanded by moving the movable half of the mold. The final density is controlled by the expansion ratio of the mold.

4. Since plastic cannot expand at low temperature even when blowing agents are present, the thickness of the skin layer at the surface can be controlled by lowering the temperature of the plastic near the mold surface. The thickness is thus controlled by cooling the mold surface, and by varying the elapsed time before the mold expansion.

For the ideal manufacturing process, the design equation that we are looking for should be of the form

$$\begin{Bmatrix} DP_1 \\ DP_2 \\ DP_3 \end{Bmatrix} = \begin{Bmatrix} a \\ \rho \\ b \end{Bmatrix} = \begin{bmatrix} \times & 0 & 0 \\ 0 & \times & 0 \\ 0 & 0 & \times \end{bmatrix} \begin{Bmatrix} PV_1 \\ PV_2 \\ PV_3 \end{Bmatrix} \qquad (6.11)$$

That is, we are looking for a process design that will yield PV_1, PV_2, and PV_3 such that the design matrix is a diagonal matrix. If we let the expansion of the mold be PV_1 (denoted by ε), the injection velocity PV_2 (denoted by V), and the temperature of the mold surface be PV_3 (denoted by T), the design equation for the proposed manufacturing process may be written as

$$\begin{Bmatrix} a \\ \rho \\ b \end{Bmatrix} = \begin{bmatrix} \times & 0 & 0 \\ \times & \times & 0 \\ 0 & \times & \times \end{bmatrix} \begin{Bmatrix} \varepsilon \\ V \\ T \end{Bmatrix} \quad (6.12)$$

The design matrix for the manufacturing process states that the proposed design is a decoupled process. Instead of using T as PV_3, the cooling time t could have been used to vary b. However, the cycle time consideration dictated that t be fixed and temperature be varied.

After the proposed design was implemented, good products were made. However, in some cases the density of the foam near the edge was lower due to larger bubbles, because the pressure at the flow front is always atmospheric. In order to solve this problem, an additional FR must be added to Eq. 6.12. How would you solve this problem?

6.7 Mixalloy Process

Statement of the Problem

A goal of metallurgy is to produce metals and alloys with desired properties through the control of microstructures and chemical compositions. To attain this goal, thermodynamics and kinetics of microstructural transformations have been studied extensively in the past. However, the design and processing of alloys still remain an art, although microstructural development during processing is understood well. This is largely due to coupling between the chemical composition and the processing conditions. In many conventional processes the microstructure and chemical compositions are inherently coupled by the constraints imposed on the process by the laws of nature (Suh, 1982, 1984).

In conventional metal processing, liquid metals are first cast into ingots that are then subsequently subjected to many intermediate processes before the final product is made. Ingots are hot- and cold-worked to improve the properties as well as to change the shape of the metal. The mechanical properties are enhanced through microstructural changes when dendrites and second-phase particles break up. After machining the worked metal to near-net shape, sometimes the parts are heat-treated to obtain the final desired properties through phase transformation, precipitation of second-phase particles, and annealing.

To a large extent these conventional metal-processing techniques are dictated by equilibrium thermodynamics and the kinetics of phase transformation. The composition of the solid is selected on the basis of the solubility of various elements and the stability of solutions. The number of phases in equilibrium is controlled by equilibrium thermodynamics. The microstructure is controlled by manipulating the rate of phase transformation and by subjecting the metal to mechanical deformation. The kinetics of phase transformation is, in turn, affected by the rate of temperature change and the chemical composition.

Therefore, any change in the chemical composition and/or process conditions results in major changes in all final properties of the metal. For example, it is difficult to change just one final characteristic (such as ductility) of metals by varying just one process (i.e., design) parameter, and/or chemistry of the metal without affecting the rest of the characteristics (such as hardness, electrical conductivity, and the modules). Accordingly, these characteristics are generally coupled in conventional metallurgical processes.

A metallurgical system which does not allow the changes in any one of the properties without altering other properties when any one of the input variables is changed, is a coupled system. The properties are controlled by the grain size, the volume fraction of the phases, and the chemical composition of each phase, each of which cannot be varied independently without affecting others. In this respect, most conventional metal processing techniques are coupled processes.

The design problem before us is to develop an uncoupled ideal metal microstructure and an uncoupled metallurgical process. As discussed in the next section, a solution for the metal structure is to create an alloy that has a large number of very small, hard, insoluble particles (of the order of 500 Å) with a short mean free path (again of the order of 500 Å) inside a metal matrix. Such a microstructure will yield high strength, provide hot hardness, and give up high creep resistance. However, such a microstructure is difficult to achieve through conventional processes, as mentioned earlier.

One way of producing such a structure in copper alloys is through internal oxidation. In this process, copper with a small amount of dissolved aluminum is exposed to an oxygen environment, which then diffuses into the copper. When oxygen atoms react with aluminum atoms, they form small aluminum oxide (Al_2O_3) particles inside the copper matrix. Such a Cu–Al_2O_3 alloy is called the dispersion-hardened alloy.

The internal oxidation process is limited to special alloy systems, and only a limited volume fraction of hard particles can be introduced this way. Another possible way of creating such a structure is through powder metallurgy, where fine particles of different compositions are thoroughly mixed and compacted. However, such a powder metallurgy process cannot produce a finely dispersed structure without extensive mechanical working, which is then called mechanical alloying. The task here is to design a new method of producing dispersion-hardened metal alloys without the use of a lengthy diffusion process.

Ideal Two-phase Structure

Before discussing the specific process, we first consider the ideal structure for a dispersion-hardened alloy and define a few terms. The *ideal* two-phase structure of Cu–Al_2O_3 alloy is illustrated schematically in Fig. 6.18. The microstructure shows a pure copper matrix phase without any alloying elements, very small (100 Å) aluminum oxide particles that are bonded strongly to the matrix phase, and the spacing between the oxide particles of a few hundred to 1,000 Å. In such a structure the yield strength in shear, τ, is governed by spacing of the hard particles, l, and the shear modulus, G, of the matrix phase, since the shear stress required to extrude dislocations by the Orowan mechanism is given by

$$\tau = 2Gb/l \tag{6.13}$$

where b is the Burgers vector, whose magnitude is about 5 Å. In the case of iron matrix, G is 11×10^6 p.s.i.

Furthermore, since aluminum oxide does not dissolve in the matrix even at the melting point of the matrix, the material can have high strength and good creep resistance even at these extremely high service temperatures. It can be shown theoretically and experimentally (Argon, 1976; Su, 1980) that when the particles are smaller than a critical size (approximately 200–300 Å), cracks cannot nucleate around these particles, because the energy criterion for crack nucleation cannot be satisfied. Therefore, this ideal material has not only high strength, but also excellent toughness. It should be noted that the hard particles must not be too small, since dislocations will then cut through them; that is, they should be sufficiently large to block dislocations, but not large enough to become crack-nucleation sites.

Furthermore, since the matrix phase can be a single element or a solution, the matrix phase can control certain properties such as electrical conductivity and flow strength. When the matrix consists of a pure single element, the electrical conductivity is nearly that of the pure element, whereas in commercial alloys conductivity is reduced substantially, due to

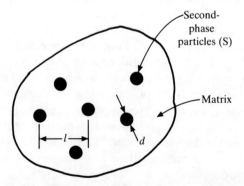

Figure 6.18. Ideal structure of two-phase metals.

the use of substitutional alloying elements. By adding a small amount of solute, the yield strength can be increased.

When it is desired to make a high-strength metal without the use of hard particles (or even with the use of hard particles), we can increase the strength by making very small grains, so as to promote dislocation pile-up at grain boundaries. It can be shown that the yield strength, σ_y, of metals, is related to the grain size by

$$\sigma_y = \sigma_0 + Kd^{-1/2} \tag{6.14}$$

where σ_0 and K are material constants, and d is the grain size. Unfortunately, conventional metallurgical processes cannot easily produce small grain size of $1\,\mu\text{m}$ or less, especially in the case of face-centered cubic (f.c.c.) metals. If d can be reduced by a factor of 100 (for example, from 100 to $1\,\mu\text{m}$), the yield strength can be increased by nearly a factor of 10.

The Mixalloy Process

In view of the shortcomings of conventional metallurgical processes, a new process, called the Mixalloy Process, was created (Suh, 1981a,b). This new invention is in some ways a result of the use of analogy in creative design discussed in Chapter 1. We investigated the impingement mixing process for the reactive processing of polymers. This knowledge on mixing triggered the basic concept for the Mixalloy Process. The Mixalloy Process for making the ideal metal structure consists of the following two basic steps:

1. Mixing and casting of liquid components to create a fine mixture of two solid phases with very small grains.
2. Chemical reaction between the elements dissolved in two liquid phases during or subsequent to the mixing and casting process.

There can be many variations and permutations of these basic steps (Suh, 1981a,b; Suh et al., 1982). The second step necessarily follows the first step, but the process does not have to involve the second step in many applications.

Step 1. The mixing and casting process of the mixalloys is shown in Fig. 6.19. The Mixalloy Process consists of two of more molten-liquid reservoirs, the same number of conduits (pipes) through which the molten liquids are pumped into the mixing chamber, the same number of nozzles for injection of the liquid streams into the mixing chamber, and a chilled mold in which the thoroughly mixed molten metals are cast. The liquid jets flowing into the mixing chamber assume turbulent motion due to their high Reynolds number as they emerge from the nozzle. These turbulent jet streams impinge against each other in the mixing chamber, creating a homogenous mixture on both the macro- and the microscale.

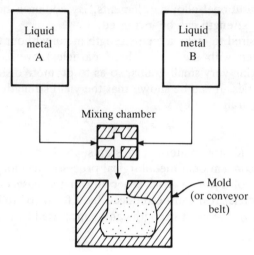

Figure 6.19. Mixalloy process.

The size, l_m, of turbulent eddies (which may be thought of as rigid-body motion of liquid particles) decreases with the increase in the Reynolds number, Re, based on the diameter of the nozzle, D, as (Tucker and Suh, 1980)

$$\frac{l_m}{D} = A \, \mathrm{Re}^{-3/4} = A\left(\frac{\rho V D}{\mu}\right)^{-3/4} \tag{6.15}$$

where V is the velocity of the liquid, μ is the viscosity, ρ is the density, and A is a proportionality constant. Since, for liquid metals, the viscosity, μ, is very low and the density, ρ, is very large, and since A is of the order of unity, small eddies of the turbulent flow can be created very easily at low head pressures. For example, in processing lead and tin, a 40-p.s.i. head pressure can easily raise the Reynolds number to about 20,000, making the eddy size approximately 1 μm. The experimental results agree quite well with the theoretical prediction given by Eq. 6.15 (Suh et al., 1982; Tucker and Suh, 1980).

Now consider the case of two liquid streams, A and B, impinging against each other in the mixing chamber. Due to the large-scale mixing motion, the eddies of A and B mix uniformly in the mixing chamber. Although the turbulent motion decays rapidly, the eddies maintain their size for a sufficient time, if the time constant for coalescence is larger than that for the large-scale flow of the liquids out of the mixing chamber. When this liquid stream flows into the chilled mold or is cast on a chilled plate, the eddy size is retained in the frozen structure of the metal. The eddies near the center of the mold may coalesce during the solidification process. The growth of the grains can be controlled by altering the heat transfer rate through the change in the thickness of the part and mold temperature.

When the liquid stream A has a significantly higher melting point than

B, the initial temperatures of A and B may be chosen so that the eddies of the metal with the high melting point (i.e., metal A) may be made to freeze almost instantaneously upon contact with the eddies of the opposing stream (i.e., metal B), which is at a lower temperature. The solidification of the lower-melting phase that surrounds the frozen particles of higher-melting phase can then take place in the mold (see Appendix 6.A).

If A and B can form intermetallic compounds, the interface of A and B eddies will consist of intermetallic compounds. If they are immiscible, a compound consisting of fine grains of immiscible metal phases can also be created through the use of the Mixalloy Process.

The mixalloys can be used after completing only this first step, since they have enhanced mechanical properties, due to their small grain size. We can also make very different metals (and/or nonmetals) with unique metallurgical microstructures and chemical properties, since any combination of elements can be used to make new alloys and compounds, including immiscible systems. However, more interesting results can be obtained by inducing chemical reactions between the constituents in Phase A and Phase B, which is the second step in the Mixalloy Process.

Step 2. Consider, as an example, the process of manufacturing very hard copper which is electrically almost as conductive as pure copper (which is very soft). Such a metal may replace expensive beryllium copper in certain applications. (Beryllium copper is used extensively in the electronics industry, due to its high hardness, strong resistance to stress relaxation, and reasonable electrical conductivity, which is typically 20–50% of pure cooper.) This can be done by creating the ideal structure, shown in Fig. 6.18, through the use of slightly different alloys in Streams A and B.

For this purpose let Stream A be Cu–3% Cu_2O solution and Stream B be Cu–2% Al solution (note that these solutions have a lower melting point than pure copper). Then, after impingement mixing, the two phases will be uniformly distributed in the solidified state, as shown in Fig. 6.19. If the chemical reaction between oxygen and aluminum has not taken place in the mixing chamber, we can heat the solid to a high temperature (say to half of the melting temperature of copper) to promote mass diffusion; aluminum from Phase A will diffuse into B and oxygen from B will diffuse into A, aluminum diffusing faster.

When aluminum meets oxygen, they will react to form aluminum oxide, since the free energy of formation of aluminum oxide is lower than that of cuprous oxide. Once Al_2O_3 nucleates, these nucleated oxide particles will grow larger with the additional reaction and diffusion of Al_2O_3 to the nucleation sites. The final size of the oxide particles will depend on the number of particles nucleated, which depends on the size of turbulent eddies. As a result of the reaction we obtain a solid with pure copper matrix strengthened by small aluminum oxide particles per the reaction

$$Cu/Cu_2O + Cu/Al \rightarrow Cu + Al_2O_3 \qquad (6.16)$$

The FRs of the ideal copper alloy may be written as

FR_1 = High strength, τ.
FR_2 = High toughness, K.
FR_3 = High conductivity, σ.

The design equation for the ideal microstructure of metals shown in Fig. 6.18 may then be written as

$$\begin{Bmatrix} FR_1 \\ FR_2 \\ FR_3 \end{Bmatrix} = \begin{Bmatrix} \tau \\ K \\ \sigma \end{Bmatrix} = \begin{bmatrix} \times & 0 & 0 \\ 0 & \times & 0 \\ 0 & 0 & \times \end{bmatrix} \begin{Bmatrix} l \\ d \\ C \end{Bmatrix} \quad (6.17)$$

where C denotes the purity of the matrix phase, and l, d, and C are DP_1, DP_2, and DP_3, respectively. The particle spacing l does clearly control the strength in view of Eq. 6.13. When d is larger than a critical size, the toughness K is adversely affected. The electrical conductivity of copper decreases precipitously when a small amount of solute atoms are dissolved in the matrix phase.

The design equation for the Mixalloy Process may be expressed as

$$\begin{Bmatrix} DP_1 \\ DP_2 \\ DP_3 \end{Bmatrix} = \begin{Bmatrix} l \\ d \\ C \end{Bmatrix} = \begin{bmatrix} \times & 0 & 0 \\ \times & \times & 0 \\ \times & 0 & \times \end{bmatrix} \begin{Bmatrix} V_f \\ t \\ M \end{Bmatrix} \quad (6.18)$$

where V_f is the volume fraction of aluminum (or oxygen) in the molten liquid, t is the time available for chemical reaction and coalescence, and M the mix quality. V_f, t, and M are PV_1, PV_2, and PV_3, respectively. The assumptions made in writing Eq. 6.18 are: the larger the volume fraction of reactants is, the smaller the mean free spacing; the longer the reaction time is, the larger the particle by coalescence; and when the mixing is poor, the matrix will have solutes dissolved in the matrix. The volume fraction also affects the particle size by accelerating the coalescence process and the electrical conductivity by the presence of particles. These expressions are obtained through qualitative arguments. When the tolerances for the FRs are specified, we can invoke Theorem 8 in order to determine precisely which element of the design matrix can be neglected without committing serious errors.

Equation 6.18 is a design equation that relates the microstructures to the Mixalloy Process, whereas Eq. 6.17 is the design equation for the property–microstructure relationship. The substitution of Eq. 6.18 into Eq. 6.17 yields a design equation which relates the designed material properties directly to the processing parameters. It is interesting to note that the volume fraction of oxides does affect the toughness and electrical conductivity, according to the resultant design equation for the property–processing parameter relationship (see Problem 6.5). This is in accordance with Theorem 9 (the theorem of design for manufacturability).

Figure 6.20 shows the microstructure of dispersion-hardened copper

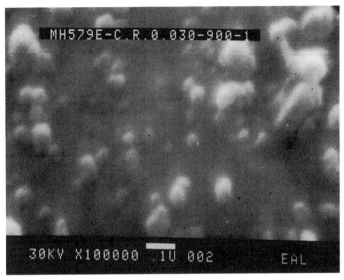

Figure 6.20. Microstructure of dispersion-hardened copper alloys made by the Mixalloy Process. (Courtesy of Sutek Corporation, Hudson, MA.)

alloy produced by Sutek Corporation of Hudson, MA, using the Mixalloy Process. The small particles are hard ceramic particles. These copper alloys made by the Mixalloy Process (MXT-5) have outstanding hot hardness, as shown in Fig. 6.21, and also very good electrical conductivity, as shown in Fig. 6.22. The Mixalloy Process is, indeed, a unique process.

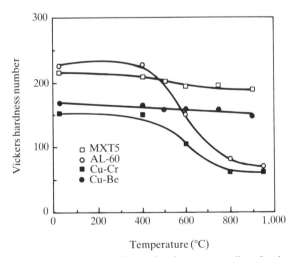

Figure 6.21. The room temperature hardness of various copper allys after heat treatment for 1 hour at the temperatures indicated. MXT5 is a dispersion-hardened copper alloy made by the Mixalloy Process, AL-60 is a dispersion-hardened alloy with Al_2O_3 made by the internal oxidation process, Cu–Be is a beryllium–copper alloy, and Cu–Cr is a copper–chromium alloy. (Courtesy of Sutek Corporation, Hudson, MA.)

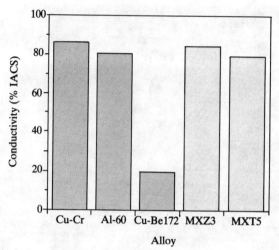

Figure 6.22. Electrical conductivities of various copper alloys. Al-60 is a dispersion-hardened copper alloy with Al_2O_3 made by internal oxidation, Cu–Be 172 is a copper–beryllium, MXZ3 and MXT5 are dispersion-hardened copper alloys made by Mixalloy Corporation. (Courtesy of Sutek Corporation, Hudson, MA.)

Concluding Observations

The Mixalloy Process is conceptually a simple process. However, the actual development of the process for industrial use took extensive research and development work, because of the difficulties involved in handling liquid metals at high temperatures; that is, the lower-level FRs and DPs presented many problems during the implementation stage, which had to be solved to make the process commercially viable. This is typical of many successful development projects. The important point to note in terms of the design axioms is that the design axioms apply at all levels of the FR, DP, and PV hierarchies.

6.8 Method and Apparatus for Mixing Solid Particles

Statement of the Problem

Many processes require the mixing of solid particles of different materials, particularly when such particles are relatively small; for example, of powder sizes in a range of about 1 μm to about 1 mm. Such mixtures are required in mixing dry materials to form pills or other drug forms, in mixing polymeric materials, in mixing ceramics, and in mixing additives to materials. The dry mixing of fine powders is a very difficult process, as they tend to segregate as a function of particle size; to agglomerate, forming clumps; and to compact into a cake. Powders are sometimes mixed in a liquid to separate individual particles from each other and facilitate stirring action for macro- and microscale mixing.

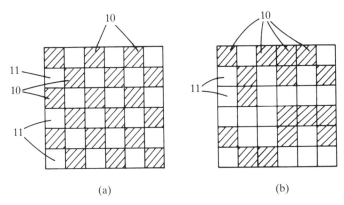

Figure 6.23. Two-component mixture of equal voluem fraction: (a) perfect mixture; (b) random mixture.

In many applications a "perfect" mixture is desired. A "perfect" mixture is defined as one in which each component is evenly distributed throughout the mixture, so that with reference to the smallest sample of interest, the ratio of particle components in every such sample is the same as the ratio of components in the entire mixture. Figure 6.23a shows the distribution of two components of equal volume fraction in a perfect mixture. However, in the absence of any "internal" attractive forces between the particles, thermodynamics does not favor the formation of the perfect mixture. The Gibbs free energy of a mixture, ΔG, may be expressed as

$$\Delta G = \Delta H - T \Delta S \qquad (6.19)$$

ΔH is the change in the enthalpy of the system due to mixing, ΔS is the change of the entropy of the system due to mixing, and T is the absolute temperature of mixing. Mixing occurs when ΔG decreases due to mixing. In the absence any change in enthalpy (i.e., $\Delta H = 0$), the mixing is dominated by the entropic term, ΔS. ΔS is the maximum for a "random" mixture, which is larger than ΔS for a perfect mixture. When $\Delta H < T(\Delta S_p - \Delta S_r)$, where the subscripts p and r denote perfect and random mixtures, respectively, a perfect mixture can result from the mixing operation.

In mixing powders, ΔH is typically nearly equal to zero and, therefore, the mixture tends to be "random." A random mixture is a mixture in which the probability of finding any particle of a specified type is the same at all points in the mixture. Figure 6.23b shows the distribution of two components of equal volume fraction in a random mixture. For a random mixture, the number of particles of one type in a plurality of samples of the same size follows the binomial distribution. In all mixing operations, the best mixture that can be hoped for in terms of the quality of mixing is the random mixture. Often the quality of mixing is not as good as the random mixture.

224   The Principles of Design

The mixing process to be designed here requires that the mixture be *significantly* better than the random mixture, the ultimate being the perfect mixture. The process must be inexpensive and capable of mixing a large volume of materials.

Solution

Figure 6.24 shows the proposed solution to the problem of mixing fine powders (Suh and Tucker, 1977), which was developed prior to the development of the design axioms. Two different powders, A and B, are stored in separate containers, through which air flows. The airstream carries the particles and conveys them to channels that direct the flow of particles past corona discharge devices. The corona devices have high-voltage corona point electrodes and ground electrodes. One of the electrodes is supplied with a positive voltage with respect to ground, and the other electrode is supplied with a negative voltage with respect to ground. The corona discharge across the electrodes ionizes the air particles. The ionized air particles combine with the particles A and B as they pass between the electrodes, so as to impart a positive and negative charge on the particles, respectively. The corona power supplies provide voltages that produce electric fields of about 5–15 kV/cm.

Since the particles in each stream are charged with the same charge, the stream spreads as each stream leaves the region of each corona discharge device. As they enter a mixing chamber, the streams of oppositely charged

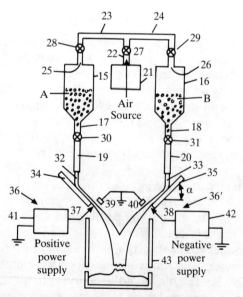

Figure 6.24. Apparatus to mixing powder by means of electrostatic charging. Powder A, which is charged positively, combines with Powder B, which is charged negatively. (From Suh and Tucker, 1977.)

particles attract each other, so particles of one charge tend to pair up with particles of the other charge as both streams are conveyed down through the mixing chamber.

In hindsight, the FR of this mixing process is:

$$FR = \text{Combine one Particle A with one Particle B}.$$

The physical solution devised (i.e., DP) is to use opposite electrostatic charges to provide internal forces among particles. At the next level of the FR hierarchy, the global FR can be decomposed into the following seven FRs:

FR_1 = Transport Powder A to the mixing area.
FR_2 = Transport Powder B to the mixing area.
FR_3 = Meter the flow rate of Powder A.
FR_4 = Meter the flow rate of Powder B.
FR_5 = Charge Powder A positively.
FR_6 = Charge Powder B negatively.
FR_7 = Combine Powders A and B.

In order to satisfy Axiom 1, we must seek DPs that yield a diagonal design matrix. The DPs chosen to satisfy Axiom 1 and the FRs are:

DP_1 = Flow rate of air through Chamber A (No. 15 in Fig. 6.24) = Q_A.
DP_2 = Flow rate of air through Chamber B (No. 16) = Q_B.
DP_3 = Area of valve opening (No. 30) = A_A.
DP_4 = Area of valve opening (No. 31) = A_B.
DP_5 = Positive voltage at the corona discharge = V_A.
DP_6 = Negative voltage at the corona discharge = V_B.
DP_7 = Angle of the chennels that direct the flow = α.

Then, the design equation for the proposed process may be written as

$$\begin{Bmatrix} FR_1 \\ FR_2 \\ FR_3 \\ FR_4 \\ FR_5 \\ FR_6 \\ FR_7 \end{Bmatrix} = \begin{bmatrix} \times & 0 & 0 & 0 & 0 & 0 & 0 \\ 0 & \times & 0 & 0 & 0 & 0 & 0 \\ \times & 0 & \times & 0 & 0 & 0 & 0 \\ 0 & \times & 0 & \times & 0 & 0 & 0 \\ 0 & 0 & 0 & 0 & \times & 0 & 0 \\ 0 & 0 & 0 & 0 & 0 & \times & 0 \\ 0 & 0 & 0 & 0 & \times & \times & \times \end{bmatrix} \begin{Bmatrix} Q_A \\ Q_B \\ A_A \\ A_B \\ V_A \\ V_B \\ \alpha \end{Bmatrix} \quad (6.20)$$

Experiments were done by constructing the apparatus shown in Fig. 6.24. Before the experimental results can be discussed, a quantative measure for mixing needs to be developed. In evaluating the quality of a mixture, a quantitative measure can be determined by counting the number of particles of one type in a plurality of separate samples, each having a total of n particles. If we take a sample of a theoretical random mixture and count the number of particles of one type, the result will be different from sample to sample, since the composition of random mixtures is not exactly identical everywhere. It can be shown that the standard deviation, σ, for a random mixture of the number of particles of one type among samples each containing n particles is

$$\sigma^2 = (1-a)a/n \qquad (6.21)$$

where a is the fraction of that type of particle in the random mixture. This theoretical standard deviation for a random mixture can be compared with the measured statistical standard deviation, S. In a completely "unmixed" state, S will be at a maximum, and as the mixture becomes closer to a perfect mixture, S decreases and becomes zero for a perfect mixture. The square of the statistical standard deviation, S, is computed and compared with the square of the standard deviation σ expected from a random mixture. A mixing index M can then be defined as

$$M = S^2/\sigma^2 \qquad (6.22)$$

If $M = 1$, the mixture is defined as the random mixture. If $M > 1$, the mixture is worse than random. A perfect mixture is the one with $M = 0$.

The experimental results showed that the mixing index M varied from about 0.44 to 0.65 when the particles were charged, whereas M was equal to 2 when particles were not charged. Thus, the electric charge improves the mix quality considerably.

Concluding Observations

The preliminary experimental results indicate that the powder-mixing apparatus is more effective than complicated mechanical mixing devices. However, it is not a perfect device, in that the mixing index M, although small, is greater than zero. This problem was caused by the lower-level FRs and DPs, since they did not fully satisfy the higher-level FRs. For example, the use of valve opening in metering flow rates might not have been the best approach. In order to improve the performance of this device, the detailed lower-level designs must be rigorously pursued.

6.9 MIT RIM Machine

Analysis of the Functional Independence

In Example 3.4, reaction injection molding (RIM) of polyurethane parts is discussed. Here we consider another RIM machine, called the MIT RIM machine. This machine was developed prior to the age of the design axioms. Unlike the design of the RIM machine discussed in Example 3.4, the MIT RIM machine is a coupled design, although at the time the machine was designed we thought intuitively that the design was a good one. Later, as we tried to apply Axiom 1 to the MIT RIM machine, we found (to our dismay!) that the machine is a coupled machine. As a consequence, the design discussed in Example 3.4 was developed (Rinderle and Suh, 1982).

The FRs of the RIM machine are, as discussed in Example 3.4,

FR_1 = Deliver liquid at high flow rate = Q.
FR_2 = Deliver an adequately mixed liquid = Z.
FR_3 = Deliver properly metered liquids = M.

Figure 6.25 shows the design of the MIT RIM machine (Suh et al., 1977; Suh and Tucker, 1979). It consists of two reservoirs for polyol and diisocyanate, two small pumps that charge two accumulators, the mechanically coupled hydraulic motor–pump for metering (as used in Example 3.4), and the mixing chamber. The accumulators are pressurized after they are filled with liquids. The liquids in the accumulators can be discharged rapidly into metering and mixing systems to deliver the liquid to the mold. In this system the flow rate, Q depends on both the accumulator pressure P and the orifice diameter D. An approximate relationship between Q and these DPs may be written as

$$Q = K_1\omega = K_2 P^{1/2} D^2 \tag{6.23}$$

where K_1 and K_2 are scaling constants, ω and is the motor–pump speed.

Similarly, the mix quality, Z, can be related to P and D, under the assumption (which is a good one; see Tucker and Suh, 1980) that the mix quality is proportional to the Reynolds number, based on the diameter of the nozzle to the $-3/4$ power. Then z may be expressed as (Tucker, 1978)

$$Z = K_3(P^{-3/2}D)^{1/4} \tag{6.24}$$

where K_3 is a scaling constant. The metering ratio M is controlled by the gear ratio of the mechanically coupled gear pumps.

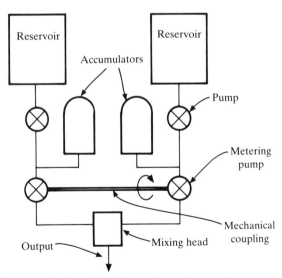

Figure 6.25. The MIT RIM machine with metering system.

Then, the design equation for the MIT RIM machine may be written as

$$\begin{Bmatrix} dQ \\ dZ \\ dM \end{Bmatrix} = \begin{bmatrix} \times & \times & 0 \\ \times & \times & 0 \\ 0 & 0 & \times \end{bmatrix} \begin{Bmatrix} dP \\ dD \\ dG \end{Bmatrix} \qquad (6.25)$$

In this system the FRs for flow rate Q and mix quality Z are coupled.

Rinderle and Suh (1982) analyzed the degree of coupling of such an RIM system by considering only the flow rate and the quality of mixing, since G does not affect Q and Z. The design equation for changes in the flow rate and the quality of mixing as a function of the changes in the DPs (i.e., the pressure and the diameter of the nozzle) may be written as

$$\begin{Bmatrix} dQ \\ dZ \end{Bmatrix} = \begin{bmatrix} \dfrac{\partial Q}{\partial P} & \dfrac{\partial Q}{\partial D} \\ \dfrac{\partial Z}{\partial P} & \dfrac{\partial Z}{\partial D} \end{bmatrix} \begin{Bmatrix} dP \\ dD \end{Bmatrix} = \begin{bmatrix} \tfrac{1}{2}K_2 D^2 P^{-1/2} & 2K_2 D P^{1/2} \\ -\tfrac{3}{8}K_3 D^{1/4} P^{-11/8} & \tfrac{1}{4}K_3 D^{-3/4} P^{-3/8} \end{bmatrix} \begin{Bmatrix} dP \\ dD \end{Bmatrix}$$

$$(6.26)$$

Normalizing Eq. 6.26 by dividing dQ by $K_2 P^{1/2} D^2$ and dZ by $K_3 P^{-3/8} D^{1/4}$, we can express the design equation as

$$\begin{Bmatrix} d\hat{Q} \\ d\hat{Z} \end{Bmatrix} = \begin{bmatrix} \dfrac{1}{2}\dfrac{1}{P} & \dfrac{2}{D} \\ -\dfrac{3}{8}\dfrac{1}{P} & \dfrac{1}{4D} \end{bmatrix} \begin{Bmatrix} dP \\ dD \end{Bmatrix} \qquad (6.27)$$

where the dimensionless flow rate $\hat{Q}$ and the dimensionless mix quality $\hat{Z}$ are given by

$$\hat{Q} = \frac{Q}{K_2 P^{1/2} D^2} \qquad \hat{Z} = \frac{Z}{K_3 P^{-3/8} D^{1/4}} \qquad (6.28)$$

If we scale the design by letting $P = 1$ and $D = 1$ (note that the scale and dimensions of DPs do not affect the measures of FR independence, as discussed in Secs. 4.5 and 4.6), the design matrix given in Eq. 6.27 becomes

$$\mathbf{DM} = \begin{bmatrix} \tfrac{1}{2} & 2 \\ -\tfrac{3}{8} & \tfrac{1}{4} \end{bmatrix} \qquad (6.29)$$

The curvilinear set of flow and mix quality isograms for the RIM machine are shown plotted in the physical space in Fig. 6.26 (Rinderle and Suh, 1982). The figure clearly shows that it is a coupled design, because the isograms for flow and mixing are not orthogonal to each other, and because they are not parallel to DP-axes, P and D. The plot has been scaled (per Eq. 6.29) so that the mix quality and flow rate equal unity when both the pressure and the orifice diameter are unity, and so that a 100% change in pressure has the same magnitude as a 100% change in orifice diameter.

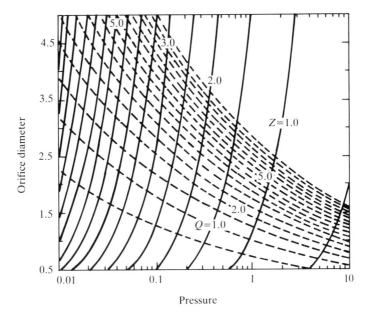

Figure 6.26. FR isograms for the MIT RIM machine. Broken curves show the flow performance (0.5 to 8.0 by 0.5). Solid curves indicate the mix quality (0.5 to 8.0 by 0.5).

To meet or exceed the FRs, the design point must lie to the right of the line marked "mix = 1" and above the line where "flow = 1."

The figure shows how the performance of the machine varies as the DPs are changed. The coupling measures R and S for the design matrix given by Eq. 6.29 are

$$\text{Reangularity, } R = 0.69$$
$$\text{Semangularity, } S = 0.1$$

These measures do not depend on the values of FRs or DPs, because the measures reangularity and semangularity are normalized. These results indicate that in certain regimes it is better to use D to control Q and P for Z.

The design equations are very simple, and can be inverted to obtain isograms of the DPs in the functional space, as shown in Fig. 6.27. The angular relationships between DP isograms and FR coordinates on the graph determine the extent to which each function can be controlled independently by a change in a single DP. A favorable design point can be chosen by examining the curvilinear DP isograms. From these figures it appears that the most favorable design point is near the normalization point where the pressure and orifice diameter are unity.

Concluding Observations

This case study is a lesson in the vulnerability of our thought processes. It is shown that our intuition and instinct, regardless of how we view our own

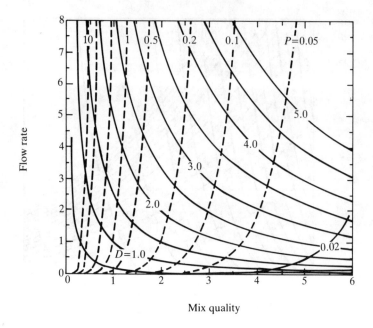

Figure 6.27. DP isograms for the MIT RIM machine. Curves of constant orifice diameter are broken (0.5 to 5.0 by 0.5). Curves of constant pressure are solid (0.01 to 10 on logarithmic scale).

ability, can misguide the decision-making process. What seemed like a perfectly acceptable solution to the RIM problem (i.e., the addition of the mechanically interconnected hydraulic motor–pump) could still not generate a solution that is consistent with Axiom 1. The original design could not be decoupled by the addition of the metering unit, since the original design matrix did not satisfy the requirement specified by Theorem 2.

6.10 Microcellular Plastics

Introduction

One of the promising new developments in polymer processing is the advent of microcellular plastics. In this section a case study on the design of a manufacturing process for microcellular plastics is presented. This case study was presented by Kumar (1988) as part of his doctoral thesis at MIT. The objective was to synthesize a process to produce plastic parts with a given geometry and with a microcellular structure (Kumar and Suh, 1988a,b). Conventional means to impart deformation were found to be detrimental to the structure. Deformation had to be integrated in the foaming process, such that the requirements on both the part geometry and the microstructure were satisfied. The basic idea was to decouple the cell nucleation and growth processes from deformation.

Unlike other case studies presented so far, this case study describes experiments conducted to obtain critical information in designing an uncoupled process. When the physical processes involved are extremely complex, we have to perform critical experiments, since a realistic mathematical analysis that can predict the causality between the FRs and DPs a priori without resorting to experiments may be difficult to achieve. Both experimental and analytical methods must be employed to find the effect of varying a DP on FRs.

Statement of the Problem

Microcellular thermoplastics refer to plastic foams with cell diameters of the order of 10 μm or less. This idea for the material was originally conceived by the author in response to a request from an industrial sponsor who asked us to innovate a means of reducing the amount of plastic used in mass-produced items without compromising their mechanical properties.

The original rationale for the microcellular plastic was that if voids smaller than the critical flaw size pre-existing in polymers can be nucleated, then the material cost could be reduced without compromising the mechanical properties. Having conceptually designed the material that satisfies the perceived needs and FRs, the next step is the design of the manufacturing process. Such a process has been developed using a thermodynamic instability phenomenon to achieve the cell nucleation (Martini et al., 1982, 1984). Polystyrene foams with cell diameters in the range 2–25 μm (hence the word "microcellular"), and void fractions of 0.05–0.80 were produced. This process also provides the possibility to foam thin-walled plastic parts with thickness in the range 1–2 mm, which cannot be foamed by conventional means, due to excessive loss of strength. In this section we describe the synthesis of a process for producing parts with a given geometry and microcellular structure.

The basic procedure involves saturating the polystyrene sheet with a high-pressure gas such as nitrogen and carbon dioxide at room temperature. When the specimen supersaturated with nitrogen is heated above the glass transition temperature, and when the pressure is suddenly released, a very large number of bubbles nucleate, due to the thermodynamic instability of the dissolved gas. Because of the extremely large number of cells (of the order of several hundred million cells per cubic centimeter) that can be nucleated by this process, and because of the limited amount of dissolved gas, small bubbles are generated compared with those formed by conventional processes (Thorne, 1979). Thus, it is possible to obtain cell diameters of the order of 10 μm. An example of microcellular foam produced by this method is shown in Fig. 6.28.

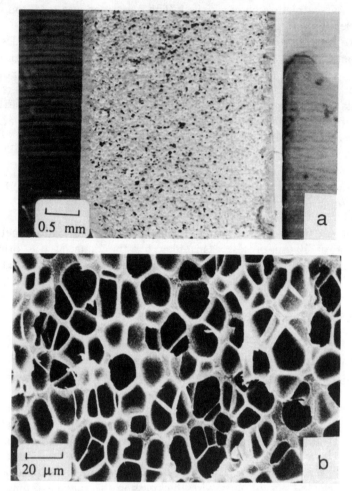

Figure 6.28. Examples of microcellular foam. (a) Overall view across thickness; (b) close-up, showing structural uniformity.

When such a foamed sheet is deformed using conventional thermoforming techniques, the cells become grossly distorted and sometimes destroyed (see Fig. 6.29). To circumvent this problem, we may form the part first (say, by injection molding), and then use the microcellular process to foam it. However, when the plastic is heated above its glass transition temperature for foaming, this process distorts the part shape, as shown in Fig. 6.30, due to the relaxation of the residual stresses and molecular orientation introduced during the original forming process. In the process just discussed, the FRs of producing the microstructure and the geometry are coupled. A coupled process is difficult to implement and control.

The goal of this case study is to develop a process whereby a desired final part geometry can be attained, while preserving the microcellular

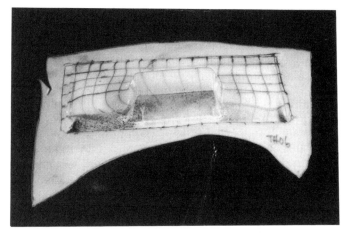

Figure 6.29. Photograph of a foamed sheet that was later thermoformed. Note the gross distortion in the cell structure.

structure. We seek to decouple the process so that the microstructure and the geometry can be controlled independently by defining the conditions under which the polymer must be processed.

Design of the Process

The basic process design objective is: "starting with a polystyrene sheet saturated with nitrogen, produce a part with a given geometry and microcellular structure."

Based on the perceived needs presented to the MIT researchers by its

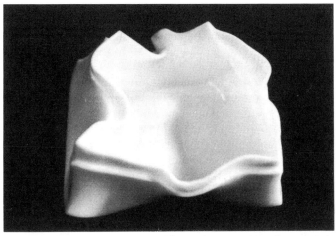

Figure 6.30. Photograph of a injection molded polystyrene box that was microcellularly foamed. Note the distortion in geometry.

sponsor, the FRs for the product were defined as:

FR_1 = Reduce materials consumption.
FR_2 = Enhance or maintain the mechanical properties of the original material.
FR_3 = Form three-dimensional geometrical shapes

In order to satisfy the above three FRs for the final product, the microcellular plastic concept was conceived and presented to the sponsor. The critical parameters are that cell sizes do not exceed a targer value, that the part contains a certain void fraction, and that it satisfies the part geometry requirement. To achieve these goals, we will need to control independently the number of cells nucleated and the cell size, in addition to geometry. The DPs for the product (and thus, the FRs for this process) can therefore be stated as:

DP_1 = Cell density (number of cell nucleated per unit volume).
DP_2 = Cell size.
DP_3 = "Thermoforming" of the plastic sheet.

In general, to create the microcellular structure we would like to obtain a high cell density and a small cell size. The pressure at which polystyrene is saturated with nitrogen affects directly the number of cells nucleated (i.e., the higher the saturation pressure is, the higher the cell density) due to greater thermodynamic instability. Cell size, on the other hand, can be controlled in part by the time and temperature to which the plastic is exposed during the forming cycle, when excess gas is present. Thus, the key PVs are:

PV_1 = Saturation pressure for nitrogen.
PV_2 = Time–temperature exposure.
PV_3 = Deformation.

Here "deformation" is used in a general sense. It is assumed that an appropriate method of deformation to achieve the desired part can be used.

The design equation may now be written as

$$\begin{Bmatrix} DP_1 \\ DP_2 \\ DP_3 \end{Bmatrix} = \begin{bmatrix} A_{11} & A_{12} & A_{13} \\ A_{21} & A_{22} & A_{23} \\ A_{31} & A_{32} & A_{33} \end{bmatrix} \begin{Bmatrix} PV_1 \\ PV_2 \\ PV_3 \end{Bmatrix} \qquad (6.30)$$

According to Theorem 9 (design for manufacturability), the design matrix [**A**] of Eq. 6.30 must be either a diagonal or a triangular matrix for the process design to satisfy Axiom 1. To get some insight into the design equation, let us establish the design matrix for a particular process idea. We set $A_{ij} = 0$ if PV_j has no significant effect on FR_i in terms of the perceived tolerance on FRs.

Consider the simple case in which we first produce a microcellular structure in the sheet by suddenly releasing the pressure when the plastic sheet supersaturated with gas reaches a softening temperature, and then impose deformation on it by using a thermoformer (i.e., a machine that heats the plastic and deforms it using vacuum or mechanical force). As mentioned previously, this procedure results in grossly distorted cells in regions of high deformation gradient. The realization of part geometry (FR_3) is thus coupled with cell density (FR_1) and cell size (FR_2), which are adversely affected by deformation. In this case the independence of the FRs is not maintained, violating the first axiom.

The design equation for this process may be written as

$$\begin{Bmatrix} DP_1 \\ DP_2 \\ DP_3 \end{Bmatrix} = \begin{Bmatrix} \text{Cell density} \\ \text{Cell size} \\ \text{Thermoforming} \end{Bmatrix}$$

$$= \begin{bmatrix} A_{11} & A_{12} & \times \\ A_{21} & A_{22} & \times \\ A_{31} & A_{32} & A_{33} \end{bmatrix} \begin{Bmatrix} \text{Saturation pressure} \\ \text{Time-temp. exposure} \\ \text{Deformation} \end{Bmatrix} \quad (6.31)$$

In the design matrix × denotes the coupling between the corresponding FRs and DPs.

Determination of A_{ij}

Let us examine the other off-diagonal coefficients in the design matrix one by one.

(i) $A_{12} = \partial(FR_1)/\partial(PV_2)$

A_{12} represents the effect of time–temperature exposure on cell nucleation. Bubble nucleation in plastics is a rate process; accordingly, the usual Arrhenius-type dependence of cell density on temperature is expected. If a higher temperature is employed, for example, to attain a higher void fraction, we may expect a higher cell density (see Appendix 6B). Similarly, foaming for an extended length of time to achieve a certain growth may induce some secondary nucleation. In both cases the effect of time–temperature exposure on cell nucleation will affect the cell density. Thus, coefficient $A_{12} \neq 0$.

(ii) $A_{21} = \partial(FR_2)/\partial(PV_1)$

This represents the effect of saturation on cell size. High saturation pressures are needed to obtain high cell nucleation rates (see Appendix 6C). For a given concentration of gas, the bubble size will be smaller when the number of cells nucleated is large. On the other hand, more gas will dissolve in the polymer at a higher saturation pressure, and may cause the bubble size to exceed the target value. Thus, there is a possibility of coupling between the FRs of high cell density and small cell size. We will need to address this issue in our process design. At this point we shall leave A_{21} as undetermined.

(iii) $A_{31} = \partial(DP_3)/\partial(PV_1)$

Since saturation pressure (PV_1) has no effect on the "thermoforming" of the sheet (DP_3), and thus on part geometry (FR_3), $A_{31} = 0$.

(iv) $A_{32} = \partial(DP_3)/\partial(PV_2)$

Again, since time–temperature exposure has no effect on part geometry, $A_{32} = 0$. Equation 6.31 can now be written as

$$\begin{Bmatrix} DP_1 \\ DP_2 \\ DP_3 \end{Bmatrix} = \begin{bmatrix} A_{11} & A_{12} & \times \\ A_{21} & A_{22} & \times \\ 0 & 0 & A_{33} \end{bmatrix} \begin{Bmatrix} PV_1 \\ PV_2 \\ PV_3 \end{Bmatrix} \tag{6.32}$$

Equation 6.32 suggests that if $A_{21} = 0$, then we may be able to decouple the FRs by sequential application of the PVs. So, if we vary PV_3 first, then PV_2, and finally PV_1, we can control the FRs independently. This sequence of PVs is, however, physically not possible, as it requires PV_2 (cell growth) to precede PV_1 (cell nucleation). We have to determine a new set of process conditions that will eliminate the coupling shown in Eq. 6.32.

Ideally, in order fully to decouple deformation from nucleation and growth of cells, we would like to accomplish deformation first, followed by nucleation and growth of bubbles. To achieve this, we will need to suppress nucleation during the deformation step. This could possibly be done by maintaining sufficiently high pressure, either mechanical or hydrostatic, during deformation. Maintaining the pressure would eliminate the thermodynamic instability that causes cell nucleation. In this scheme, cell nucleation will have to occur (following deformation) under external pressure. We will need to investigate how nucleation under external pressure affects cell density. Furthermore, since deformation may induce molecular orientation, we will need to establish the effect that such molecular orientation may have on cell nucleation, a primary FR.

To summarize, our design strategy calls for an investigation of the effect of saturation pressure on cell size, in order to determine the nature of coupling between cell density and cell size. Furthermore, in order to evaluate a proposed scheme to decouple deformation and cell nucleation, experimental investigation of the effect of deformation on cell nucleation and growth is required. The process design will be modified until a diagonal, or an acceptable triangular design matrix has been obtained.

Experiment

All experiments were conducted on DOW XP 6065 polystyrene, with average molecular weight of 200,000. Circular disks 2 in. (50.8 mm) in diameter and 1/16 in. (1.55 mm) thick were injection molded from the resin. These disks were saturated with nitrogen in a pressure vessel and foamed using the microcellular process. The microstructure was studied on a scanning electron microscope.

To determine the cell density, a micrograph showing 100–200 bubbles was obtained and the exact number of bubbles was determined. Figure 6.28 is typical of a micrograph used to determine the cell density. Assuming isotropic distribution of bubbles, the number of cells per cubic centimeter of foam, N_f, was determined. The average cell diameter was usually obtained from a second micrograph at a higher magnification. The void fraction, V_f, in the foam was estimated from

$$V_f = (\pi/6)D^3 N_f \tag{6.33}$$

The number of cells nucleated per cubic centimeter of original, unfoamed polymer, N_0, was determined from

$$N_0 = N_f(1 - V_f) \tag{6.34}$$

and is reported here as the cell density.

Effect of saturation pressure on cell size. To investigate the effect of gas saturation pressure on cell size, polystyrene disks were saturated at different nitrogen pressures, and were allowed to foam in the glycerin bath at 115°C for a sufficiently long time to develop the limiting (maximum) void fraction. The average cell size obtained from this experiment is plotted in Fig. 6.31a as a function of the nitrogen saturation pressure. The cell nucleation density, obtained in a separate experiment, is plotted in Fig. 6.31b. Note that the cell density has been plotted on a logarithmic scale.

The average cell size is seen to decrease linearly with increasing saturation pressure. The cell nucleation density, on the other hand, is seen to increase exponentially with the nitrogen saturation pressure. The average bubble volume is equal to the available gas volume divided by the number of bubbles nucleated. As the saturation pressure is raised, the total volume of voids increases in proportion to the saturation pressure. However, due to the exponential increase in the number of cells, the average volume occupied by an individual cell actually decreases as the saturation pressure is raised (see Appendix 6C). Therefore, the saturation pressure has a much more significant effect on cell density than on cell size. Consequently, the coefficient A_{21} in Eq. 6.30 can be taken to be zero since the variation in cell size over the operating pressure range is negligible.

Effect of external pressure on cell nucleation. In order to uncouple cell nucleation from deformation, it was proposed first to get the part geometry and then to obtain the microstructure. Since we needed to heat the supersaturated polystyrene sheet above the glass transition temperature before we could deform it, we had somehow to suppress cell nucleation during the deformation step. It was thought that if sufficient external pressure was maintained during deformation, then cell nucleation could be effectively suppressed, since the external pressure would reduce the driving force for cell nucleation. After the deformation is complete, the external pressure would be reduced, allowing cells to nucleate.

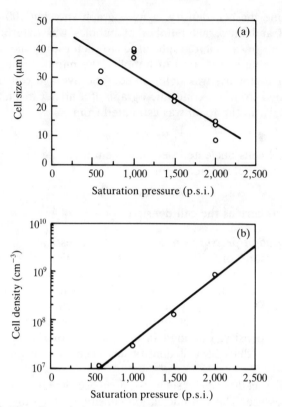

Figure 6.31. (a) The effect of nitrogen saturation pressure on the cell size (1000 p.s.i. = 6.895 MN/m^2). (b) The effect of nitrogen saturation pressure on the cell nucleation density.

Under this scheme the cells should not nucleate while under external pressure. We need to insure that cells will not nucleate under these conditions until the external pressure is removed, for otherwise deformation and cell nucleation would remain coupled, defeating the original purpose.

The set-up shown in Fig. 6.32 was used to investigate the effect of external pressure on cell nucleation. A mold with a cavity was placed in a

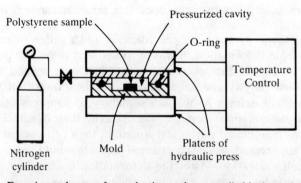

Figure 6.32. Experimental set-up for nucleation under controlled hydrostatic pressure.

hydraulic press. The mold was connected to a pressurized nitrogen cylinder. The cavity pressure could thus be controlled. The platens of the press were connected to a temperature control unit. The cavity temperature was monitored independently by a thermocouple installed under the bottom cavity surface.

A polystyrene disk saturated at 2,000 p.s.i. (13.8 MPa) was placed in the mold, and the cavity was pressurized to the desired external pressure. The sample was heated to 115°C for a few minutes, and then cooled to room temperature. The external pressure was then brought to atmospheric and the sample was examined. This experiment was repeated at different cavity pressures.

The results are shown in Fig. 6.33. The open squares are data from the hydrostatic external pressure described earlier. The solid triangles are data from a second experiment, where the saturated polystyrene disk was placed directly on the platen of the hydraulic press and compressed between the platens to a desired external pressure. The objective was to create external pressure by direct mechanical contact. The disk was subjected to the same thermal cycle as described earlier. Data from this experiment are referred to as "squeeze pressure" data.

We note that there is no difference in the cell density data from the two experiments. *External pressure is seen to have no effect at all on cell density.* This means that we can not suppress cell nucleation by maintaining an external pressure. This is a most surprising result. Homogenous nucleation theory predicts a decrease in cell nucleation density of some 8 orders of magnitude as the external pressure is increased from zero to 1,000 p.s.i. (6.9 MPa). Although this theory is not strictly applicable, we nevertheless expected a significant drop in cell density as we increased the external pressure. Clearly, there is a big discrepancy between the theory and experiment.

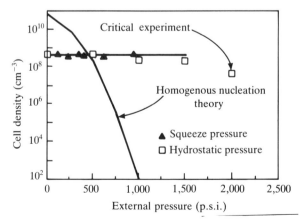

Figure 6.33. The effect of external pressure on the cell nucleation density (1000 p.s.i. = 6.895 MN/m^2).

In order to investigate this discrepancy, we performed the following critical experiment. A polystyrene disk was saturated at 2,000 p.s.i. and placed in the mold (see Fig. 6.32). The cavity was pressurized to 2,000 p.s.i. (13.8 MPa) and then heated to 115°C as for the hydrostatic external pressure experiments. Since in this experiment an external pressure equal to the saturation pressure was maintained, we expected no bubbles at all, as there is no supersaturation of nitrogen to cause cell nucleation.

Much to our surprise, the disk turned white, showing that bubbles had nucleated. A cell density of 44 million per cubic centimeter was determined, showing substantial nucleation. These data have also been plotted in Fig. 6.33 as "critical experiments". What is the driving force for cell nucleation in these experiments?.

A decrease in the solubility of nitrogen with increasing temperature would explain cell nucleation in the sample, as it was heated from room temperature to the glass transition temperature and beyond. Indeed, a review of the literature showed that the solubility of nitrogen in polystyrene decreased with temperature (Durrill and Griskey, 1969).

We measured the solubility of nitrogen in polystyrene as a function of the temperature. Polystyrene samples were saturated with 1,500 p.s.i. (10.3 MPa) nitrogen at different temperatures, using the set-up shown in Fig. 6.32. The samples were weighed on a Mettler (Model H51AR) balance capable of detecting changes in weight up to 10 μg. The solubility data are shown in Fig. 6.34. We have plotted Henry's constant, measured (in cubic centimeters of nitrogen per gram of polystyrene per atmosphere at STP) as a function of temperature. We see that solubility drops by nearly 40% as the temperature increases from room temperature to the glass transition temperature.

This drop in solubility explains the significant cell nucleation observed in the critical experiment. These experimental results are consistent with

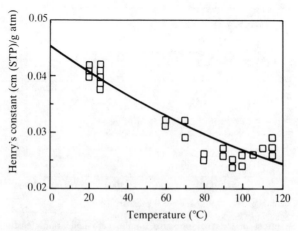

Figure 6.34. The effect of temperature on nitrogen solubility.

the thermodynamic state of matter, which is typically a function of two thermodynamic properties. In this case the solubility varies as a function of pressure and temperature. The effects of both pressure and temperature on solubility are significant.

This new understanding of cell nucleation in polystyrene led us to reassess the process strategy. It became clear that cell nucleation cannot be suppressed by employing external pressure. Therefore, we physically cannot deform the plastic before cell nucleation. On the other hand, the cells will nucleate as soon as we reach the glass transition temperature, regardless of any external pressure. We could therefore employ temperature as the means of precipitating cell nucleation through phase separation. This idea leads to the rather simple process described in the next section.

Synthesis of the Final Product

Our basic goal was to realize a given part geometry without harming the microstructure. We considered the idea of deforming soon after cell nucleation, when the cells are still small. The majority of cell growth will be attained after deformation in a heated mold at a suitable temperature. This process is shown schematically in Fig. 6.35. The idea to nucleate the bubbles first, then apply deformation, and then grow the bubbles in a heated mold, appeared feasible, sinc deformation can be accomplished in a short time relative to that needed to get significant cell growth.

We studied cell growth as a function of time. A number of polystyrene disks were saturated at 2,000 p.s.i. (13.8 MPa) and foamed at 115°C for different lengths of time. The results are shown in Fig. 6.36. We see that at this temperature it takes some 25 seconds to reach an average cell size of 10 μm, and some 100 seconds to approach the limiting final cell size. By contrast, the deformation can be imposed in 1 or 2 seconds. It is therefore possible to attain the part geometry before the cells become too large.

Note that, in the process proposed in Fig. 6.35, there is a possibility that

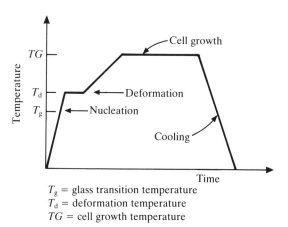

T_g = glass transition temperature
T_d = deformation temperature
TG = cell growth temperature

Figure 6.35. Schematic of the proposed process.

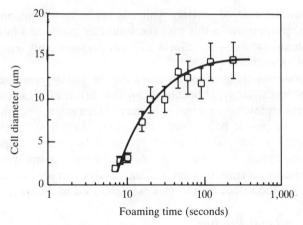

Figure 6.36. Growth in cell size as a function of time.

the cells may become nonspherical under deformation. However, subsequent cell growth will tend to restore the spherical shape. If cells are allowed to grow to form a honeycomb-like structure, then the initial deviation from a spherical shape will be inconsequential.

A conventional thermoforming machine was used to test the feasibility of the process. A commercial grade high-impact polystyrene sheet was saturated with nitrogen at 2,000 p.s.i. (13.8 MPa) and placed in the thermoformer. The sheet was heated in an infrared oven, and formed into a container, using a heated mold in a plug-assist vacuum-forming process. A sample from the bottom of the container was studied under the scanning electron microscope. Figure 6.37 shows the micrographs from these samples. It is seen that a microcellular structure has been successfully produced in the container. This demonstrates the feasibility of the process concept.

Let us examine the nondiagonal elements of the design matrix for this process. Coefficient A_{12} represents the effect of time–temperature exposure on cell nucleation density. In our process, cell nucleation occurs before deformation, as the sheet is heated to the glass transition temperature. The cell growth temperature, TG, therefore has no effect on cell density, and we can set $A_{12} = 0$.

Coefficient A_{13}, the effect of deformation on cell density, is clearly zero, since cell nucleation precedes deformation. This is also seen in Fig. 6.33, where the cell density was found to be independent of the external pressure. Coefficient A_{21} represents the effect of saturation pressure on the cell size. We have shown earlier that $A_{21} = 0$.

In our process, cell growth occurs predominantly after deformation. Therefore, we may set $A_{32} = 0$, signifying that deformation (PV_3) has no significant effect on the final cell size distribution. However, it should be noted that the possibility for deformation to affect the shape of the cells

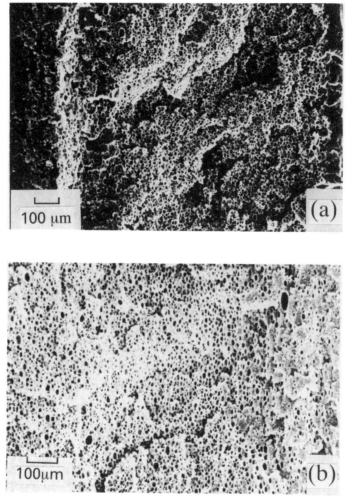

Figure 6.37. Scanning electron micrographs of a container with microcellular structure. (a) Sample from the bottom; (b) sample from the container wall.

still exists, and process conditions will need to be adjusted to keep the shape within acceptable limits.

The saturation pressure (PV_1) and time–temperature exposure for cell growth have no bearing on thermoforming; therefore $A_{31} = A_{32} = 0$.

With all of the off-diagonal terms equal to zero for this process, we have obtained a diagonal design matrix, which was the objective of axiomatic design.

Concluding Observations

We have presented a case study that illustrates the application of the axiomatic approach to the design of a manufacturing process to produce microcellular parts from polystyrene. This case study involved critical experiments to determine the elements of the design matrix. We were able to synthesize a process in which deformation and microstructure are uncoupled. The process has a diagonal design matrix, showing that the FRs can be controlled independently in the process. Finally, the process concept is demonstrated on a conventional thermoforming machine. This process is based on the recognition that the amount of gas dissolved in the polymer is a function of both the temperature and the pressure (as expected from thermodynamics), and that sudden thermodynamic instability can be induced to cause phase separation. A combined evaluation of the process, applying basic laws of nature and the design axioms, yielded an uncoupled process.

6.11 Summary

Several case studies related to manufacturing processes, materials processing, and machines have been presented in this chapter. The case studies illustrate how Axioms 1 and 2 can be used in solving real problems. Furthermore, the case studies illustrate the relationship between the design of products and the design of processes. The design of the product is described by a relationship between the FRs and the DPs of the product, whereas the design of the manufacturing process is given in terms of DPs of the product and the PVs of the manufacturing process. In the latter case DPs are equivalent to FRs of the manufacturing process, per Theorem 9 (design for manufacturability).

These case studies reveal certain common truths that could be derived from the axioms and the corollaries. One of these is stated as Theorem 15, which states:

> Theorem 15 (Design–Manufacturing Interface)
> When the manufacturing system compromises the independence of the FRs of the product, either the design of the product must be modified or a new manufacturing process must be designed and/or used to maintain the independence of the FRs of the product.

This theorem is related to Theorem 9. It implies that the product must be so designed that it can be manufactured by a given set of machines and equipment.

The case studies indicate that many design decisions can be made qualitatively if and when we understand the basic engineering, physics, and chemistry involved. Only in cases where the decision is less certain do we need to resort to an elaborate analysis at the beginning. However, as design proceeds to lower levels of the FR and DP hierarchies, more-precise analysis will be required to optimize the solution. Such detailed analyses are not presented here, so as not to confuse the design issues. The references cited provide additional details.

References

Agron, A.S., "Formation of Cavities from Non-deformable Second-phase Particles in Low Temperature Ductile Fracture," *Journal of Engineering Materials and Technology, Transactions of A.S.M.E.* **98**:60–68, 1976.

Colton, J.S., and Suh, N.P., "Nucleation of Microcellular Foam: Theory and Practice," *SPE Technical Papers* **XXXII**:45–47, 1986.

Durrill, P.L., and Griskey, R.G., "Diffusion and Solution of Gases into Thermally Softened or Molten Polymers," *AICHE Journal* 106–110, January, 1969.

Kumar, V., "Process Synthesis for Manufacturing Microcellular Plastic Parts: A Case Study in Axiomatic Design," Ph.D. Thesis, MIT, 1988.

Kumar, V., and Suh, N.P., "Structure of Microcellular Thermoplastic Foam," *Proceedings of the S.P.E. ANTEC '88*, Atlanta, April, 1988, pp. 715–718.

Kumar, V., and Suh, N.P., "Process Synthesis for Manufacturing Microcellular Thermoplastic Parts: A Case Study in Axiomatic Design," *Proceedings of the A.S.M.E. Manufacturing International '88*, Atlanta, April 1988, Vol. 1, 29–38.

Martini, J.E., Waldman, F.A., and Suh, N.P., "The Production and Analysis of Microcellular Thermoplastic Foam," *S.P.E. Technical Papers* **XXVIII**:674–676, 1982.

Martini-Vvedensky, J.E., Suh, N.P., and Waldman, F.A., U.S. Patent 4,473,665, 1984.

McCree, J., "Vented Compression Molding; A New Method for Molding Particular Materials," S.M. Thesis, MIT, 1981.

McCree, J., and Erwin, L., "Vented Compression Molding," *Journal of Engineering for Industry* **106**(2):103–106, May 1984.

Rinderle, J.R., and Suh, N.P., "Measures of Functional Coupling in Design," *Journal of Engineering for Industry, Transactions of A.S.M.E.* **104**:383–388, 1982.

Stelson, K.A., "The Adaptive Control of Brakeforming Using In-process Measuring for the Identification of Workpiece Material Characteristics," Ph.D. Thesis, MIT, 1981.

Su, K.-Y. "Void Nucleation in Particulate Filled Polymeric Materials and Its Implications on Friction and Wear Properties," Ph.D. Thesis, MIT, 1980.

Suh, N.P., USM Corporation Report, EX 18422, 1961 (unpublished).

Suh, N.P., U.S. Patent 4,278,622, July 14, 1981a.

Suh, N.P., U.S. Patent 4,279,843, July 21, 1981b.

Suh, N.P., "Orthonormal Processing of Metals, Part I: Concept and Theory," *Journal of Engineering for Industry, Transactions of A.S.M.E.* **104**:327–331, 1982.

Suh, N.P., "Development of the Science Base for the Manufacturing Field Through the Axiomatic Approach," *Journal of Robotics and Computer-Integrated Manufacturing* **1**(3/4):397–415, 1984.

Suh, N.P., and Tucker III, C.L., U.S. Patent 4,034,966, July 12, 1977.

Suh, N.P., and Tucker III, C.L., U.S. Patent 4,170,319, October 9, 1979.

Suh, N.P., Malguarnera, S.C., and Anderson, F.H., U.S. Patent 4,019,652, April 26, 1977.

Suh, N.P., Tsuda, H., Moon, M., and Saka, N., "Orthonormal Processing of Metals, Part II: Mixalloying Process," *Journal of Engineering for Industry* **104**:332–338, 1982.

Thorne, J., "Principles of Thermoplastic Structural Foam Molding: A Review," *Science and Technology of Polymer Process*, Suh, N.P., and Sung, N., eds, 77–131. MIT Press, Cambridge, MA, 1979.

Tucker III, C.L., "Reaction Injection Molding of Reinforced Polymer Parts," Ph.D. Thesis, MIT 1978.

Tucker III, C.L., and Suh, N.P., "Mixing for Reaction Injection Molding I. Impingement Mixing of Liquids," *Polymer Engineering and Science* **20**:875–886, 1980.

246

Problems

6.1. In discussing the manufacture of the multilense plate in Section 6.2, it is stated that a possible solution is to heat the mold to insure an isothermal flow of plastic during the filling process and to cool the mold upon completion of the filling process. This is an acceptable solution in Section 6.2 because the cycle time is not specified. If we now add a constraint that the cycle time must be less than 30 seconds, how should the mold be modified to accomplish the isothermal filling and rapid cooling? Justify your solution by developing the design matrix.

6.2. Develop an alternate process that can mold the thermal insulation layer of the Space Shuttle Project discussed in Section 6.4.

6.3. In the USM foam molding process, the cell size along the flow front was larger due to the low pressure at the plastic–air interface. Develop a new process that will yield a uniform cell size even at the flow front. First state the FRs clearly and design a process. State your solution in the form of a design equation.

6.4. Prove theorem 7.

6.5. Derive the design equation for the property–processing parameter relationship for the Mixalloy Process. What are the implications of the resultant design equation?

6.6. It was found that the solid particles formed through a chemical reaction in the Mixalloy Process segregate from the matrix phase metal when the part is thick. Design an alternate process or modify the Mixalloy Process to eliminate this segregation problem.

6.7. The powder-mixing device shown in Fig. 6.24 yields a mixture that is considerably better than a random mixture but is not a perfect mixture. Design a better powder-mixing device and develop the design equation.

6.8. From appropriate design equations, plot the isograms shown in Figs. 6.26 and 6.27.

6.9. Design an RIM machine that does not violate Axiom 1. Justify your design by writing down the design equation.

6.10. Ceramic blocks made of TiC–Al_2O_3 composite are coated with a thin alumina (Al_2O_3) layer by chemical vapor deposition in order to enhance their electrical properties. The blocks must be sliced into small wafers as

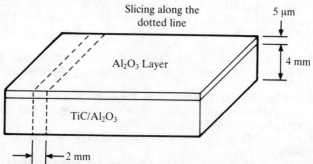

shown in the figure without delaminating the thin alumina layer during the slicing operation. The wafers must be produced at a low cost. Design a manufacturing process.
6.11. A defense contractor is interested in spreading fine copper wires of length 1 ft in a small region in space as a means of reflecting electromagnetic signals. When they carried the copper wires in a tube into space using a rocket and exploded it in space, the copper wires drifted away from each other in a very short time. On the other hand, when they did not use such a strong force, the copper wires stuck together and did not spread well in space. Propose a solution to this problem.

6A: MIXALLOY PROCESS

6A.1 Concept

The concept of the Mixalloy Process is illustrated in Fig. 6.19. The basic system consists of two (or more) molten metal streams emerging from nozzles and impinging upon each other in the mixing chamber. The liquid streams undergo violent turbulent motion, due to the large ratio of inertial-to-viscous forces. The turbulent motion of the liquids consists of small eddies that mix intimately in the mixing chamber, due to the large momentum of the liquid streams.

When the liquid metals have different melting points, the heat transfer between the eddies of these metals causes the metal in the higher melting point eddy to freeze. Freezing may occur last in the high-shear zone between the eddies, because nucleation cannot occur due to the liquid motion until sufficient supercooling takes place.

The mixture of the partially frozen slurry (liquid with solid particles) is then injected into a mold or cast onto a chilled conveyer belt and rapidly solidified. Thermal energy is continuously removed from the mixture of the molten metals in the mixing chamber and the runner, so as to supercool the metal mixture and make a slurry. The final distribution of the higher melting and the lower melting phase is controlled by the ratio of liquid stream flow rates.

From the instant when the liquid metal emerges from the nozzle and mixes with the liquids from other nozzles, a number of interesting phenomena take place in the liquid metals, which can be used for independent control of the chemical composition and microstructure. Each of these phenomena is described below.

6A.2 Theoretical Considerations

Mixing and Eddy Size

When the liquid streams emerge from the nozzle, the fluid can be either laminar or turbulent. The laminar–turbulent flow transition occurs when the Reynolds number, based on the nozzle diameter, exceeds about 50 (Tucker and Suh, 1978). With directly opposed nozzles, the critical Reynolds number for the laminar–turbulent transition does not depend on the nozzle diameter. Turbulent flow consists of eddies that may be thought of as small liquid particles (or vortices) undergoing rigid-body motion (both rotational and translational) with intense shear at the inerface between the adjoining eddies. It has been shown that the size of the eddy decreases with the increase in the Reynolds number (Davis, 1972); that is, as the inertia of the fluid is increased with respect to fluid viscosity, progressively smaller eddies are created. Based on the concept of Kolmogoroff's microscale (Kolmogoroff, 1941) and the dimensional analysis of the fluid motion in the mixing chamber, the size of the eddy, l, can be related to the Reynolds number (Tucker and Suh, 1978; Tucker, 1979) as:

$$\frac{l}{D} = k\left(\frac{\rho V D}{\mu}\right)^{-3/4} = k\,\mathrm{Re}^{-3/4} \qquad (6\mathrm{A}.1)$$

where D is the nozzle diameter, and ρ, μ, and V are the density, the viscosity, and the velocity of the liquid at the nozzle exit, respectively, and k is a proportionality constant. For example, lead at 643 K emerging from a 6.35-mm diameter nozzle under a head pressure of 2.76 bar has an eddy size of the order of 0.7 μm, assuming that $k = 1$. The general relationship between the Reynolds number and the eddy size is given in Fig. 6A.1, assuming that $k = 1$.

The fluid motion in the mixing chamber is very complex. In addition to the small-scale mixing of the eddies, a large-scale mixing of the liquid

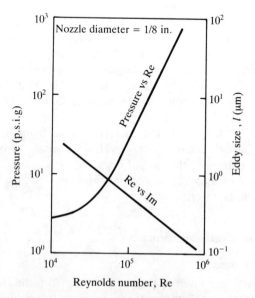

Figure 6A.1. Pressure and eddy size as functions of the Reynolds number.

streams emerging from the nozzles also occurs. A careful investigation of the liquid motion in the impingement chamber shows that both the large-scale mixing of liquid components emerging from the nozzles and the small-scale mixing of eddies on a microscale occur readily due to the large momentum carried by each liquid stream when the Reynolds number is high (Kolmogoroff, 1941). Therefore, within one chamber diameter downstream of the impingement point, all turbulent flows achieve large-scale distribution of fluids.

Heat Transfer and Solidification

Case 1: Metals with a large difference in melting points. When the impinging metal streams have a large difference in their melting points, the higher-melting component can be frozen as solid particles in the liquid matrix of the lower-melting component; freezing occurs almost instantaneously. For example, when the eddy size is about $0.7\,\mu\text{m}$, the characteristic thermal diffusion time is about 3×10^{-6} second. Even if the eddy size is as large as $70\,\mu\text{m}$, the diffusion time is still of the order of 1 ms.

The complete freezing of the lower-melting component can take place in the mold. However, even this phase may be supercooled into a slurry in the mixing chamber without freezing the entire mixture, since nucleation cannot occur readily without much supercooling, due to the turbulent fluid motion in the region of intense shear.

The necessary condition for freezing the metal with the higher melting point can be established through heat balance. To freeze the metal with higher melting point, the initial temperature of the low-melting metal should be set so that the condition

$$\Delta T_2 > \frac{m_1 c_1}{m_2 c_2} \Delta T_1 + \frac{m_1 \lambda_1}{m_2 c_2}$$
$$\Delta T_2 = T_{m1} - T_2 \qquad \Delta T_1 = T_1 - T_{m1} \tag{6A.2}$$

is satisfied where T_m and T are the melting and the initial temperatures, respectively; m and c are the mass flow rate and heat capacity, respectively; and λ is the heat of fusion. Subscripts 1 and 2 refer to the metals with high and low melting points, respectively. Since ΔT_2 cannot be greater than $\Delta T_m = T_{m1} - T_{m2}$, the maximum allowable initial temperatures of high-melting point metal $(T_1)_{\max}$ is given by

$$(T_1)_{\max} = \Delta T_m \left(\frac{m_2 c_2}{m_1 c_1}\right) - \frac{\lambda_1}{c_1} + T_{m1} \tag{6A.3}$$

The minimum melting point difference required for a pair of metals to form a slurry immediately after impingement is obtained by letting $T_1 = T_{m1}$ as

$$(\Delta T_m)_{\min} = m_1 \lambda_1 / m_2 c_2 \tag{6A.4}$$

Equation 6A.4 states that the smaller the heat of fusion of the high-melting metal (or the greater the heat capacity of the low-melting metal) is, the smaller the required melting point difference to freeze the higher-melting metal. The minimum melting point difference required depends on the ratio of mass flow rates.

It can be easily appreciated that the resulting microstructure will have grain sizes smaller than or equal to the eddy size, because there can be more than one nucleation site in each eddy. Even the grain size of the component with low melting temperature cannot be larger than the mean free spacing of the component with higher melting point, which has already solidified. Therefore, the volume fraction of each component may determine the grain size of the matrix phase of the resultant solid.

The melting point differences can be made large by choosing metals with large differences in their melting points, or by forming a solution through the addition of alloying elements into the metal with low melting point in order to depress the liquidus temperature. However, this addition of alloying elements will tie the chemical composition to the process condition, to a limited degree, making it a quasi-coupled system.

Case 2: Metals with a small difference in melting points. When the difference in the melting points is small, the control of the size of each phase is more difficult since a precise control of the temperature of the mixture is necessary. In this case the difference in the nucleation rate between the interior of eddies and the periphery of the eddy due to the shearing motion at the boundary must be made use of by removing heat through the wall of the mixing chamber. Since the interior of the eddy is undergoing a rigid-body motion while the periphery is sheared extensively, solidification must proceed from the center of each eddy when the temperature of the eddy is nearly isothermal.

Then, resultant slurries must be injected into a mold (or cast onto a conveyor belt) before complete freezing occurs in the mixing chamber and the runner. For this process to be possible, the temperature of the liquid metal must be nearly equilibrated along the radial direction, so that the temperature is a function of x and not of r (see Fig. 6.A2). This will prevent the mixture from forming a frozen layer by solidifying on the wall of the mixing chamber.

Under the assumption that the temperature of the slurry is a function of only the axial distance x from the impingement point, the temperature decay of the mixture in the mixing changer with a constant wall temperature T_w can be shown to be

$$T = (T_0 - T_w) \exp\left(-\frac{2h}{\rho c V}\frac{x}{r}\right) + T_w \qquad (6A.5)$$

where T is the average temperature of the impinging liquids at $x = 0$ (given by Eq. A1 of Suh, 1982); h is the heat transfer coefficient; V is the axial velocity of the liquid; ρ and c are average density and heat capacity of the

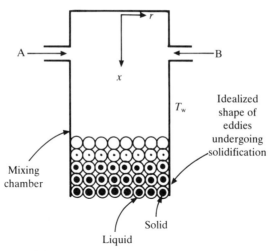

Figure 6A.2. Conceptual picture of how small grains may form in casting metals with a small difference in melting points.

metals, respectively; and r is the radius of the mixing chamber. The heat transfer coefficient h is given by the empirical relationship (Giedt, 1957)

$$hr/k = Nu = 0.625(\text{Re Pr})^{0.4} \qquad (6A.6)$$

where k is the thermal conductivity of the liquid, Pr the Prandtl number, Nu the Nusselt number, and Re the Reynolds number. The mixing chamber geometry (L_c/r) can be determined from Eq. 6.A5 by assuming that the desired temperature of the mixture is the average of the two melting points; that is,

$$T = \tfrac{1}{2}(T_{m1} + T_{m2}) \qquad (6A.7)$$

Microstructure, Chemical Composition, and Other Metallurgical Considerations

The final grain and particle sizes are controlled primarily by the Reynolds number of the impinging liquid streams. This requires that the heat transfer between the eddies be properly controlled by setting the initial temperatures of the liquids appropriately, per Eq. 6A.2. In the case of the metals with nearly identical melting points, heat must be rapidly extracted from the metal throught the wall of the mixing chamber.

Because the eddy size can be reduced by increasing the Reynolds number, very small grain sizes can be generated by the Mixalloy Process. Ordinarily it is nearly impossible to make metals with submicrometer grains by conventional means. In the Mixalloy Process the grain size is nearly independent of the kinetics of solidification if the melting points of metals mixed are significantly different. Otherwise, the mixed liquids must be solidified before the liquids dissolve in each other.

Any chemical composition of various phases is possible using the Mixalloy Process, since the phases are put into the structure by mechanical means, provided that they can be frozen before dissolution can occur. In this case the process is not governed by equlibrium thermodynamics. This eliminates the need to add alloying elements for the purpose of controlling the microstructure. In conventional metallurgical process the alloying elements are often used to control the thermodynamics and the kinetics of phase transformation, and thus the microstructure. Since the process does not couple these two requirements, we are now in a position to generate many different kinds of metal alloys and composites that could not have been made in the past, including immiscible systems.

This process also offers other advantages. Because two different liquids meet at the interface of eddies, intermetallic compounds with desirable properties can be generated at these interfaces by the judicious choice of metallic elements. Materials with high hardness and toughness may be made in this way. For example, if three liquid streams consisting of three different kinds of metals are impinged against each other, the metals can be so chosen that a pair of these metals forms a hard intermetallic or ceramic compound, while the remaining two pairs improve impact toughness by being mutually soluble. These kinds of metals may have very interesting tribological properties. There are virtually unlimited metal combinations that could be produced and investigated.

Another interesting possibility is the manufacture of powders of controlled size (Suh, 1981). The material to be made into powder can be impinged against a material with low melting point and frozen. The frozen particles can then be separated from the slurry by mechanical means. This overcomes the need to crush cast slabs mechanically and to sieve the powder in order to collect particles of the proper size. It should also be mentioned that by forming an intermetallic or hard layer on the surface of particles with tough inner core, new kinds of abrasives and polishing compound can be manufactured by this Mixalloy Process.

Dispersion-hardened alloys can also be made by introducing two elements of a compound into these two liquid streams, one in each stream. During mixing of the liquid streams, they react, and form hard, small particles. When these small particles are unifromly distributed throughout the matrix, the hardness of the alloy will increase, in inverse proportion to the distance between the paths.

References

Davis, J.T., *Turbulence Phenomena*. Academic Press, NY, 1972.
Giedt, W.H., *Principles of Engineering Heat Transfer*. Van Nostrand, NY, 1957, p. 186.
Kolmogoroff, A.N., "The Local Structure of Turbulence in Incompressible Fluid for Very Large Reynolds Numbers," *C.R. Academy of Sciences, U.S.S.R.* **30**:301 1941. (Translation available in *Turbulence: Classic Papers on Statistical Theory,* Friedander, S. K., and Topper, L., eds. Interscience, NY, 1961.)

Suh, N.P., "Process for Making Uniform Size Particles," U.S. Patent 4,279,843, 1981.

Suh, N.P., "Orthonormal Processing of Metals. Part 1: Concept and Theory," *Journal of Engineering for Industry* **104**:327–331, 1982.

Tucker, C.L., "Reaction Injection Molding of Reinforced Polymer Parts," Ph.D. Thesis, MIT, 1979.

Tucker, C.L., and Suh, N.P., "Impingement Mixing—A Fluid Mechanical Approach," *Proceedings of 36th Annual Technical Meeting, Society of Plastics Engineers*, 1978, p. 158.

6B: EFFECT OF TEMPERATURE AND SATURATION PRESSURE ON CELL DENSITY

Bubbles in polymers may nucleate either homogenously or heterogenously. The specific mechanism depends upon whether a second phase is present in the polymer due to an insoluble additive or a nucleating agent. Homogeneous nucleation occurs when the dissolved gas molecules come together for a sufficiently long time to produce a stable bubble nucleus. The homogenous nucleation rate is (Colton and Suh, 1986)

$$N_{\text{HOM}} = C_0 f_0 \exp(-\Delta G^*/kT) \tag{6B.1}$$

where

C_0 = concentration of gas molecules
f_0 = frequency factor for gas molecules joining the nucleus
k = Boltzmann's constant
T = temperature, in K
ΔG^* = activation energy

The activation energy for homogenous nucleation is

$$\Delta G = \frac{16\pi\gamma^3}{3(p_s - p_0)^2} \tag{6B.2}$$

where γ is the surface energy of the polymer, p_s is the gas saturation pressure, and p_0 the environmental pressure.

The effect of temperature and gas saturation pressure on the cell nucleation density can be seen qualitatively from Eqs. 6B.1 and 6B.2, respectively. Equation 6B.1 shows that the higher the temperature is, the higher the nucleation rate will be. Equation 6B.2 shows that a higher saturation pressure leads to a lower activation energy barrier for the nucleation rate, as is evident from Eq. 6B.1. Thus, a higher gas saturation pressure leads to a higher cell density.

Reference

Cotton, J.S., and Suh, N.P., "Nucleation of Microcellular Foam: Theory and Practice," *SPE Technical Papers* **XXXII**:45–47, 1986.

6C: EFFECT OF SATURATION PRESSURE ON CELL SIZE

Consider 1 g of polymer that is saturated with a gas at temperature T_s and a pressure p_s. The amount of gas dissolved in the polymer is given by Henry's law (Durrill and Griskey, 1969):

$$C = Hp_s \tag{6C.1}$$

where

C = gas concentration at saturation, cm^3 (STP)/g of polymer
H = Henry's law constant for the gas–polymer system, cm^3 (STP)/g atm
p_s = gas saturation pressure, atm

The volume that the dissolved gas would occupy at saturation conditions is

$$v_s = C(p_0/p_s)(T_s/T_0) \tag{6C.2}$$

where T_0 and p_0 refer to the standard conditions. Upon substituting for C from Eq. 6C.1,

$$v_0 = H(T_s/T_0) \tag{6C.3}$$

where we have used the fact that $p_0 = 1$ atm. Note that if the saturation occurs at T_0 (i.e., 20°C), then the volume occupied by the dissolved gas at saturation per gram of polymer is simply equal to the Henry's law constant expressed in cubic centimeters.

Now assume that the saturated plastic is foamed at a constant temperature T_2 at atmospheric pressure. Assuming ideal gas behavior, the volume occupied by the dissolved gas at foaming temperature is

$$v_s = Hp_s(T_2/T_0) \tag{6C.4}$$

Equation 6C.4 shows that the dissolved gas volume at foaming temperature increases linearly with the gas saturation pressure. However, the bubble size depends on the number of bubbles present.

Let N_0 be the number of bubbles nucleated per cubic centimeter of unfoamed plastic. If d_0 is the density of the unfoamed plastic, then the number of bubbles nucleated in 1 g of plastic, N_0', is

$$N_0' = N_0/d_0 \tag{6C.5}$$

Let us assume that all the bubbles have the same diameter, D. Then, equating the volume occupied by bubbles to the right-hand side of Eq. 6C.4 gives

$$D^3 = \left(\frac{6Hd_0}{\pi N_0}\right)\left(\frac{T_2}{T_0}\right)p_s \tag{6C.6}$$

The data in Fig. 6.32b show that the cell density N_0 increases exponentially with saturation pressure; that is,

$$N_0 = A \exp(p_s/B) \tag{6C.7}$$

where A and B are empirical constants. Substitution for N_0 in Eq. 6C.6 gives

$$D^3 = \left(\frac{6Hd_0}{\pi A}\right)\left(\frac{T_2}{T_0}\right) p_s \exp(-p_s/B) \qquad (6C.8)$$

or

$$D = \left(\frac{6Hd_0 T_2}{\pi A T_0}\right)^{1/3} p_s^{1/3} \exp(-p_s/3B) \qquad (6C.9)$$

Equation 6C.9 shows the effect of the saturation, p_s on the average cell size, D. Due to the exponential increase in the number of cells nucleated, the cell size actually decreases as the saturation pressure is raised.

Reference

Durrill, P.L., and Griskey, R.G., "Diffusion and Solution of Gases in Thermally Softened or Molten Polymers: 2. Relation of Diffusivities and Solubilities with Temperature, Pressure and Structural Characteristics," *A.I.Ch.E. Journal* **15**:106–121, 1969.

7
CASE STUDIES: DESIGN OF PRODUCTS

7.1 Introduction

The design axioms apply to all synthesis processes: the design of manufacturing processes, machines, products, and even organizations. Chapter 6 presents a number of manufacturing-related case studies, involving the design and analysis of products, manufacturing operations, materials processing, and manufacturing machinery. In this chapter six case studies concerning the design of products are chosen for analysis, in order to illustrate the use of the design axioms in a somewhat different context, although the distinction between this chapter and Chapter 6 is somewhat arbitrary. These case studies involve primarily the use of the Independence Axiom, although the Information Axiom is also applied wherever appropriate in making design decisions. In Chapter 8 case studies are presented to show how the information content (thus Axiom 2) can be used in selecting the best design among several feasible designs. Chapter 9 presents case studies that involve organizational and system issues; it shows that the design axioms may apply equally well to the design of organizations and systems.

Since we are surrounded by products, it is easy for us to pick any one product and analyze its design from the axiomatic point of view. The analysis can be at different hierarchial levels, ranging from the overall evaluation of the ultimate FRs of the entire product (e.g., the design of airplanes), to specific design issues involving such problems as part integration and the use of symmetry. In this chapter six case studies are presented: the creation of a conductive grinding wheel, the creation of coated carbide tools, the analysis and redesign of simple engine lathe components, the generation of a laser architecture, analysis of a robot arm called the "direct drive" arm, and the design of automobile wheelcovers. These case studies are chosen at random.

A classic example of design failure is a product called the "flying machine," built early in the twentieth century. In those days the inventors tried what seemed obvious—imitating birds by using flapping wings so that they could achieve what birds do so naturally. Although their aspiration and determination (i.e., "If birds can fly, so can we!") were admirable, their attempt was an unmitigated failure. Later, the Wright brothers came along with their fixed-wing design, an innovation that opened up new technological and scientific frontiers.

The question that many people might have asked over the past eight decades, in view of the failure of the "bird-like flying machine," is, "What was wrong with the idea of wanting to fly like birds? After all, birds do indeed fly!"

The answer can be provided in terms of the design axioms, more specifically in terms of the FRs and the complexity of birds. The basic mistake the early would be inventors made was trying to copy the geometrical shape of birds' wings without fully understanding the relationship between the various functions and the physical embodiment (i.e., DP) of birds' wings. These types of mistakes are more commonly made, even today, than people like to admit.

Birds' wings satisfy many different FRs: a bird can take off vertically or horizontally; it can cruise at varying altitudes; it can remain nearly still in mid-air; it can fly at different speeds and change directions. Besides providing the propelling force, a bird's wings fulfll many other functions that are not related to flying, and there is the added feature that they can be folded and tucked away when the bird is not flying.

When the early inventors tried simply to duplicate the geometrical shape of the wing without understanding which functions of the wing are related to flying requirements, they might have satisfied a wrong set of FRs, or they might have tried to satisfy more FRs than was necessary. Furthermore, their design might have coupled the FRs to such an extent that the crude control systems that they employed would not have met the required FRs.

The genius of the Wright brothers was that they minimized the number of FRs. Subsonic airplanes satisfy the following three principal FRs that are essential for flying: near-horizontal take-off; optimum cruising speed at a given altitude; and change in direction. The wings of these airplanes provide the vertical lift, which can be varied by changing the speed and by extending the flap. The rudder at the tail, which steers the airplane, and the propellers, which provide the propulsion, are functionally separate from the wings. This historical example reaffirms the two key points of the axiomatic approach to design: the number of FRs must be kept at a minimum, per the Information Axiom, and the independence of FRs must be maintained, per the Independence Axiom.

After the product is designed, it must be manufactured. Some people claim that simultaneous considerations must be given to manufacturing operations as the product is being designed, so as to maximize productivity. This is the so-called "simultaneous engineering" or "life cycle engineering." Although it is necessary to design products that can be manufactured effectively, it may not be necessary to consider details of manufacturing operations during the early phase of product design, or at high levels of the FR hierarchy. The first thing to do during the product design stage is to develop a product design concept that satisfies the FRs and their independence. We then have to consider whether or not a manufacturing process exists that also satisfies the DPs of the product and the independence of FRs of the product as well as constraints such as cost. In Chapter 6 many case studies are presented that involve "design for manufacturability." Theorems 9 and 15 specify the conditions (the FR–DP and the DP–PV relationships) that the product and the process designs must satisfy to enable production of the product.

At the highest level of the FR hierarchy, the design of products (e.g., airplanes) can normally proceed without regard to manufacturing processes. However, as the FRs are decomposed into lower-level FRs, and as details of component design are being developed, the manufacturability issue becomes increasingly important. As seen in Section 6.2, where the design and manufacture of the multilense plate was presented, the manufacturing operation should not couple the FRs of the product. In this case either new manufacturing processes must be sought or the design of the product must be altered (i.e., Theorem 15). In this sense, iteration is an essential feature of the design process.

7.2 Electrically Conductive Grinding Wheels

Introduction to the Case Study

Aluminum alloys with hard refractory particles are being manufactured to increase stiffness, hardness and wear resistance while still maintaining low density.

Such alloys are difficult to process by either machining or grinding, since normal cutting tools wear rapidly during machining, due to the presence of hard particles, and since grinding wheels cease to grind because the soft aluminum matrix fills up the pores between hard abrasive particles in the grinding wheel. Our design task is to develop a solution to this manufacturing problem. At the highest level of the FR hierarchy, the FR that we hope to satisfy is the processing of aluminum alloys with hard refractory particles. Having decided on the overall FR, we have to consider various options in the physical domain.

After considering all plausible solutions (such as laser machining) in the physical domain, we came to the conclusion that one way of processing such a material is *electrochemically assisted grinding,* which is a well-known materials processing technique. In such a grinding operation, d.c. power is supplied to the metal-bonded diamond grinding wheel, which rotates at high speed. Electric current is then passed to the electrically conducting workpiece through a suitable electrolyte. The workpiece is consumed in accordance with Faraday's law, while the diamonds serve to remove the products of reaction, in addition to removing the hard particles by mechanical abrasive action.

Metal-bonded diamond wheels or metal-bonded aluminum oxide grinding wheels are ideally suited for this application. Metal provides *both* bonding and electrical conduction, which may be a good use of the physical integration principle (i.e., Corollary 3, integration of physical parts). This metal bonding does not couple FRs, since the bonding requirement can be controlled separately from the conductivity requirement by varying the thickness of the metal bond. The presence of a thin metal layer is sufficient to satisfy the electrical conductivity requirement. The only problem with metal-bonded grinding wheels is the high cost of the wheel, which may be caused by using metal bonding that exceeds the minimum conductivity requirement to satisfy the strength requirement.

It was decided to produce low-cost, electrically conductive grinding wheels. This work was done prior to the advent of the design axioms. In fact, it was one of the examples that helped to inspire the design axioms, as described in Chapter 1.

FRs and Constraints of the Electrically Conductive Grinding Wheel

Having chosen the electrochemically assisted grinding approach as the process in the physical domain, we now go back to the functional domain to establish the next set of FRs. The FRs of the electrically conductive grinding wheel are established as

FR_1 = Conduct sufficient electric current uniformly at the grinding wheel–workpiece interface to remove the metal matrix electrolytically during the electrochemical grinding operation.

FR_2 = Provide abrasive action to remove hard particles from the aluminum–alumina workpiece by bonding the abrasive particles strongly to the grinding wheel.

The constraints are

C_1 = Maintain compatibility with conventional grinding operation.
C_2 = Keep cost significantly lower (about one-tenth) than the metal-bonded wheel.

Solution

A physical solution that was consistent with the FRs and constraints was designed by Nehru and Suh (1967). The solution consists of satisfying the bonding requirement, using cheaper grinding wheels that use either resin or vitreous bonds, and satisfying the electrical conductivity requirement separately through metal plating. The invention encompasses the impregnation of inexpensive, nonconductive or semiconductive commercially available porous grinding wheels (made of aluminum oxide or silicon carbide) with an electrically conductive substance. Wall portions of the interstices and foramina of such porous abrasive bodies are coated with an electrically conductive material by first catalyzing them and then forcing an electroless plating solution through the pores (see Fig. 7.1).

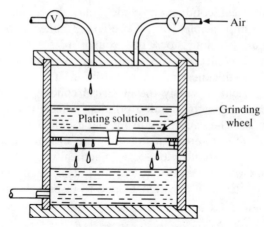

Figure 7.1. Apparatus for electroless plating of grinding wheels.

By utilizing electroless plating solutions, a metal can be deposited on an originally nonmetallic surface that has been catalyzed. The process relies on the reduction of metallic ions on a metallic surface. When the surface is nonmetallic, the surface must be catalyzed, such as with palladium. This is done by exposing the surface to an aqueous solution of palladium chloride and hydrochloric acid. Metals such as copper, nickel, cobalt, and gold are deposited on these catalyzed surfaces electrolessly from solutions containing these metal ions, using hypophosphite as the reducing agent. Electroless deposition is ideally suited for this application, because the deposition occurs at any exposed surfaces, even at re-entrant corners, where electrolytic deposition cannot occur due to the electric field effect.

Resin-bonded aluminum oxide grinding wheels were first coated with copper, followed by nickel, using electroless solutions. Copper is highly conductive, whereas nickel is more resistant to corrosion. The compositions of these solutions are given in Table 7.1. The deposition procedure involved a slight pressure differential between the upper part of the plating chamber above the grinding wheel, where the electroless plating solution was originally poured, and the bottom part of the chamber below the grinding wheel (see Fig. 7.1).

Before we could commence the plating operation, the wheels were first cleaned and catalyzed. The cleaning process was accomplished by passing a 30–50% HCl solution through the wheel. The wheel was then catalyzed by passing a suitable catalyzing solution through the wheel. The catalytic solution was made by combining a palladium chloride solution with about 0.02 g of palladium and 20 mL of HCl per liter of water. The wheel was exposed to the catalytic soluton for about 20–25 minutes. The wheel was then washed. The palladium ions on the grinding wheel were then reduced to metallic state by passing an "accelerator" solution which consists of 150 g of sodium hypophosphite ($NaH_2PO_2 \cdot H_2O$) per liter of water.

The copper plating solution was then passed through the grinding wheel

Case Studies: Design of Products 261

TABLE 7.1 Electroless Plating Solution Composition

Copper	
Water	1 L
Potassium sodium tartrate	94.1 g
Copper sulfate	28.4 g
Sodium hydroxide	37.4 g
Sodium carbonate	31.2 g
37% formaldehyde	30 cm^3
Nickel	
Water	1 L
Nickel chloride (NiCl$_2$·6H$_2$O)	16.04 g
Sodium hypophosphite (NaH$_2$PO$_2$·H$_2$O)	23.86 g
Glycine (aminoacetic acid) (NH$_2$CH$_2$COOH)	5.07
Malic acid (C$_4$H$_6$O$_5$)	4 × 10^{-3} g
Pb^{2+} (PbCl$_2$)	
pH (adjusted with NaOH)	6.5
Bath temperature	210°F at 1 atm

for about 1 hour. The differential pressure was needed to plate inside crevices of the grinding wheel. After completing the plating process with copper, the wheel was cleaned and plated with nickel, in order to prevent the rapid corrosion of copper. Unlike the copper deposition, which is done at room temperature, nickel deposition was done at 210°F because of its very slow deposition rate. The wheel was exposed for about 1 hour to the hot nickel solution, whose pH was maintained at about 6.5 by constant addition of sodium hydroxide solution. At a pH lower than 5.5 the reactions deplete the nickel from the plate, whereas at higher pH values of about 7 the nickel spontaneously reduces in the solution. In order to build up the nickel layer, the electrolessly plated wheel was exposed to electrolytic nickel solution, since the nickel layer was very thin.

This case study indicates that, during the process of developing a low cost, electrically conductive grinding wheel, we found that we had to reiterate the design process by adding a third FR, which was required to provide the corrosion resistance. The design equation may be written as

$$\begin{Bmatrix} FR_1 \\ FR_2 \\ FR_3 \end{Bmatrix} = \begin{Bmatrix} \text{Electrical conduction} \\ \text{Bonding of abrasive particles} \\ \text{Corrosion resistance} \end{Bmatrix}$$

$$= \begin{bmatrix} \times & 0 & \otimes \\ 0 & \times & 0 \\ 0 & 0 & \times \end{bmatrix} \begin{Bmatrix} \text{Copper layer} \\ \text{Resin bond} \\ \text{Nickel layer} \end{Bmatrix} \quad (7.1)$$

The design is a nearly uncoupled design, since the thin nickel layer does not affect the electrical conductivity, and thus is negligible, per Theorem 7.

Concluding Observations

The case study described in this section illustrates, once again, the process of design: we must start out in the functional domain by first establishing the FRs at the highest FR hierarchical level, followed by the creation of solutions in the physical domain. In the physical domain we consider all plausible solutions (which is the essence of the morphological technique discussed in Chapter 1). After selecting the most promising physical solution that satisfies the FRs, we move back to the functional domain to establish the next level of FRs in the FR hierarchy. This zig-zagging must continue to the nth level of the DP hierarchy until we obtain a satisfactory solution.

Although it is not done in this case study, when there are two or more competing solutions we must use the Information Axiom to select the best design from among the solutions that satisfy the FRs and the Independence Axiom.

At first glance the metal-bonded grinding wheel appeared to be a good solution in terms of the Independence Axiom. Only when we considered the cost of the metal-bonded wheel was it necessary to design other solutions.

In this case study we treated the cost as a constraint. In this formulation, as long as the manufacturing cost was within the constraint, we would have accepted the solution as being acceptable if the FRs were satisfied. When we treat cost as an FR rather than a constraint, the information content (I) associated with cost must be computed, using the definition adopted in Chapter 4, and then added to the total information content. The best solution is then an uncoupled design that satisfies the FRs with the least total information.

7.3 Coated Carbide Tools

Statement of the Problem

In cutting steel, the life of cutting tools limits the maximum cutting speed. During the past 100 years, the cutting speed has increased by more than an order of magnitude, thanks to the introduction of better cutting-tool materials. Early in the twentieth centruy high-carbon steel was used, which cut mild steel at a maximum speed of 100 ft/min. However, this was not a good material, because of its brittleness and rapid wear rates. Since then, with the advent of high-speed steel (HSS) tools, the cutting speed was increased to about 450 ft/min. HSS tool materials are tough, and have reasonable wear resistance at high cutting speeds. They are still being used.

When the tungsten carbide–cobalt cermet tool was introduced in the mid-1920s, the cutting speed jumped to 600 ft/min. However, this increased cutting speed was obtained at the expense of toughness. This speed was further increased to about 800 ft/min when complex carbides consisting of titanium carbide, tantalum carbide, tungsten carbide, and cobalt were introduced. Again the toughness decreased when the amount of complex carbide increased substantially. The cutting speed reached the range of 2,000 ft/min with the inroduction of aluminum oxide tools at a significant loss of toughness. The toughness of these alumina-based tools has increased in recent years with the introduction of aluminum oxide–titanium carbide tools and alumina fiber-reinforced silicon carbide tools.

Notwithstanding the advances made with the introduction of better cutting-tool materials, it is difficult to satisfy the FRs of a cutting tool with a single "homogenous" material. One material cannot satisfy both the surface requirements, related to wear resistance, and the bulk requirements, related to toughness and stiffness. This is indicated by the fact that the increased cutting speed has been obtained at the loss of toughness, which is very important in intermittent cutting.

This case study involves the development of better cutting tools. The task is the creation of cutting tools that last longer than existing tools and can, by being tough, withstand impact loading during cutting.

The FRs of an ideal cutting tool are: (1) wear resistance (i.e., long life); (2) toughness; and (3) stiffness. The "homogenous" tool materials that are based on WC–Co could not satisfy the first two FRs independently, because as we change the composition of the material to increase any one of these FRs (e.g., wear resistance), we have to sacrifice the other property (e.g., toughness). The third FR (i.e., stiffness) could be satisfied independently by changing the geometry of the tool bit—the stiffness increases with the thickness of the tool bit. Therefore, by Theorem 1 (coupling due to insufficient number of DPs), a single, homogenous material can not be the ideal tool material. What we have done in the past is to use different cutting tools to obtain the maximum tool life or greater toughness for a given application, and to subordinate other requirements. For example, when the toughness is the most important consideration, we used the HSS tool, although its wear resistance was not as good as those of carbide tools.

Beginning in 1965 (the design axioms did not exist then), the author and his students began to search for ways of changing the wear resistance of tungsten carbide-based cutting tools without substantially changing the toughness (Suh, 1977). One of the techniques investigated at the time was the coating of tungsten carbide tools with materials that are more wear-resistant. The idea was to make carbide tools last as long as aluminum oxide (alumina) tools, without sacrificing their better toughness. The idea of coating cutting tools was a novel idea then, although it is done routinely now. The project necessitated an in-depth understanding of the wear process (Suh, 1986), which is reviewed next.

Description of Tungsten Carbide Tools and Their Properties

Tungsten carbide–cobalt cermet tools are made by bonding small tungsten carbide particles (typically 1–3 μm in diameter), with cobalt as the binder.

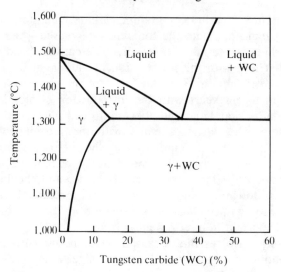

Figure 7.2. Approximate pseudo-binary tungsten carbide–cobalt system. (From Sandford, E. J., and Trent, E. M., "Symposium on Power Metallurgy," Iron and Steel Institute, Special Report No. 38, 1947.)

The bond strength between WC and cobalt is very strong, because tungsten carbide (melting point 2,870°C) and cobalt form an eutectic at 1,320°C, as shown in Fig. 7.2. At the eutective temperature, which is the lowest melting temperature possible, the liquid solution consisting of tungsten carbide and cobalt undergoes solidification, yielding a fine, two-phase microstructure that consists of the WC-rich phase and the cobalt-rich phase.

The cermet is made by means of powder metallurgy techniques. The tungsten carbide powder with cobalt powder are ball-milled together for 24–72 hours, either in alcohol or dry, and then cold-pressed to a desired shape. They are then presintered to give them the green strength at about 810°C. The pre-sintered tungsten carbide–cobalt cermet has the consistency of chalk, which can easily be shaped into the final geometry. The pre-sintered carbide is then sintered further at a temperature slightly above the eutetic temperature, in order to consolidate the powder to a fully dense state, due to the formation of the liquid eutetic phase between the carbide particles. These tools are then ground to the final dimensions.

These tools wear by two mechanisms: abrasion of the surface layer, presumably by hard particles in the workpiece, and dissolution of the tool material in the workpiece (Suh, 1986). The abrasion resistance is controlled by the hardness of the surface, whereas the free energy change and kinetics govern the rate of wear by dissolution. The hardness depends on the hardness and volume fraction of WC and cobalt. When the cobalt content is very small, the hardness is primarily controlled by the hardness of WC. The tougness of WC–Co cermet depends sensitively on the amount of cobalt.

Case Studies: Design of Products

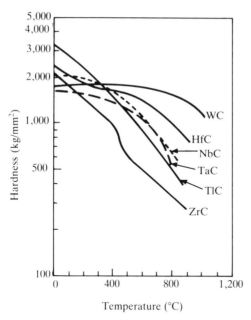

Figure 7.3. Hardness of various carbides. (Data from Toth, L. E., *Transition Metal Carbides and Nitrides*. Academic Press, NY, 1971.)

The hot hardness of tungsten carbide is among the highest, as shown in Fig. 7.3, but its free energy of formation is also very high, compared with other carbides, nitrides, and oxides (see Fig. 7.4). The high free energy of formation means that the material may be less stable chemically, although its abrasion resistance is outstanding because of its hot hardness. Furthermore, the solubility of tungsten carbide in iron is so high that it readily forms solutions with iron (Kramer and Suh, 1980), resulting in high wear rates of these tungsten carbide tools when cutting steel. Therefore, the shortcoming of tungsten carbide–cobalt cermet is its chemical stability, despite the fact that it is hard and tough (among carbide cermets and oxide tools it is one of the toughest, although it is not as tough as HSS tools).

In order to control the toughness without affecting the wear resistance too much, the volume fraction of cobalt is varied from 6 to 10%. However, the wear resistance is affected by the change in the cobalt content. To improve wear resistance, complex carbides (TiC–TaC–WC) are used, but the amount of TiC and TaC must be small, so as not to affect the toughness of the tool.

The design equation for the original WC–Co cermet may be written as

$$\begin{Bmatrix} FR_1 \\ FR_2 \\ FR_3 \end{Bmatrix} = \begin{Bmatrix} \text{Wear resistance} \\ \text{Toughness} \\ \text{Stiffness} \end{Bmatrix} = \begin{bmatrix} \times & \otimes & 0 \\ \times & \times & 0 \\ \times & \times & \times \end{bmatrix} \begin{Bmatrix} \text{WC content} \\ \text{Cobalt content} \\ \text{Geometry} \end{Bmatrix} \quad (7.2)$$

where DP_1, DP_2, and DP_3 are, respectively, WC content, cobalt content,

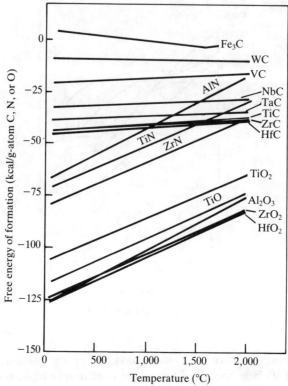

Figure 7.4. Free energy foundation of various hard materials. (Data from *Janaf Thermochemical Data*. The Dow Chemical Company, Midland, MI, 1965; Elliot, J. F., and Gleiser, M., *Thermochemistry for Steelmaking*, Vol. 1. Addison–Wesley, Reading, MA, 1960.)

and geometry. The WC–Co cermet tool is a coupled system if the design goal is to improve the wear resistance and toughness significantly. The basic problem is that WC is not the best material to have for wear resistance at the surface, due to its poor chemical stability, although it has an excellent abrasion resistance; but we cannot replace it with other wear-resistant ceramic materials because there are no suitable binders for them that can provide the toughness.

Design of an Ideal Tool

From the foregoing discussion it is clear that the wear resistance of the cutting tool can be further decomposed into: (1) the abrasion resistance of the surface layer; and (2) the chemical stability of the material. Then, the FRs are

FR_1 = Abrasion resistance.
FR_2 = Chemical stability.
FR_3 = Toughness.
FR_4 = Stiffness.

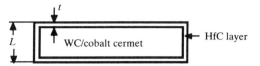

Figure 7.5. HfC-coated WC–Co cermet tool.

FR_1 and FR_2 are surface properties, whereas FR_3 and FR_4 are primary bulk properties.

In order to overcome the high wear rates of tungsten carbide in cutting steel, it is proposed that a thin layer (about 5 μm thick—see Fig. 7.5) of hafnium carbide be coated to prevent rapid dissolution of the tool material in iron. (Hafnium carbide is chemically much more stable than, and almost as hard as, WC—see Figs 7.3 and 7.4.) Then the design equation may be written as

$$\begin{Bmatrix} FR_1 \\ FR_2 \\ FR_3 \\ FR_4 \end{Bmatrix} = \begin{Bmatrix} \text{Abrasion resistance} \\ \text{Chemical stability} \\ \text{Toughness} \\ \text{Stiffness} \end{Bmatrix}$$

$$= \begin{bmatrix} \times & 0 & 0 & 0 \\ 0 & \times & 0 & 0 \\ 0 & 0 & \times & 0 \\ 0 & 0 & 0 & \times \end{bmatrix} \begin{Bmatrix} DP_1 = \text{Hardness of the coating} \\ DP_2 = \text{Free energy of formation} \\ DP_3 = \text{Substrate toughness} \\ DP_4 = \text{Thickness of the tool, } L \end{Bmatrix} \quad (7.3)$$

The design equation shows that the coated tool satisfies Axiom 1. The design matrix is diagonal. The ideal coating thickness is found to be about 5 μm. When it is thicker that this, it lowers the toughness due to the cracks that initiate at the surface in the pure HfC layer and propogate into the bulk. The effect of the coating on the wear rate is significant, as shown in Fig. 7.6. The hafnium carbide-coated tool lasts for about an order of magnitude longer than the WC–Co tool (Kramer, 1979; Kramer and Suh, 1980; Suh, 1986). The coating can be applied by chemical vapor deposition, or physical deposition techniques such as sputtering (Suh, 1977).

Concluding Observations

This case study shows that, by recognizing the shortcoming of tungsten carbide cermet tools (it violated Axiom 1), we could develop an uncoupled product in the form of coated carbide tools. It took about 10 years, after the research was initiated, to develop commercial products fully. The most important step in the development process was the recognition of the problem, but in those days it took much trial and error to define the problem, since we did not have the benefit of the design axioms.

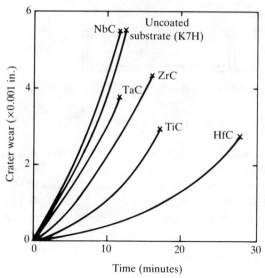

Figure 7.6. Test results of crater wear for various carbide coatings. The coating thickness is approximately 5 μm. The cutting time was about 12 minutes on AISI 4340–700 SFPM, 0.005 in./rev., 0.050 in. depth.

7.4 Analysis of and Alternative Designs for Engine Lathe Components

General Description of the Case Study

During the early phase of the axiomatics research at MIT, one of the graduate students, William Tice, undertook a case study of simple engine lathers under the supervision of Professor Nathan H. Cook (Tice, 1980). They analyzed the design features of two different kinds of metal-cutting engine lathes: a small hobby lathe ("Atlas" lathe) and a prototype South Bend engine lathe for small lot production. (Specifications for the lathes are given in Table 7.2.) The major objective of their study was to demonstrate the validity or the invalidity of the axioms by analyzing the components of the lathes. The material presented in this section is taken from Tice (1980).

The research methodology used by Tice and Cook is represented by the flowchart shown in Fig. 7.7. They randomly chose several components of the lathes for investigation. Their first task was to generate a set of FRs through an analysis of the specific function of the component, although there was no way in which they could be certain if their FRs were the same as those used by the original designer of the lathes. Similarly, they specified constraints based primarily on geometry and operational features of the lathes. Once the set of FRs and constraints were chosen, they searched for alternate ideas that could satisfy the same set of FRs and constraints.

Case Studies: Design of Products 269

TABLE 7.2 Lathe Specification

	Atlas	South Bend
Swing over bed (in.)	6	12.125
Swing over cross-slide (in.)	3.375	6.5
Distance between centers (in.)	19.25	25
Thread range (threads/in.)	8–96	4–480
Longitudinal feed range (in./rev)	0.0024–0.0078	0.0007–0.0836
Cross-feed range (in./rev)	—	0.0002–0.0314
Spindle speeds	8	Infinite
Spindle speed range (r.p.m.)	55–2,300	60–2,400
Cross-slide travel (in.)	4.75	7.25
Compound rest travel (in.)	1.75	2.5625
Tailstock spindle travel (in.)	1.25	3.125
Tailstock taper offset (in.)	0.5625	0.500
Motor drive power (HP)	1/4–1/3	2

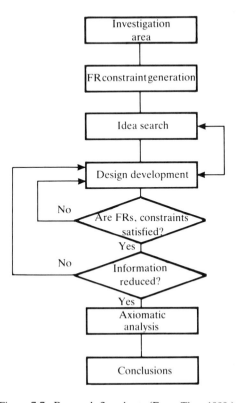

Figure 7.7. Research flowchart. (From Tice, 1980.)

This is the creative part of the design process, where the design axioms play only a minor role. Design concepts were generated by brainstorming, and the selected ideas were checked later for feasibility. An absence of any new solutions to the problem was taken to mean that the existing design is close to optimal. At this time the design development phase began. By design development, they meant the actual conversion of the concept into specific geometries with dimensions and materials, along with the analysis of forces that the device will encounter.

They then undertook "FR, C Analysis" to determine if the alternative design satisfies the FRs and constraints. This analysis involved theoretical calculations, or prototype and model testing. The new design was tested to see whether it would perform at the levels indicated by the FRs, and the independence of the FRs was checked. The constraints were satisfied by proper selection of components. If the FRs and constraints were not satisfied by the proposed design, then it was either changed or abandoned. They then analyzed for information content, using their own ad hoc measures, which incorporate weighting functions (however, this part is not presented here).

The final step in their case study was the axiomatic *analysis*. The designs were examined to determine whether any corollaries were followed, and which of the axioms applied to the improvement; many times a corollary was used even though no conscious attempt at its application had been made. They also investigated whether the designs contradicted any of the axioms or corollaries; this was done by comparing the axiomatic evaluation of the design alternative with an intuitive one.

Conclusions on the success of the redesign effort were drawn. In the cases where the alternatives were functionally equivalent, reduction of information content was used as the criterion; in cases where a coupling was present, the design was improved by eliminating the coupling, which normally also reduced the information content. They encountered uncertainties regarding the success of a design when the original set of FRs and constraints was not known exactly, and when the difference in the information content was small.

Four specific examples of their case study are reviewed here: tailstock spindle, headstock gearing, tailstock base clamp system, and carriage control. Figure 7.8 shows the general location of these components on an engine lathe.

Tailstock Spindle

FRs and constraints of the spindle. The tailstock spindle is one of the three major subsystems of the tailstock. It can function as a tool or center holder, has axial feed capabilities, and can hold a drill on a boring bar, which can then be fed into the end of the rotating workpiece. The spindle and the tool attached to it must not rotate. When either boring or drilling a

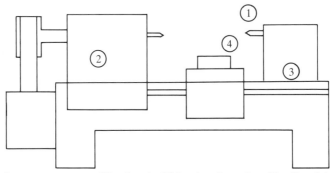

Figure 7.8. Improvement areas: (1) tailstock, (2) headstock gearing, (3) tailstock base clamp system, and (4) carriage control. (From Tice, 1980.)

hole, the cutting action is provided by the velocity difference between the workpiece and the tool. When it is used as a center holder to support one end of a workpiece, the rotation of the spindel is again undesirable, due to possible interaction with the feed device. Sometimes a "live" center is used to isolate the rotation of the spindle from the feed mechanism, but these lathes did not have this feature.

The FRs of the spindle are determined to as follows:

FR_1 = Support a tool or center holder with a #2 Morse taper bore.
FR_2 = Lock the spindle to prevent rotation.
FR_3 = Translate axially in tailstock housing.
FR_4 = Withstand 100-lb axial compressive force.

The constraints are

C_1 = The device must be compatible with the spindle feed mechanism (3/8-16 left-hand power screw).
C_2 = Spindle maximum size:. 3/4 in. outside diameter, $3\frac{1}{2}$ in.
C_3 = Clamping and/or rotation must require user force of approximately 10 lb.
C_4 = The lifetime must be comparable with the lathe lifetime.

Alternative design description. The spindle was examined, and it was decided that the information content could be reduced by integrating FR_2 and FR_4 in one physical piece. In the original design these two FRs are satisfied by two separate devices as shown in Fig. 7.9. The pin in Fig. 7.9 acts as the rotation limiter to satisfy FR_2. A combination of a bolt, clamping pieces, and handle (not shown) acts as the axial motion clamp to satisfy FR_4.

In the alternate design, FR_2 and FR_4 were integrated into one part, using a pin, as shown in Fig. 7.10. This pin serves the same purpose as the clamp and pin of Fig. 7.9. The clamping function is obtained by turning the handle and threaded rod clockwise into the threaded tailstock housing, which pushes the bottom face of the pin onto the floor of the keyway

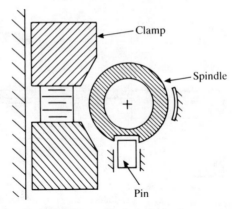

Figure 7.9. Original spindle design. (From Tice, 1980.)

milled into the cylindrical spindle. The clamping force is a function of the coefficient of friction between the pin and the keyway and the normal force generated by tightening the handle. The new design satisfies Corollary 3 (integration of physical parts).

The spindle rotation is prevented independently by the action of the sides of the pin, which contact the keyway walls whenever a torque is applied to the spindle. Tice used a brass pin and threaded rod concept in the design, in order to prevent excessive wear that would violate the lifetime constraint. Using a brass pin to prevent the wear of the rod and spindle, the helical motion of the threaded rod is decoupled from the clamping action. For this reason Tice used a brass pin and threaded rod concept in the design. The design was finalized by determining the size of the parts, based on strength calculations for the new design.

Results. The alternate design satisfies the FRs and constraints of the tailstock spindle. The new design maintains the independence of FRs, and thus satisfies Axiom 1. Since the number of parts required is less than before, and since no special tolerance requirements are presented by the new design, per Corollary 6 (largest tolerance) and Axiom 2, according to axiomatics this new design represents an improved solution to the tailstock spindle problem. The nature of the improvement conforms to the axioms, and specifically to Corollary 3 (integration of physical parts).

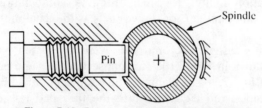

Figure 7.10. Proposed spindle. (From Tice, 1980.)

South Bend Headstock Gearing

The headstock gearing of the South Bend lathe is shown in Fig. 7.11. Through reduction of parts, there is a resultant decrease in information content. The improved design still satisfies all FRs of the original design.

Design description. The headstock gearing is a subsystem of the main spindle drive which provides rotational power to the workpiece. For each cutting operation, the optimal cutting speed depends on the material of the workpiece and cutting tool, the workpiece radius, and the type of coolant used, among other variables. Because of the large variation on these parameters, a wide range of headstock spindle speeds is required. The spindle speed of the South Bend lathe is from 60 to 2,400 r.p.m.

FRs and constraints of the headstock gearing. The FRs of the headstock gearing are

FR_1 = The device must provide an infinite number of spindle rotational speeds in the range 60–2,400 r.p.m.
FR_2 = The components must withstand torques, forces generated by the 2-HP motor drive for the lifetime of the lathe.

The constraints of the headstock gearing are

C_1 = Must employ existing 1:4 to 2.2:1 variable-speed transmission.
C_2 = Must use 1,725 r.p.m. 2-HP motor drive.
C_3 = Must fit into existing headstock casting.
C_4 = Maximum gear ratios of 1:4 reduction and 2:1 gain.
C_5 = Speed range selection must be a one-hand procedure.

Figure 7.11. South Bend headstock gearing.

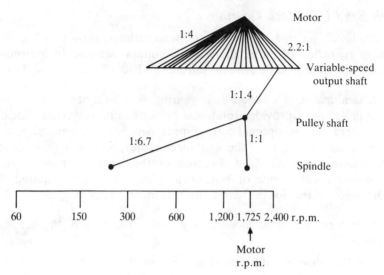

Figure 7.12. Structural diagram of the original gearing. (From Tice, 1980.)

Original design. Figure 7.12 is the structural diagram for the original South Bend spindle deive. The drive consists of the 1,725 r.p.m. motor, the variable-speed transmission, a pulley and V-belt reduction, and a two-speed gearbox. The V-belt pulleys and the gearbox were modified to reduce the information content.

The original two-speed gearbox, as shown in Fig. 7.13, consists of three shafts and five gears. Gear selection is accomplished by sliding the gears on shaft B to one of two positions by means of a control knob. It uses a double gear reduction to obtain the 1:1 and 1:6.6 ratios. With some minor modifications to the V-belt pulleys, these ratios could be obtained by a single reduction. It appears that the original headstock gearing has excess information content because of the hardware needed for the double reduction.

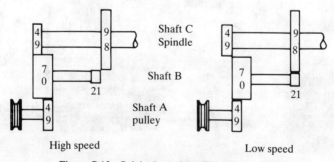

Figure 7.13. Original gearing. (From Tice, 1980.)

Case Studies: Design of Products 275

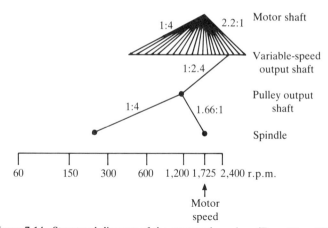

Figure 7.14. Structural diagram of the proposed gearing. (From Tice, 1980.)

Alternative design description. A new ratio organization has enabled the total system to be simplified. Figure 7.14 presents the proposed structural diagram. It shows that the V-belt pulley speed reduction has been increased from 1:1.4 to 1:2.4 and that the gearbox ratios required to obtain the same speed ranges as before are 1.66:1 and 1:4. As a consequence of this ratio reorganization, the gearbox ratios can each be obtained by using a single pair of gears (see Fig. 7.15). The proposed gearbox has only four gears and two shafts. The gears on shaft A' slide axially and transmit power from the shaft through a key. In addition to these changes, the direction of rotation of the electric motor may have to be reversed if necessary.

Results. The new design satisfies the same FRs and constraints as the original headstock gearing. According to axiomatics, the new design is better because it has less information content, due to the reduction of parts without an increase in the tolerance requirement, per Corollary 6 (largest tolerance). As an added benefit, the gearing would be more efficient than before. Since only two gears are in contact instead of the three or four that were in contact before, there will be less frictional loss, noise, and backlash error.

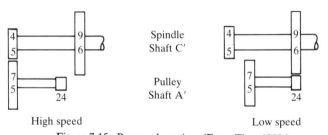

Figure 7.15. Proposed gearing. (From Tice, 1980.)

Atlas Tailstock Base Clamp

Tice and Cook also investigated the tailstock base clamp for the Atlas lathe. They found that the original design coupled the FRs and was more complicated than necessary.

Design description. The tailstock base clamp locks the tailstock housing and base to the lathe bed so that the tailstock can function as a tool or a workpiece support. The clamp design must be compatible with the offset positioning capability of the tailstock housings.

FRs and constraints. The FRs are determined to be

FR_1 = Must be able to clamp tailstock to lathe bed to withstand 100 lb of axial force.
FR_2 = Tailstock housing must be able to be positioned up to 9/16 in. to either side of center.

The constraints are

C_1 = Must be consistent in size with the present tailstock base and housing.
C_2 = Clamping control must be obtained by a manually applied force of no more than 15 lb.
C_3 = The design must have a lifetime equal to the lathe lifetime.

Existing design. The existing clamp is shown in Fig. 7.16. It consists of four components: a bolt, a washer, a nut, and a clamping piece. The clamp is locked by turning the nut clockwise with a wrench, which draws the bolt and clamp upward until the top of the clamp contacts the underside of the ways.

The device couples FR_1 and FR_2 the clamp and the tailstock alignment. On the bottom of the tailstock base there is an adjustable gib, which acts to align the tailstock in the ways (see Fig. 7.17). The gib interferes with the

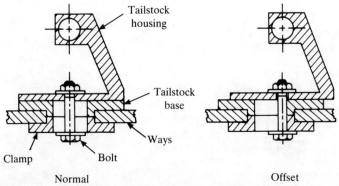

Figure 7.16. Atlas tailstock base clamp. (From Tice, 1980.)

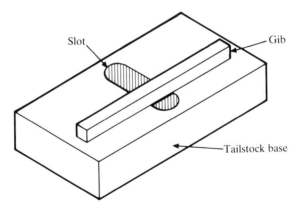

Figure 7.17. Coupling example. (From Tice, 1980.)

clamp bolt when the tailstock is in position. The clamping bolt must be free to move relative to the base, in order to provide the offset capability. The gib interferes with this requirement on one side of center, preventing the full 9/16-in. offset because the bolt comes into contact with the gib.

Alternative description. An alternate design that simplifies the clamp and eliminates the coupling is shown in Fig. 7.18. This design consists of a bolt, a washer, and a threaded clamping piece. To operate the clamp, the head of the bolt is turned, drawing the clamp upwards. Coupling is eliminated by putting a slot in the tailstock housing, rather than in the base and the clamp. The gib does not interfere with the tailstock in any way, since there is now no motion of the clamping bolt when the housing is offset. This alternate design offers an additional benefit; now the clamping force is evenly distributed between the two ways, because the bolt is always centrally located, whereas previously the clamping force would be unevenly distributed when the housing was offset.

Results. The new design is an uncoupled system and has less information content (see Tice, 1980). The new design represents an improved device,

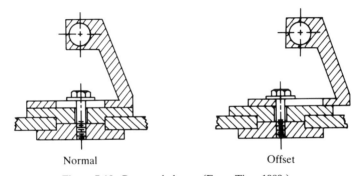

Figure 7.18. Proposed clamp. (From Tice, 1980.)

per Axiom 1. This conclusion is supported by the intuitive reasoning that the design performs as intended and costs less to manufacture.

South Bend Lead Screw–Feed Rod

So far only the components that violated the design axioms have been analyzed. However, as expected, there were also many well-designed components in these lathes, design characteristics which demonstrate the principle of integration.

The South Bend lead screw–feed rod is shown in Fig. 7.19. It is a standard lead screw with a keyway machined along its length. The threads, along with a pair of half-nuts, power the carriage for threading operations. The key in the keyway and the friction-drive power the carriage when a power feed is required.

The relevant FRs are

FR_1 = Provide power transmission device between gearbox and carriage for threading operations.
FR_2 = Provide power transmission device between gearbox and carriage for power-feed operations.

One of the constraints is that wear of the lead screw threads be minimal. FR_1 and FR_2 are independent FRs, since the threading operation requires a far more precise correspondence between the workpiece and tool motion, and because the required transmission ratios are significantly different for feeds and threads.

A common method of satisfying these two FRs is to use a lead screw and half-nuts for FR_1 and a keyed feed rod for FR_2. The South Bend designers have integrated the two FRs into one device, a lead screw with a keyway. The keyway is small enough not to interfere with the power transmission between the threads and the half-nuts. When the keyway is used for power transmission, the threads are not used, and so experience no wear. This is an example of a successful integration of functions into one part.

Intuitively the South Bend design seems better because of its simplicity. This is reflected in a lower information content. It can be concluded that the integrated feed–lead screw is a better device according to Corollary 3, since the information content is less and the sane FRs are satisfied independently.

Figure 7.19. South Bend lead screw–feed rod. (From Tice, 1980.)

Case Studies: Design of Products   279

Concluding Observations on the Lathe Design Case Study

The case study conducted by Tice and Cook dealt with relatively simple lathes. They showed that the design could be improved per the design axioms. They also found that some designs could not be improved further because they satisfied the axioms, and that no better alternate ideas than those embodied in the lathe could be developed. These conclusions were based on the specific set of FRs that they developed for the components that they studied. In other, more-complicated lathes these FRs may not necessarily be applicable for similar parts. In this case the resultant designs that satisfy this new set of FRs may be different from those presented in this section, per Theorem 5 (need for new design) (see Chapter 3). The important message of this case study is that the design axioms facilitate the evaluation of existing designs and enable quick identification of poor designs. Once the problem is identified based on the axioms, better designs can be developed.

7.5 Laser Architecture

Statement of the Problem

As part of the laser fusion project at the National Laboratory, a powerful laser is being developed to initiate a fusion reaction by bombarding the "fuel" with an intense laser beam. In designing the laser system, the cost of the laser is one of the major considerations, and a large component of that cost is related to the laser amplification architecture adapted by the laboratory, wherein a 1.06 μm laser beam is amplified by passing it through a series of laser glass plates.

The principle of the amplifier is illustrated in Fig. 7.20. The laser glass is

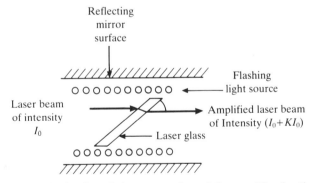

Figure 7.20. Schematic drawing of the construction of the amplifier for the laser beam. Energy is stored in the laser glass by using the flashing light source. The light beam stored in the laser glass travels in the glass back and forth. When a laser beam enters the laser glass at Brewster's angle (α), the laser beam enters the glass without reflection from the surface. The laser beam exits the glass amplified. The beam intensity increases by KI_0, where K is a materials constant and I_0 is the intensity of the entering beam.

an energy-storage device. This glass is enclosed in a clean chamber containing a series of flashing fluorescent light sources. The light emitted by the fluorescent light sources is stored in the glass. When a coherent laser beam is incident into the laser glass at Brewster's angle α, the laser beam enters the glass without any reflection of the beam at the surface. Some of the stored energy then leaves the glass with the laser beam as the laser beam is transmitted through the thickness of the glass.

If the amplitude of the incident laser beam is I_0, the transmitted amplified laser beam has the intensity $I_0 + KI_0$. If the laser beam passes through the laser glass plates n times in the amplifier, the intensity becomes $I_0(1 + K)^n$. The maximum intensity of the laser beam, I_{max}, that can be tolerated by the laser glass without shattering determines the maximum amplification ratio of the amplifier.

The design objective is to develop the amplifier architecture so as to minimize the manufacturing cost of the amplifier, which is largely determined by the number of laser glass plates used in the amplifier.

The FRs of the amplifier may be stated as

FR_1 = The incident laser beam must be amplified by $R = I_{max}/I_0$.
FR_2 = Each laser glass disk should be subjected to maximum intensity of laser beam, I_{max}.

The current architecture is shown in Fig. 7.21. It consists of a pattern of laser glass plates. In this design the incident 1.06 μm laser beam is continuously amplified as it travels along the amplifier, which passes through each laser glass plate only once. This laser architecture can easily satisfy FR_1, but not FR_2. The task here is to devise a lower-cost laser architecture, since only the last laser glass plate is effectively used to amplify the intensity of the beam from $I_0(1 + K)^{n-1}$ to $I_0(1 + K)^n$. For maximum efficiency, the laser glass should be used at its maximum capacity, I_{max}, by subjecting each laser glass plate to the most intense laser beam that the glass can withstand. We need to devise a new design solution that satisfies FR_2 as well as FR_1 without compromising the functional independence of each FR.

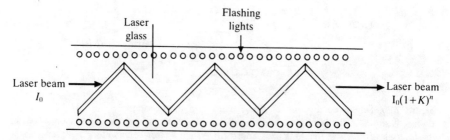

Figure 7.21. The current architecture. The laser beam is amplified by a series of laser glasses. The intensity of the incident laser beam on the first laser glass is I_0, whereas the last laser glass is subjected to intensity $I_0(1 + K)^{n-1}$.

Solution

The shortcoming of the current design is that the laser glass is not used at its maximum efficiency, since the amplification of the laser beam depends on the magnitude of the incident laser beam. One clear way to increase the efficiency of the laser system is to design a system in which each laser beam extracts more of the energy pumped into the laser disks (i.e., glass) than is extracted in current designs. Smith and Finger (1985) proposed a laser architecture different from the original design, based on axiomatic thinking. Their solution is presented here.

The solution that they sought to develop was a multipass laser architecture that would subject each glass disk to the same maximum intensity. One such solution is to have a laser beam travel down a series of laser disks and then return through the same disks in the opposite direction. In order to achieve this amplifier effect, an optical switch is needed to perform a dual function: to act as a mirror to reflect the laser beam, and to transmit the beam at other times when its intensity exceeds a certain limit. The use of such a switch is not a practical consideration, since such a device is not available.

Smith and Finger could not develop a design that satisfied FR_2 fully. Therefore, they relaxed the FR by replacing FR_2 with the following: FR_2' = all beams should pass through the same number of disks, and every disk should be used twice in a complementary manner; that is, each disk should have two light beams going through at two different incident angles (both still at Brewster's angle), such that the total intensity of the laser beam passing through the $(j+1)$th laser disk is the sum of the two intensities

$$I = I_0(1 + K)^j + I_0(1 + K)^{n-j} \qquad (7.4)$$

In this configuration a disk that is used first by one beam should be used last for another; a disk that is used second by one beam should be used second to last for another beam, etc. The laser beam is most intense for the glass plates subjected to the nth amplification. The laser beam is less intense for the glass plates located away from ends.

One simple design that satisfies this condition is illustrated in Fig. 7.22. In this design a series of mirrors is used to reflect the laser beams. The use of such a large number of mirrors may increase the cost significantly, and may require the use of a much larger space. Then the next question is: "Is

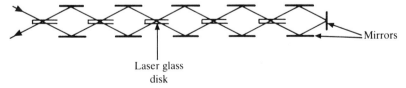

Figure 7.22. A simple laser architecture that does not require an optical switch, but uses many mirrors.

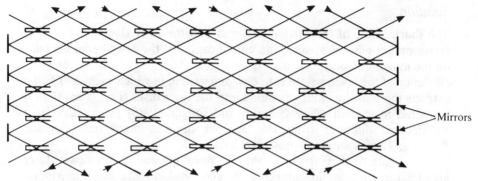

Figure 7.23. A laser architecture that minimizes the use of mirrors. Two separate beams pass through each glass plate.

there a way to make use of the dual-pass disk concept in a manner not requiring a significant increase in the number of mirrors, use of space, etc.?" One solution is to let one beam pass through a series of disks in one direction and then let another beam pass through the same series in the opposite direction.

Figure 7.23 shows an arrangement of the laser glass disks that satisfies the requirements of transfering the most energy from each disk to the beam energy and a minimum number of mirrors. An additional set of mirrors (not shown in the figure) may be required to focus the two beams coming out of the amplifier on the same target. However, the figure does illustrate the basic concept that every glass disk is subjected to a laser beam of nearly maximum intensity.

Concluding Observations

One of the important elements in developing the new laser amplifier architecture was the recognition that one of the FRs established by Smith and Finger was not satisfied by the original architecture used at the National Laboratory. In some ways this illustrates the process of refining FRs through an iteration of the conceptual design. The original design was acceptable if the cost of manufacturing the amplifier was not an important requirement. However, one of the two most important constraints of this laser architecture was the cost of manufacturing the amplifiers. Having identified the problem with the original design, they could not develop a design that fully utilized the capability of each glass plate. Therefore, they relaxed their FRs. For this modified set of FRs, the new architecture is a decoupled design, since the number of laser glass disks can be determined after the intensity of the incoming laser beam is known. The design equation may be written as

$$\begin{Bmatrix} FR_1 \\ FR_2 \end{Bmatrix} = \begin{bmatrix} A_{11} & A_{12} \\ A_{11} & A_{22} \end{bmatrix} \begin{Bmatrix} DP_1 \\ DP_2 \end{Bmatrix}$$

$$\begin{Bmatrix} R \\ I'_{max} \end{Bmatrix} = \begin{bmatrix} \times & 0 \\ \times & \times \end{bmatrix} \begin{Bmatrix} N \\ I_0 \end{Bmatrix}$$

(7.5)

Case Studies: Design of Products   283

One of the basic questions that the reader should address is: "Are there design solutions that satisfy the original set of FRs that Jack and Susan established but could not satisfy?"

7.6 Direct-drive Robot Arm

Introduction

A robot may be defined as a programmable manipulator. Robots have become important means of moving objects for many applications. In manufacturing, robots are used for spot welding, assembly, painting, and inspection. In many hazardous applications, robots handle nuclear, toxic, and high-temperature materials. The most common robot design is the articulated arm, which consists of a series of linkages joined together at hinges. In a conventional robot the arm is driven using a series of transmission mechanisms such as gears, lead screws, and chains.

One of the difficulties in controlling this type of mechanical arm is caused by the fact that the multiple links are coupled and their behavior is highly nonlinear. The control of such a robot is difficult, due to the uncertainty in arm dynamics and such unknown quantities as friction, deflection, backlash, compressibility, and wear of mechanical components (i.e., gears, lead screw, steel belt, servovalve, and structural parts). For these reasons the decoupling of arm dynamics in these robots is not an easy task.

In order to avoid the shortcomings of the conventional mechanical robot arm, Asada et al. (1983) designed a direct-drive arm in which the joint axes are directly coupled to rotors of high-torque electric motors, thus avoiding the need for transmission mechanisms between the motors and their loads. Conventional mechanical arms are driven through gears, chains, and lead screws. The designers of the robot initially thought that the simple mechanism of a direct-drive arm would allow a clear and precise modelling of arm dynamics for accurate positioning, and for compensation of interactive and nonlinear torques in high-speed manipulation. They also expected such a robot to have high stiffness, no backlash, and low friction. Their direct-drive arm is shown schematically in Fig. 7.24.

When they designed and constructed the direct-drive arm, they realized that the robot was not as precise as was originally expected. What was wrong with the original reasoning, if anything? How is the shortcoming to be overcome? Could we have noticed the problem before proceeding with the design and construction of the direct-drive arm? These questions are answered in this section, based on the work of Asada and Youcef-Toumi (1984).

Description of the Direct-drive Arm and its Problems

The overall view of the direct-drive arm developed by Asada et al. (1983) is shown in Fig. 7.24. The arm has six degrees of freedom, all of which are

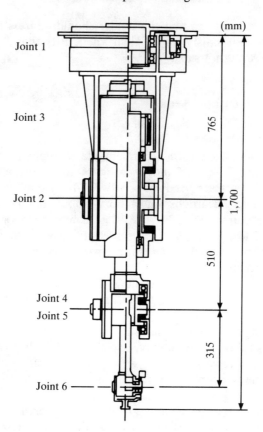

Figure 7.24. Overall schematic drawing of the direct-drive arm. (From Asada et al., 1983.)

articulated direct-drive joints. Joint 1 is a rotational joint about a vertical axis; Joint 2 is a rotational joint about a horizontal axis; Joints 3 and 4 rotate the forearm, and Joints 5 and 6 perform a rotational and a bending motion of the wrist part. The total length of the arm is 1.7 m. The movable range of Joints 1 and 5 is 330 degrees. The remaining joints can move 180 degrees. The maximum payload is 6 kg, including the weight of the gripper.

At each joint, high-performance d.c. torque motors were used. The motor, which consists of a rotor, a stator, and a brush ring, is installed directly at the joint housing. The rotor was attached to the hollow shaft, and the stator and the brush ring to the case. To develop the large torque required to turn the joint shaft directly, motors with large diameters were used. At Joint 1 the diameter of the motor and the peak torque were 56 cm and 204 N m, respectively. Joint 2 has two motors, one on each side of the upper arm, 30 cm in diameter and of 136 N m peak torque.

The motors at Joints 4, 5, and 6 had to be lightweight to reduce the load on the upper joints, so they used samarium cobalt motors. The motor at

Joint 4 was 23 cm in diameter with 54 N m peak torque, and the motors at Joints 5 and 6 were 8 cm in diameter with 6.8 N m peak torque. An optical shaft encoder was installed at each joint to measure the joint angle and its angular velocity.

In the above design, gearings are eliminated. Therefore, the robot has no backlash, low friction, and high mechanical stiffness. It also has major shortcomings. Since the drive system has no means of amplifying the motor output torque mechanically, it cannot exert a large force for a long time without overworking the motor. Also, the motor itself is a load for the next motor down the serial linkage, thereby rapidly increasing the drive torque required. This results in a heavy arm weight and a relatively small load capacity. Another drawback is that direct-drive motors are too sensitive, and lead to changes and disturbances, thus reducing accuracy.

The problem caused by the lack of gearing can be seen from the dynamic equation for the rotor. For the ith joint of the arm driven through gearing with gear ratio k_i, the equation of motion becomes (Asada and Youcef-Toumi, 1984)

$$\left(h_{ri} + \frac{h_{ai}}{k_i^2}\right)\ddot{\theta}_i + \frac{\tau_{ci} + \tau_{ni}}{k_i} = \tau_i \tag{7.6}$$

where h_{ri} is the inertia of the motor rotor, including the gearing (an invariant); h_{ai} is the arm's inertia reflected to the ith joint axis; τ_{ci} is the interactive inertia torque; τ_{ni} is the nonlinear torque reflected to the ith joint axis; $\ddot{\theta}_i$ is the angular acceleration of the ith joint; and τ_i is the torque exerted on the ith joint.

Although h_{ai} is configuration-dependent, its contribution to inertia is reduced by a factor of k_i^2 when reflected at the motor axis. τ_{ci} and τ_{ni} are also reduced by k_i, which implies that the arm dynamics are less important when the actuator has a large gear ratio. Since the direct-drive arm does not have large k_i, the varying inertia, interactive, and nonlinear forces are reflected directly to the motor axes to make the direct-drive most sensitive to the arm dynamics, complicating the control problem (Asada and Youcef-Toumi, 1984).

The design task here is to develop means for overcoming these difficulties of the direct-drive arm. In order to simplify the discussion, we consider the simple two-degree-of-freedom serial drive arm shown in Fig. 7.25 as the direct-drive arm for analysis.

Solution

The FRs of a robot arm may be stated to be the following:

FR$_1$ = The overall stiffness, K (i.e., resistance to deflection when the load is applied at the end effector).
FR$_2$ = The overall accuracy in positioning the end effector, δ.
FR$_3$ = Rapid acceleration of Arm 1 (high torque/low inertia), $\ddot{\theta}_1$.
FR$_3$ = Rapid acceleration of Arm 2, $\ddot{\theta}_2$.

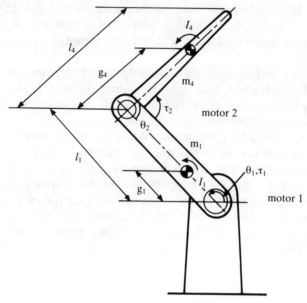

Figure 7.25. Two-degree-of-freedom robot arm. (From Asada and Youcef-Toumi, 1984.)

In the case of the two-degree-of-freedom serial drive arm shown in Fig. 7.25, the DPs are:

DP_1 = Stiffness of the motor 1 (torque exerted by the rotor of the motor 1 divided by rotation) = τ_1/θ_1.

DP_2 = Stiffness of the motor 2 (i.e., torque exerted by the rotor of the motor 2 divided by rotation) = τ_2/θ_2.

DP_3 = Inertia of arm 1 = $(H_{ij})_1$.

DP_4 = Inertia of arm 2 = $(H_{ij})_2$.

The design equation may be written as

$$\begin{Bmatrix} K \\ \delta \\ \ddot{\theta}_1 \\ \ddot{\theta}_2 \end{Bmatrix} = \begin{bmatrix} \times & \times & 0 & 0 \\ \times & \times & \times & \times \\ \times & \times & \times & \times \\ \times & \times & \times & \times \end{bmatrix} \begin{Bmatrix} \tau_1/\theta_1 \\ \tau_2/\theta_2 \\ (H_{ij})_1 \\ (H_{ij})_2 \end{Bmatrix} \quad (7.7)$$

In writing Eq. 7.7, an assumption is made that the stiffness is primarily governed by the deflection at the joints, rather than the bending of the arms. Furthermore, there is interaction between the two motors. The rotors of the motor must exert a high torque at low speeds, primarily to counterbalance the torque generated by the externally applied load, probably for a long time. They must also exert a very high torque in order to be able to accelerate the arm rapidly. The accuracy of the arm is affected by the rotational deflection at the joints, the deflection of the arms, and the inertia of the arms and the rotors. According to Eq. 7.7, the robot arm shown in Fig. 7.25 is clearly a coupled system.

In order to be able to use the robot represented by Eq. 7.7, we must be willing to accept the following possible actions: (1) lower the stiffness and accuracy requirements of the arms; (2) limit the acceleration to a low value; or (3) even eliminate some of the specified FRs, such as stiffness, accuracy, and acceleration. Any one of these three actions is equivalent to redefining the design problem. In order to determine the best trade-off in redefining the FRs, analysis of the arm shown in Fig. 7.25 may be useful.

Equation 7.7 can be justified in part by actually writing the equation of motion of the arm. The equation of motion of a direct-drive arm is given by (Asada and Youcef-Toumi, 1984)

$$H_{ii}\ddot{\theta}_i + \sum_{j \neq i} H_{ij}\ddot{\theta}_j + \sum_j \sum_k \left(\frac{\partial H_{ij}}{\partial \theta_j} - \frac{1}{2}\frac{\partial H_{jk}}{\partial \theta_i}\right)\dot{\theta}_j\dot{\theta}_k + \tau_{gi} = \tau_i \qquad (7.8)$$

where θ_i is the angle of the ith motor, τ_i is the torque of the ith motor, H_{ij} is the ij element of the arm inertia tensor reflected to the motor axes, and τ_{gi} is the gravity torque. The first term on the left-hand side of Eq. 7.8 represents the inertia torque, which depends on the arm configuration; the second term represents the interactive inertia torque; and the third term represents the nonlinear velocity torque due to centrifugal and Coriolis acceleration. The two-degree-of-freedom direct-drive arm shown in Fig. 7.25 is highly nonlinear, since the inertia tensor is given by

$$\mathbf{H} = \begin{bmatrix} H_{11} & H_{12} \\ H_{12} & H_{22} \end{bmatrix}$$

where

$$\begin{aligned} H_{11} &= I_1 + m_1 g_1^2 + I_4 + m_4(l_1^2 + g_4^2 - 2l_1 g_4 \cos \theta_2) \\ H_{22} &= I_4 + m_4 g_4^2 \\ H_{12} &= I_4 + m_4 g_4^2 - m_4 l_1 g_4 \cos \theta_2 \end{aligned} \qquad (7.9)$$

Off-diagonal element H_{12} governs the interaction between the two motors. Both H_{12} and H_{11} are functions of θ_2, and thus are dependent on the configuration.

In order to overcome the shortcomings of the arm shown in Fig. 7.25, Asada and Youcef-Toumi (1984) proposed an alternate arm design shown in Fig. 7.26. In this design the motor at Joint 5 is eliminated, and two rotors are placed at the base (Joints 1 and 2). These two motors at the base drive the two input links, and cause a two-dimensional motion at the tip of the arm. The weight of one motor is not a load on the other. Also, the reaction torque of one motor does not directly exert load on the other motor, since the motors are fixed on the base. In order to drive link 4, a parallel linkage (links 3 and 2) is used. The elements of the inertia tensor of this new arm is given by

$$\begin{aligned} H_{11} &= I_1 + m_1 g_1^2 + I_3 + m_3 g_3^2 + m_4 l_1^2 \\ H_{22} &= I_4 + m_4 g_4^2 + I_2 + m_2 g_2^2 + m_3 l_2^2 \\ H_{12} &= (m_3 l_2 g_3 - m_4 l_1 g_4) \cos(\theta_2 - \theta_1) \end{aligned} \qquad (7.10)$$

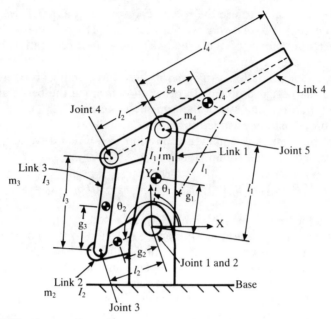

Figure 7.26. Alternate direct-drive arm designed by Asada and Youcef-Toumi (1984).

These expressions for the moments of inertia were obtained by first immobilizing motor 2 in order to compute the moment of inertia about the axis of motor 1, which is H_{11}. Similarly, motor 1 was immobilized to obtain H_{22}. H_{11} and H_{22} in this new design are independent of the positions of the links, and thus are invariants. Again, the off-diagonal element H_{12} represents the interaction between the two motors. H_{12} can be made to be zero in this design, thus eliminating the interaction between the two motors completely by controlling the mass ratio of link 3 and 4, m_4/m_3; the ratio of mass outer distances of the two links, g_4/g_3; and/or the length l_2 so that the coefficient $(m_3 l_2 g_3 - m_4 l_1 g_4)$ becomes zero. The condition for making H_{12} zero may be written as

$$\frac{m_4 g_4}{m_3 g_3} = \frac{l_2}{l_1} \tag{7.11}$$

Asada and Youcef-Toumi (1984) claimed that the elimination of H_{12} in the design shown in Fig. 7.26 is a major advantage of the new design. If we define the FRs of this new design to be the same as those of the original design (Fig. 7.25), then the FRs may be written again as

FR$_1$ = The overall stiffness, K.
FR$_2$ = The overall accuracy in positioning the end effector, δ.
FR$_3$ = Rapid acceleration of the links, $\ddot{\theta}_1$.
FR$_4$ = Rapid acceleration of the links, $\ddot{\theta}_2$.

Case Studies: Design of Products 289

The DPs of the design shown in Fig. 7.26 may be written as

DP_1 = Stiffness of motor 1 = τ_1/θ_1.
DP_2 = Stiffness of motor 2 = τ_2/θ_2.
DP_3 = Inertia reflected on motor 1.
DP_4 = Inertia reflected on motor 2.

The design equation may then be written as

$$\begin{Bmatrix} FR_1 \\ FR_2 \\ FR_3 \\ FR_4 \end{Bmatrix} = \begin{bmatrix} \times & \times & 0 & 0 \\ \times & \times & \otimes & \otimes \\ 0 & 0 & \times & \times \\ 0 & 0 & \times & \times \end{bmatrix} \begin{Bmatrix} DP_1 \\ DP_2 \\ DP_3 \\ DP_4 \end{Bmatrix} \qquad (7.12)$$

According to this design equation, even the new design is a coupled system. The basic problem is that the stiffness of the system is still in part controlled by the rotor of the system. The reason why FR_2 (i.e., the accuracy) depends on all DPs is because the motors must have high torque capability to provide accuracy, and because the inertia of the system, as reflected at the rotor, affects the position accuracy. The reason for the crossed circle (i.e., $\otimes$) is that it may be possible to lighten the inertia of the links in order to improve the accuracy.

The foregoing analysis of the direct-drive arm indicates that the design suffers from two basic problems. The problems stem from the fact that the rotor of the motors control the stiffness, and that the motions along the Cartesian coordinates (i.e., two orthogonal directions) are coupled by the articulated arm. It should be possible to overcome these shortcomings (see Problem 7.2).

7.7 Design of an Automotive Wheelcover

Statement of the Problem

Wheelcovers are put on the rim of automobile wheels, primarily for decorative reasons. In addition to the decorative and aesthetic requirements, the wheelcover must satisfy two FRs with which we are concerned here. They are: the wheelcover must stay on the wheel when the wheel goes over a bump or a pothole, and it must also be easy to remove when the tire has to be changed. Oh (1988), of General Motors Corporation, investigated the design of a wheelcover that is easy to manufacture and at the same time satisfies the two FRs. This section applies the design axioms to Oh's work.

According to the General Motors study, the percentage of dissatisfied customers varies as a function of the retention force N, which is plotted in Fig. 7.27. The retention force above which the wheelcover is difficult to remove varies from 30 N, at which it is easily removed, to 60 N, at which it is completely unremovable and therefore 100% unsatisfactory. On the

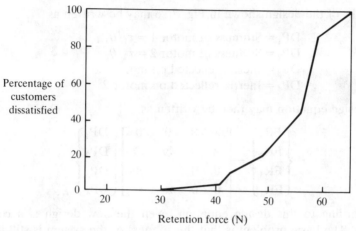

Figure 7.27. Percentage of customers dissatisfied as a function of the retention force that must be overcome in removing the wheelcover. (From Oh, 1988.)

other hand, the customers become unhappy when the retention force is so low that the wheelcover comes off the wheel. The dissatisfaction level increases as the retention force is made less than a critical value, as shown in Fig. 7.28. These two competing requirements yield a target value of the retention force that the designer of the wheelcover and the manufacturing process must satisfy. Any departure from the target value will incur some degree of customer dissatisfaction, possibly leading to the loss of market share. The superposition of these two requirements is shown in Fig. 7.29, which shows that the target value is about 34 N.

The design task is to develop a retention mechanism that satisfies the customer and enables mass-production at a low cost. Oh and his colleagues

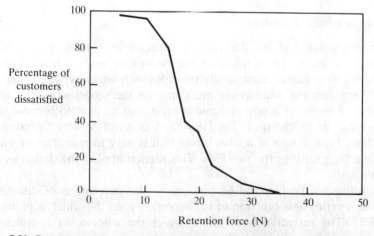

Figure 7.28. Percentage of customers dissatisfied when wheelcovers fall of plotted as a function of the retention force of the wheelcover. (From Oh, 1988.)

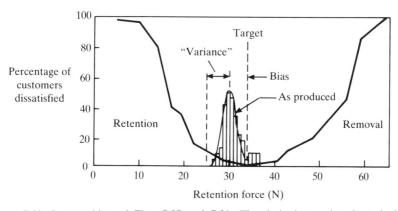

Figure 7.29. Superposition of Figs. 7.27 and 7.28. The desired retention force is 34 N (indicated as TARGET) and the actual force due to the variation in manufacturing processes. (From Oh, 1988.)

decided to use three metal clips, each with two prongs, spaced around the circumference to form a circle. The diameter of this circle, called the clip diameter, is larger than the diameter of the rim on the wheel. When the cover is pressed on the wheel, the clip acts like a spring and clicks onto the rim. For a given spring constant of the clip, the bigger the difference between the rim diameter and clip diameter is, the larger the retention force. The specific design task is to analyze this design by applying the design axioms, corollaries and theorems.

Analysis of the Proposed Design

In the proposed design the retention force is the key FR and may be stated as:

FR_1 = Provide a retention force of 34 ± 4 N.

The constrain is the low manufacturing cost. There are two possible DPs:

DP_1 = Difference between the clip diameter and the rim diameter.
DP_2 = Spring constant of the clip.

Since there are two DPs and only one FR, according to Theorem 3, we may have a redundant design. In this case we can fix, at least for the time being, the spring constant and let the difference between the clip diameter and the rim diameter be the only DP. In any case the proposed design satisfies the Independence Axiom.

Now let us apply the Information Axiom to this design problem. If we plot the probability distribution as a function of the FR, the design range and the system range may not overlap, as shown in Fig. 7.30. As discussed in Chapter 5, the bias b must be reduced to zero and the variance (which is proportional to v) must be minimized to reduce the information content. This can be done by moving the system range to the right. However, the

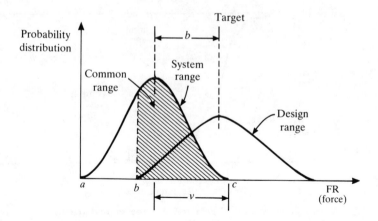

Figure 7.30. The probability distribution of the FR, force. Since the retention force is the product of the diameter difference, D, and the spring constant, K, the FR is equal to KD. Therefore, the system range can be moved to the right by chosing K as the DP and by increasing the spring constant. Another possible DP is the difference between the clip diameter and the rim diameter, but the cost of changing this DP is high.

manufacturing process may be such that the tolerances of clip diameter and the rim diameter cannot be brought under better control to shift the FR to the right without incurring major expenses. Therefore, it may not be possible to shift the system range. On the other hand, if General Motors ships out the product as it is, then they will have so many unhappy customers that they may eventually lose their market share. The question is: "How do we make the design range and the system range overlap, or at least increase the common range so as to minimize the number of unhappy customers?" Another way of putting the question is: "How do we minimize the information content?"

According to Axiom 2, the better design is the one with the minimum information content. The information content was defined as the logarithm of the inverse probability, which was generalized as

$$I = \log_2\left(\frac{\text{system range}}{\text{common range}}\right) \qquad (5.12)$$

For the case shown in Fig. 7.30, this equation may be written as

$$I = \log_2\left[\int_a^c f(D)\,dD \bigg/ \int_b^c f(D)\,dD\right] = \log\left[1 \bigg/ \int_b^c f(D)\,dD\right] \qquad (7.13)$$

where $f(D)$ is the probability function for the system range.

In the case of uniform proability, Eq. 7.13 reduces to

$$I = \log\left(\frac{c-a}{c-b}\right) \qquad (7.14)$$

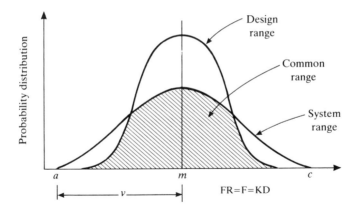

Figure 7.31. The design range is made to overlap the system range by using a softer spring, rather than by tightening up the manufacturing tolerance. The productivity can be increased further by reducing the variance, which is proportional to v.

Since the design range is fixed by the customer satisfaction requirements, we have to consider how the system range can be made to overlap the design range completely by moving it to the right. In this particular design situation there is a simple way out, because it is a redundant design with an excess DP that also affects the FR; namely, the spring constant of the clip. If we decrease the spring constant of the clip, then the system range will move toward the right, since the DP is equal to the retention force F divided by the spring constant k. This is illustrated in Fig. 7.31. This solution is much more desirable than the possibility of replacing the dies to move the system range to the right, which means increased manufacturing costs.

It is worth noting here that, as discussed in Secs. 5.7 and 5.8, the statistical quality control technique developed by Taguchi and Phadke (1984) and others which is a specific methodology that is useful in solving manufacturing problems, can also be used to solve this problem. Taguchi's method deals with various means of reducing the bias and variance by checking the sensitivity of various FRs to DPs in terms of the SN ratio. The Taguchi method is another manifestation of the design axioms.

7.8 Summary

In this chapter product design is analyzed in terms of the design axioms. The analysis was done at or close to the highest level of the design hierarchy. As in earlier chapters, only the first-order design decisions were investigated by writing the design equation (i.e., we did not worry about the use of specific bearings, etc.). Once the macroscale analysis is done, it is sometimes necessary to undertake a detailed analysis of the proposed design through modeling and numerical solutions so as to optimize the design. It may be possible to determine the conditions under which the proposed design may be a decoupled or an uncoupled design in a specific range of operation in the case of nonlinear design.

The design analysis done here for various applications can be extended to lower levels of the design hierarchy, if the higher-level analysis warrants further development of the design. In some cases, from the first-order analysis, the shortcomings of the proposed design are so obvious that continued further development may not be justifiable. The power of axiomatic design, in fact, lies in its ability to make correct design decisions and avoid costly detailed development of weak ideas.

The six examples given in this chapter illustrate the notion that the decisions made at the desing stage determine the productivity of a firm or a research project. Once wrong decisions are made during the conceptual development of a product, it may take a significant investment in time and money to find out that the original ideas had flaws. Since the design process involves iteration at all stages of the design execution, one may not be able to take correct actions at all times, including the definition of FRs to the analysis of a proposed design. What is important, however, is the need to write the *design equations,* which may quickly identify the potential problems associated with the proposed design, establish the validity of the FRs in meeting the perceived needs, and check the usefulness of the DPs chosen for generating an uncoupled or decoupled design.

The case studies presented in this chapter demonstrate that there is an unlimited number of solutions that can satisfy a given set of FRs. When two or more design solutions that satisfy Axiom 1 are presented, Axiom 2 must be applied to select the best among those acceptable solutions. In Chapter 8 a number of case studies which involve the Information Axiom are presented.

References

Asada, H., and Youcef-Toumi, K., "Analysis and Design of a Direct Drive Arm with a Five-Bar-Link Parallel Drive Mechanism," *ASME Journal of Dynamic Systems, Measurement, and Control* **106**:225–230, 1984.

Asada, H., Kanade, T., and Takeyama, I., "Control of a Direct Drive Arm," *ASME Journal of Dynamic Systems, Measurement, and Control* **105**(3):136–142, 1983.

Kramer, B.M., "An Analytical Approach to Tool Wear Prediction," Ph.D. Thesis, MIT, 1979.

Kramer, B.M., and Suh, N.P., "Tool Wear by Solution: A Quantitative Understanding," *Journal of Engineering for Industry, Transactions of A.S.M.E.,* **102**:303–339, 1980.

Nehru, A.K., and Suh, N.P., "Electrically Conductive Abrasive Bodies," U.S. Patent 3,310,390, issued March 21, 1967.

Oh, H.L., "Modeling Variation to Enhance Quality in Manufacturing," Presented at the Conference on *Uncertainty in Engineering Design,* NBS, May 10–11, 1988.

Smith, J., and Finger, S., "Laser Architectures," Report to Dr. John F. Holzrichter of Lawrence Livermore National Laboratory, Laboratory for Manufacturing and Productivity, MIT, April 26, 1985.

Suh, N.P., "Coated Carbides—Past, Present and Future," *Carbide Journal* **9**(1):3–9, 1977.

Suh, N.P., *Tribophysics.* Prentice-Hall, Englewood Cliffs, NJ, 1986.

Taguchi, G., and Phadke, M.S., "Quality Engineering Through Design Optimization," *I.E.E.E. Global Telecommunications Conference, GLOBCOM '84,* Atlanta, GA, November 26–29, 1984.

Tice, W., "The Application of Axiomatic Design Rules to an Engine Lathe Case Study," S.M. Thesis, MIT, 1980.

Problems

7.1. Design an automatic transmission for passenger vehicles that can perform the following *additional* functions: (1) operation at the "extra fuel-efficient" mode of driving; and (2) operation at the "fast acceleration" mode of driving. First specify the FRs of commercially available automatic transmissions.

7.2. Design an assembly robot that does not have the shortcomings of the articulated arm, which is a highly coupled system. Define the FRs of your robot and specify the DPs.

7.3. How could you improve the parallel arm of the Asada and Youcef-Toumi design?

7.4. Develop a laser architecture that satisfies the FRs and requires less information than the design proposed by Smith and Finger.

7.5. Design two different coffee makers for ground coffee and choose the best design.

7.6. Determine the reangularity and semangularity of a centrifugal pump.

7.7. Design a "read/write" (i.e., erasable) compact disk for information-storage devices that use a laser beam.

7.8. Design a traffic control system for major intersections in urban areas.

7.9. Design a software system that can make reservations, sell airplane tickets, and issue tickets at the airport without the intervention of airline employees.

7.10 Consider the design of human body. What are the FRs and DPs? Is the human body a coupled system? Can you improve the design?

7.11. Feller–bunchers are mobile machines consisting of a prime mover and a tree-felling attachment such as a shear (see the figure). A long boom attached to a feller-buncher is deemed beneficial for thinning trees on steep slopes where the prime mover mobility is restricted. Booms must reach 6–20 m. Thinning consists of selective removal to increase the growth of forest and obtain desired characteristics in the residual stand. Individual trees are addressed by reaching with a boom or by moving the prime mover to each tree. Define FRs and DPs. How would you improve the design?

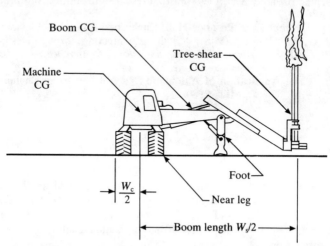

W_c = Travel corridor width
W_s = Cutting swath width

8
CASE STUDIES INVOLVING INFORMATION CONTENT

8.1 Introduction

Axiom 2 (the Information Axiom) states that among those designs that satisfy Axiom 1 (the Independence Axiom), the one that requires the least information content is the best design. The significance of this axiom and its application are illustrated in Chapter 5. The information content was defined in terms of the probability of being able to satisfy a given FR. Since probability is a dimensionless scalar quantity, information contents of all variables, regardless of their physical origin (i.e., hardness, length, cost, ect.), could be treated alike, and added together to obtain the total information content, provided that all the information is directly related to achieving a given design task represented by the FRs.

In this chapter several case studies involving information content are presented in order to augment those presented in preceding chapters (i.e., Chapters 3, 5, and 6). Many of the case studies presented in this chapter were conducted by Professor Nakazawa of Waseda University, Japan, who has continued to work on this aspect of axiomatic approach to design ever since he visited MIT for a year in 1980. During his short stay at MIT, he developed, with the author, the basic methodology and the process planning technique based on the Information Axiom (Nakazawa and Suh, 1984). The case studies presented in this chapter follow the same reasoning process as those described in Chapter 5.

Some of the case studies presented in this chapter compare the predictions made by the information content measure with those based on other criteria. These other criteria have been used in some rule-based expert systems. Many knowledge-based expert systems have been developed to aid decision making in all fields in recent years, but they are application specific and are not generalizable, whereas the information measures developed in Chapters 3 and 5 are universal. In order to illustrate this, a variety of different case studies are presented in this chapter.

The information content was variously defined in Chapter 5 as

$$I = \log_2(1/p) \qquad (5.5)$$

$$I = K \log_2(Q_2/Q_1) \qquad (5.5a)$$

$$I = \log_2(\text{range}/\text{tolerance}) \qquad (5.7)$$

$$I = -\sum p_i \log p_i \qquad (5.8)$$

where p is the probability, Q_1 is the number of equally probable a priori outcomes, Q_2 is a posteriori outcomes, and p_i are probabilities of a number of discrete events occurring in an ensemble. These definitions of the information content may be written in terms of the natural logarithm rather than logarithm to base 2, as discussed in Chapters 3 and 5.

In measuring the information content, the question we asked was, "What is the probability of satisfying an FR?" The information content was always associated with the useful information related to accomplishing a given task, as expressed in the form of FRs. These concepts are applied in the case studies presented in this chapter.

8.2 Design of a Manufacturing Process

Nakazawa (1983) conducted a laboratory scale test of Axiom 2 by determining the best manufacturing process based on the axiom and comparing it with the actual results obtained in the machine shop of Waseda University, Japan. The part to be manufactured (shown in Fig. 8.1) is made by machining a a bar stock, employing two different machining steps (i.e., surface groups) shown in Fig. 8.2. The machine shop had 24 lathes and three cylindrical grinders that could be used to manufacture the part. Surface roughness, dimensional accuracy, and cost are the FRs to be evaluated, all of which are assumed to be probabilistically independent. The technique of determining the information content is given in Sec. 5.5 (Nakazawa and Suh, 1983).

The information content of surface roughness is calculated using Eq. 5.7. The system range of the machine tools was obtained experimentally. The lower bound of the system range is chosen to be the best surface

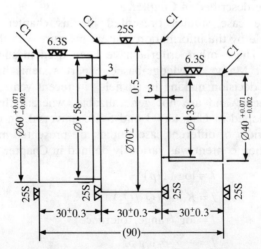

Figure 8.1. Shaft to be manufactured by machining a bar stock. Material: mild steel. (From Nakazawa, 1984.)

Case Studies Involving Information Content   299

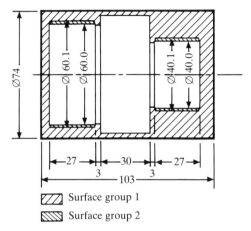

Figure 8.2. The bar is to be machined in two surface groups. Each surface group is machined by one machine. (From Nakazawa, 1984.)

roughness that can be provided by a given machine, whereas the upper bound is chosen to be 10 times the lower bound. The design range is given in Fig. 8.1, which is the upper bound.

The information content associated with dimensional accuracy can also be determined using Eq. 5.7. The system range and the designer-specified range are shown in Fig. 8.3. The designer-specified tolerance range is plotted symmetrically with respect to the vertical axis, since the tolerance is given as $\pm \Delta l$. The lower bound of the system range is equal to the minimum graduation of the cross-feed handle wheel, whereas the upper bound is taken to be 10 times the lower bound.

Equation 5.7 may then be written as

$$I = \ln \frac{l_1 + l_2}{\Delta l_3 + \Delta l_4} \qquad (8.1)$$

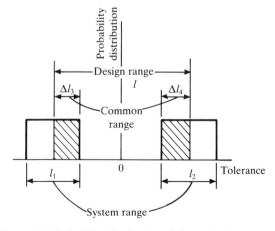

Figure 8.3. Probability distribution of dimensional accuracy.

The information content associated with cost may be determined, based on what the designer is willing to pay and what the system must charge to stay in business. In order to determine the system range for machining cost, the maximum and minimum metal removal rates, Z_{max} and Z_{min}, are obtained as

$$Z_{max} = \eta H / q_{min} \tag{8.2a}$$
$$Z_{min} = \eta H / q_{max} \tag{8.2b}$$

where H is the power of the machine tool, given in kilowatts; q_{min} and q_{max} are the minimum and the maximum energies required for cutting, respectively; and η is the mechanical efficiency of the machine tool. Figure 8.4 gives a plot of q versus the depth of cut for various materials. The maximum and the minimum machining times, T^c_{max} and T^c_{min}, are

$$T^c_{max} = V / Z_{min} \tag{8.3a}$$
$$T^c_{min} = V / Z_{max} \tag{8.3b}$$

where V is the volume to be removed. The total manufacturing time is the sum of the machining time, T^c, and loading and unloading times, T^L; that is,

$$T = T^c + T^L \tag{8.4}$$

The upper and lower bounds of the system range for cost, C_{max} and C_{min}, are determined for each machine by multiplying the maximum and the minimum times taken for manufacture by the cost per unit time. The design range is determined arbitrarily by letting the upper bound be the

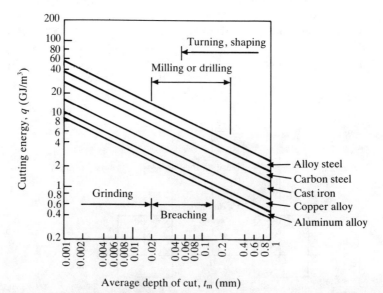

Figure 8.4. Relative energy consumed in machining various metals as a function of the depth of cut. (From Nakazawa, 1987.)

TABLE 8.1 Information Required to Generate the Surface Groups by Various Machines. L1, L2, L3, and L4 are Lathes, and G1, G2, and G3 are Grinding Machines (from Nakazawa, 1984)

	Combination			
	1	2	3	4
Surface group 1 (1st step)				
Lathe	L1	L2	L3	L4
Order	1	2	5	24
Information	1.36	1.39	1.94	4.27
Surface group 2 (2nd step)				
Cylindrical grinder	G1	G2	G3	G3
Order	1	2	3	3
Information	0.58	2.19	3.73	3.73
Total				
Order	1	21	49	72
information	1.94	3.58	5.67	8.00

mean value of C_{max} and C_{min}, and the lower bound be zero. In a mass-production situation, the cost of fixtures and/or cost of numerical control tape can be included.

The surface to be machined by one machine are grouped into one *surface group*. The information for each surface group is then calculated. The total information contents were determined for each surface group for each of 24 lathes and three cylindrical grinders. The results from four different lathes and three different cylindrical grinders are shown in Table 8.1. The information contents are listed in the order of increasing information requirements, starting from the left-most column. Combination 1 has the minimum information content, and Combination 2 has the second-lowest information content. Combination 3 is the one selected by the experienced machinist as being the best machine tool to use, and Combination 4 is the one with the maximum information content.

It is interesting to note that the experienced machinist at the Waseda workshop selected the fifth-ranked lathe by the information content analysis as the best machine tool to use for machining Surface Group 1. (This is listed in Table 8.1 as Combination 3.) For Surface Group 2 the same machinist selected the third-ranked grinder by the information content analysis, which again is listed in the third column of Table 8.1.

In order to verify the analytical results, Nakazawa had the part shown in Fig. 8.1 machined by one machinist using the machine tools listed in Table 8.1. The machinist started with Combination 1 and proceeded in the rank order listed. This gave extra credance to this method, since the machinist gained experience and thus could accomplish the job faster as the combinations with greater information contents were tried. All of the parts

TABLE 8.2 Measured Total Machining Time (seconds) (from Nakazawa, 1984)

	Combination			
	1	2	3	4
Surface group 1				
Lathe	L1	L2	L3	L4
Measured time	1,453	1,405	1,665	2,487
Surface group 2				
Cylindrical grinder	G1	G2	G3	G3
Measured time	1,158	1,205	1,433	1,481
Total	2,611	2,610	3,098	3,948

produced by these machines satisfied the FRs for dimensional accuracy and surface finish.

The total machining time is listed in Table 8.2, which includes the loading and unloading time. The actual machining times show that Combinations 1 and 2 took about the same time, but clearly Combinations 3 and 4 took much longer to produce the part than those with lower information content, notwithstanding the fact that the machinist was more experienced in producing this part when he tried Combinations 3 and 4. The machine chosen intuitively by the experienced machinist (i.e., Combination 3) took 20 % longer to produce the part, whereas the worst combination took 50% longer.

A few comments are in order here, notwithstanding the general applicability of the Nakazawa solution presented in this section. First, the information measure represented in Fig. 8.3 may not be the most rational way of calculating the information content in measuring dimensions. Instead of having two common ranges at the extreme limits of the tolerance range, the normal practice in setting up a machine should be to position the machine around the exact dimension desired. Second, it should be noted that Table 8.2 is a listing of machining time, whereas the information measure provides the optimum solution with respect to surface finish, length, and cost. The fact that the machining time is the least for the best process indicated *indirectly* that the FRs were most easily achieved when I is a minimum. Third, Nakazawa treated cost as an FR in this example. However, in this case, the cost could have been used as a constraint, since the cost is determined by and dependent on the choice of various machines. Each FR must be independent from others by definition. However, the choice of cost as an FR for the case study presented in Sec. 8.4 is consistent with the design axioms, since the selling price of a subcompact car is not related to the other FRs specified.

This case study illustrates the fact that the intuitive feeling of an experienced machinist is not necessarily reliable, and that the quantitative information content determined by the method introduced in this book is better than other measures. Nakazawa tried this method at a large Japanese machine tool company, with success.

One of the basic reasons why the information content measurement is a powerful tool is that this method can optimize solutions with respect to many variables; in fact, the number of variables is not an issue when Axiom 2 is used. In many design and manufacturing applications, the problems involve a large number of variables, and thus require the optimum solution with respect to many variables. Few other techniques can deal with such a problem.

8.3 Design of Production Schedules

Nakazawa (1984) applied the information content measure to the design of scheduling problem with success. He considered a manufacturing system with the following characteristics:

1. There are different machine tools which work on n jobs.
2. Each machine tool has ample buffer space.
3. Transportation of parts in the factory is not the limiting factor.

The overall FR of the factory scheduling is to meet the largest number of due dates (i.e., the date by which the part must be delivered) as soon as possible, in addition to satisfying the FRs on the dimensional accuracy, surface roughness and cost. Therefore, there are four FRs that must be satisfied.

This problem is solved by determining the information content associated with each of the FRs, and summing up the information contents associated with all four FRs. It is shown in Sec. 8.2 that the time required to machine a part using various machines can be substantially different. Therefore, Nakazawa simplified the problem here by determining only the information content associated with the dispatching time, and verifying the results through simulation.

The information content associated with the delivery time can be computed treating the time as a FR. The upper and lower bounds of the *system range* for the delivery time, R_i, may be expressed as

$$(R_i)_{\max} = (T_{si})_{\max} + \sum_{j=1}^{N_i} (P_{ij})_{\max} \tag{8.3a}$$

$$(R_i)_{\min} = (T_{si})_{\min} + \sum_{j=1}^{N_i} (P_{ij})_{\min} \quad i = 1, 2, \ldots, n \tag{8.3b}$$

where T_{si} is the starting time for job i, P_{ij} is the operation time of job i on machine j, including the handling time, and N_i is the total number of machine tools used for job i. The *design range* for the delivery time must be established by the designer (or by the management in this case) based on business considerations. The lower bound of the design range is chosen to be zero, and the upper bound is determined using three different ways of estimating due dates: the total work due dates (TWK), number of operation due dates (NOP), and random allowance due dates (RDM). They are calculated as follows (Conway, 1965).

1. *Total work due dates (TWK)*. The allowable time, D_a, which is the time difference between the arriving time of a job and the due date, is calculated using the rule

$$(D_a)_{max} = K \sum_{j=1}^{N_i} \bar{P}_{ij} \qquad (D_a)_{min} = 0 \qquad (8.4)$$

where $\bar{P}_{ij}$ is the mean operation time of job i, and K is a constant. In this simulation K is set to 4.

2. *Number of operations due dates (NOP)*. The allowable time, D_a, is given by

$$(D_a)_{max} = 1{,}000 N_i + 500 \text{ (minutes)} \qquad (D_a)_{min} = 0 \qquad (8.5)$$

The 500 minutes takes into consideration the start-up time.

3. *Random allowance due dates (RDM)*. The allowable time according to this method is given as

$$(D_a)_{max} = 6{,}000\text{RND} + 1{,}000 \text{ (minutes)} \qquad (D_a)_{min} = 0 \qquad (8.6)$$

where RND is a random real number between zero and unity.

In order to compare the numerical results obtained using the information content measure with other techniques, the following four dispatching rules are investigated (Bake, 1974).

1. *Shortest processing time (SPT)*. In this dispatching rule, the highest priority is given to the waiting job with the shortest operation time. Since the operation time is given by a range bound by upper and lower bounds, SPT is represented with a mean value of the maximum and minimum time.
2. *First arrival at the shop, first served (FASFS)*. The priority is given to the job that arrived at the shop earliest. Arriving time is represented by the mean value of the maximum and the minimum time when the previous operation may be completed.
3. *Earliest due dates (EDD)*. The highest priority is given to the job with the earliest due date. The due date is represented by the maximum value that is determined by the management when they accepted the job.
4. *Minimum slack time (MST)*. In this criterion, the minimum slack time is given the highest priority. The slack time, S_i, for job i as defined as

$$S_i = D_i - \sum_{j=1}^{N_i} \bar{P}_{ij} - T_p \qquad (8.7)$$

where D_i is the due date and T_p is the present time. $\bar{P}_{ij}$ is the mean value of the maximum and the minimum operating times.

The schedule that can meet as many due dates as possible is designed, using the information content measure or the dispatching values described above. Once a scheduling solution is obtained, it must be evaluated in terms of acceptable criteria. In order to be more specific, the FR related to the delivery time must be stated in terms of specific criteria. The criteria used by Nakazawa are the occurrence rate of infinite information, the mean delay time, $\bar{T}$, and the mean stay time $\bar{F}$. The mean delay time, $\bar{T}$, may be expressed as

$$\bar{T} = \left[\sum_{i=1}^{n} \max\{0, C_i - D_i\}\right]/n \qquad (8.8)$$

where C_i is the time job i is completed. Since C_i has a range, the mean value of the maximum and the minimum value is used. Max($0, C_i - D_i$) is equal to the larger value in the parentheses, either zero or $C_i - D_i$. The mean stay time, $\bar{F}$, is the time job stays in the factory, beyond the due dates and may be expressed as

$$\bar{F} = \left[\sum_{i=1}^{n} (C_i - D_i)\right]/n \qquad (8.9)$$

In this simulation the number of jobs is taken to be five and the number of machine tools to be four. Other values assumed are:

1. Number of surface groups (number of operation steps). Some integer value from one to five is randomly given.
2. Number of machine tools. To each surface group, one to four is randomly given as number of candidates for machine tools.
3. Operation starting times. Two random integer numbers between zero and 1,000 are produced, from which the upper and the lower bounds are determined.
4. Operation times. The operation time is also given randomly by a range, similar to the operation starting times.

The method for estimating the due dates is given in Eqs. 8.4–8.6.

The simulation results are shown in Tables 8.3–8.6. Table 8.3 lists the rate of occurrence of infinite information, mean delay time, and mean stay time, based on the due date computed using the TWK method. Table 8.4 lists similar results, except that the due dates are estimated using the NOP method. Table 8.6 is obtained by taking the average of the values given in Tables 8.3–8.5.

The rate of occurrence of infinite information may be the most important criterion, since the most important purpose of scheduling is "not to delay." When it is nonzero, one or more jobs will inevitably be delayed. Similarly, the mean delay time is an important measure.

The results indicate that, in terms of the rate of occurrence of infinite information, the schedules designed by the information content measure (IIM), EDD, and MST are the best. In terms of the mean delay time, the design of scheduling by the method based on Axiom 2 seems to be the best. However, the mean delay time based on other values are ad hoc, and therefore no conclusive statements can be made.

TABLE 8.3 Evaluation of Designed Schedule (TWK)

Evaluation Measure	Rate of Occurrence of Infinite Information (%)	Mean Delay Time (minutes)	Mean Stay Time (minutes)
IIM*	0	0.2	2,422.8
SPT	14	104.1	2,729.1
FASFS	14	21.9	2,408.6
EDO	0	0.2	2,370.9
MST	0	0.2	2,402.6

*IIM is the acronym for the information integration method that uses Axiom 2.

TABLE 8.4 Evaluation of Designed Schedule (NOP)

Evaluation Measure	Rate of Occurrence of Infinite Information (%)	Mean Delay Time (minutes)	Mean Stay Time (minutes)
IIM	0	6.9	2,424.9
SPT	50	338.8	2,813.1
FASFS	17	72.8	2,669.1
EDO	0	47.8	2,649.8
MST	0	35.4	2,712.3

TABLE 8.5 Evaluation of Designed Schedule (RDM)

Evaluation Measure	Rate of Occurrence of Infinite Information (%)	Mean Delay Time (minutes)	Mean Stay Time (minutes)
IIM	0	44.0	2,914.1
SPT	25	200.1	2,801.7
FASFS	25	100.5	2,568.9
EDO	0	117.5	3,259.4
MST	0	68.0	3,291.8

TABLE 8.6 Total Evaluation of Designed Schedule

Evaluation Measure	Rate of Occurrence of Infinite Information %	Mean Delay Time (minutes)	Mean Stay Time (minutes)
IIM	0	16.7	2,587.3
SPT	30	214.3	2,781.3
FASFS	19	65.1	2,548.9
EDO	0	55.1	2,760.0
MST	0	34.5	2,802.2

8.4 Selection of Subcompact Automobile

Nakazawa (1987) applied the information content measure to determine the best subcompact car from among those sold commercially in Japan. In purchasing cars, there are so many FRs that the customer wishes to satisfy that they make a decision difficult.

Six different automobiles, Cars A–F, were considered for evaluation. All of these cars were 1980 models. The data presented in the manufacturers' catalogue were used in the evaluation and analysis. The FRs evaluated were: (1) fuel efficiency; (2) acceleration characteristics; (3) noise level inside the car; (4) trunk space; and (5) price. The design ranges for these FRs were established arbitrarily by Nakazawa, based on the desired characteristics that he wished to see in a compact car.

The information content was determined by establishing the *system range* for each of the FRs and by determining the overlap between the system range and the design range (i.e., the designer-specified range). The system range for each of the vehicles is listed in Table 8.7. The design ranges that Nakazawa chose are also given in the table. Some values of the design range are given in terms of "satisfaction function," which is defined by Nakazawa as follows.

TABLE 8.7 System Range for Cars A–F and the Design Range (from Nakazawa, 1987)

(a) Automobiles evaluated
   (1) Car A   (2) Car B   (3) Car C
   (4) Car D   (5) Car E   (6) Car F

(b) Fuel efficiency (km/L)
   (1) 12.4–24.6   (2) 11.1–21.9   (3) 11.0–21.8
   (4) 11.4–22.1   (5) 8.7–16.7   (6) 15.7–26.5
   Design range = The average value of the entire system ranges = 17.0 km/L

(c) Acceleration (seconds)
   (1) 4.7–15.9   (2) 4.3–14.4   (3) 4.8–16.6
   (4) 5.0–16.9   (5) 5.0–15.2   (6) 6.1–23.1
   Design range = The smallest value of the upper limit of the system range = 14.4 seconds

(d) Noise level in the car (dB)
   (1) 45–68   (2) 49–71   (3) 45–68
   (4) 48–69   (5) 45–68   (6) 58–74
   Design range = Below 55 dB

(e) Trunk space (L)
   (1) 102   (2) 148   (3) 178
   (4) 169   (5) 178   (6) 148
   Design range = Above 0.4 of the satisfaction function

(f) Price (×1,000 yen)
   (1) 828   (2) 844   (3) 824
   (4) 770   (5) 860   (6) 919
   Design range = above 50% of the satisfaction function

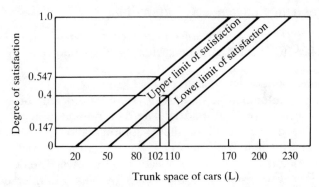

Figure 8.5. Degree of satisfaction versus trunk space of compact cars. (From Nakazawa, 1987.)

The "satisfaction function" was created in order to deal with the uncertainties involved in defining the design range and the system range in terms of a given physical parameter. For example, in defining the design range for the trunk space, it is difficult to say precisely what the design range should be. On the other hand, one may state that, "If the trunk space is in the range of 165–230 L, I will be 100% happy," or, "If the trunk space is in the range of 20–80 L, I will be completely unhappy." Between these limits, the degree of satisfaction may vary from zero to unity as shown in Fig. 8.5. The abscissa of the figure is for the trunk space, and the ordinate is the axis for satisfaction level. There is no reason why the upper and lower bound curves should be linear, but it may be a reasonable first-order approximation.

According to Fig. 8.5, the level of satisfaction can vary at a given trunk space. For example, if the trunk space is 102 L, the satisfaction level is in the range of 0.147–0.547, the difference value being 0.4. These results can be translated into probabilities, as we have done so far, by plotting the

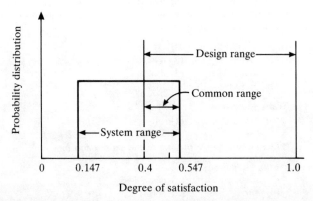

Figure 8.6. Probability distribution as a function of the degree of satisfaction for trunk space of compact cars. (From Nakazawa, 1987.)

Case Studies Involving Information Content 309

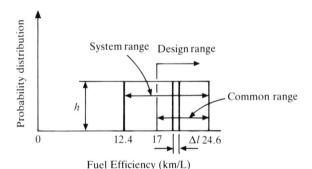

Figure 8.7. Probability distribution as a function of fuel efficiency of compact cars. (From Nakazawa, 1987.)

probability density as a function of the FR (shown in Fig. 8.6). In this plot the FR is converted from trunk space to "satisfaction level."

The information content for the trunk space can be computed for the case where the designer-specified range (or the design range for short) is from 0.4 to 1.0 in terms of the satisfaction function in the preceding paragraph. The probability distribution function for the trunk space plotted in Fig. 8.6 shows that the system range for Car A is 0.147–0.547 and the information content is $\ln(0.4/0.147) = 1.0$ nat.

The design range for fuel efficiency is arbitrarily defined by the designer. In this case it is defined as anything greater than 17 km/L, based on the average of the fuel efficiency of the six cars as shown in Fig. 8.7. Depending on one's need and financial outlook, different designers may choose a different range for the fuel efficiency. The information content for the fuel efficiency of Car A is obtained, using Eq. 5.7, as

$$I = \ln \frac{\text{system range}}{\text{tolerance}} = \ln \frac{24.6 - 12.4}{24.6 - 17.0} = 0.473 \text{ nat}$$

The FR associated with acceleration is given in terms of the time that it takes to reach 80 km/h from 40 km/h in the 4th gear of standard

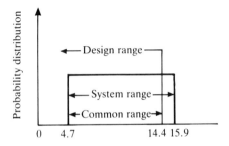

Time taken to reach 80 km/h from 40 km/h (seconds)

Figure 8.8. Probability distribution as a function of the time taken to reach 80 km/h from 40 km/h. (From Nakazawa, 1987.)

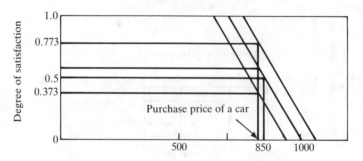

Figure 8.9. Purchase price of compact cars and the customer satisfaction (in degree of satisfaction). (From Nakazawa, 1987.)

transmission. The smallest value of the upper limit of the system range (i.e., 14.4 km/h) is chosen as the upper bound of the design range, which is plotted in Fig. 8.8. Again, for Car A the information content is

$$I = \ln \frac{15.9 - 4.7}{14.4 - 4.7} = 0.144 \text{ nat}$$

The information content for the noise level can be obtained similarly. The design range is chosen to be 0–55 dB. At a noise level of 55 dB, a normal conversation can be conducted.

The satisfaction function for purchase price is presented in Fig. 8.9. It is also plotted in terms of the probability distribution function and the level of satisfaction in Fig. 8.10 (hedonistic scale of values). It shows that if the price is less than 850,000 yen, the level of satisfaction is greater than 0.5.

Summing up the information contents determined for various FRs established as criteria for the purchase of a subcompact car, the total information content can be determined for the six automobiles under

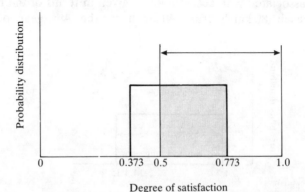

Figure 8.10. Probability distribution as a function of the degree of satisfaction based on the purchase price of compact cars. (From Nakazawa, 1987.)

TABLE 8.8 Information Content for Sub-Compact Car

	Car A	Car B	Car C	Car D	Car E	Car F
Fuel efficiency	0.473	0.790	0.811	0.741	∞	0.128
Accel. charact.	0.144	0	0.206	0.236	0.082	0.717
Noise level	0.833	1.300	0.833	1.099	0.833	∞
Trunk space	1.000	0	0	0	0	0.125
Price	0.382	0.598	0.332	0	0.873	∞
Total inf. cont.	2.832	2.688	2.182	2.076	∞	∞

consideration. Table 8.8 gives the information content for each criterion for each of the six cars and the total information content. According to this analysis of the information contents, the best car to purchase is Car D, followed by Car C. The difference in the information contents between these two cars is marginal. In this case the choice may depend on other factors that appeal to the purchaser. However, Cars E and F are not acceptable vehicles. Car F has failed to meet the criteria for noise level and price, whereas Car E has failed to satisfy the fuel efficiency requirement.

8.5 Job Scheduling by Information Content Measure and Others

Introduction

Scheduling of jobs in batch manufacturing operations is an important task that has a significant effect on productivity. In Chapter 5 and Secs. 8.2 and 8.3 it is shown how the process planning and job scheduling can be done based on the Information Axiom (Axiom 2) when these tasks must be optimized with respect to many criteria in a multivariable environment. In this section we compare the solution obtained for a job-scheduling problem by the information method and by other methods such as the mean flow time (MFT), mean tardiness (MT), and mean slack time (MST). A simple problem is chosen for this case study (Tokawa, 1988).

Statement of the Problem

The manufacturing system consists of two kinds of machine tools (M1 and M2), which are used to perform two types of machining tasks (Tasks A and B) as shown in Fig. 8.11. The processing time for each task by each machine is given as a1, a2, b1, and b2, where a and b refer to Tasks A and B, respectively, and 1 and 2 refer to machine tools M1 and M2, respectively. The process time includes all material-handling time.

Three task orders consisting of various combinations of Tasks A and B are issued simultaneously for these two machine tools to work on. Both machines can perform either Task A or Task B. In the example shown in

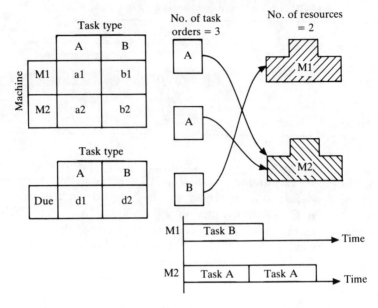

Figure 8.11. Configuration of the manufacturing system. Two machines M1 and M2 must perform various operations that consist of Task A and Task B.

Fig. 8.11, three tasks—two Task As for M2 and one Task B for M1—are to be performed by M1 and M2 in the time sequence indicated. In Fig. 8.11 the due dates are given as d1 and d2. Our assignment is to schedule the jobs for these machines so as to produce these parts in the shortest possible time. It may be assumed that the machines are available when the task orders are issued.

The schedule made by the information content measure is to be compared with three other dispatching rules: MFT, MT, and MST. MFT, MT, and MST are computed by means of the equations

$$\text{MFT} = \frac{1}{N} \sum_{1}^{N} (T_{\text{com}} - T_{\text{arr}}) \tag{8.10}$$

$$\text{MT} = \frac{1}{N} \sum_{1}^{N} \max[0, (T_{\text{com}} - T_{\text{due}})] \tag{8.11}$$

$$\text{MST} = \frac{1}{N} \sum_{1}^{N} \max[0, (T_{\text{due}} - T_{\text{com}})] \tag{8.12}$$

where N is the number of tasks.

The information content associated with time is computed using the following system, design, and common ranges:

$$\text{System range} = (T_{\text{com2}} - T_{\text{com1}}) \tag{8.13}$$

$$\text{Design range} = (T_{\text{due}} - 0) \tag{8.14}$$

Case Studies Involving Information Content

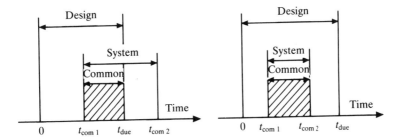

Figure 8.12. System, design, and common ranges.

$$\text{Common range} = \min[(T_{\text{due}} - T_{\text{com1}}), (T_{\text{com2}} - T_{\text{com1}})] \tag{8.15}$$

where T_{com1} and T_{com2} are the lower and the upper bounds of the part completion time as shown in Figs. 8.12 and 8.13.

A process time data base often contains only typical process times and does not contain actually surveyed process time ranges. Therefore, we assume here that the upper limit is n times the typical process time T_{com} retrieved from the process time data base, where $n > 1$. Then, the information content may be written as

$$I = \ln \frac{(nT_{\text{com}} - T_{\text{com}})}{\min[(T_{\text{due}} - T_{\text{com}}), (nT_{\text{com}} - T_{\text{com}})]} \tag{8.16}$$

where $n > 1$ and $T_{\text{due}} > T_{\text{com}}$. When $T_{\text{due}} < T_{\text{com}}$, it is assumed that a new delivery date will be negotiated with the customer.

Simulation Results

Tokawa (1988) performed simulations of the job-scheduling problem. He conducted a large number of simulations, out of which four sample data

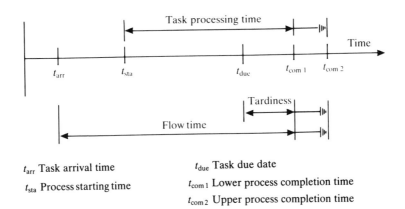

t_{arr} Task arrival time
t_{sta} Process starting time
t_{due} Task due date
t_{com1} Lower process completion time
t_{com2} Upper process completion time

Figure 8.13. Time line for processing a task.

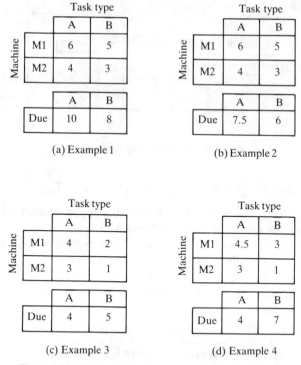

Figure 8.14. Process and due time table of simulation data.

are shown which illustrate the performance of the previously mentioned dispatching rules. Figure 8.14 shows these sample data.

In Example 1 (Fig. 8.14), Task A takes 6 time units (e.g., hours) on machine M1. The three other figures in the upper matrix read in the same manner. On the other hand, the lower table shows the due dates. The due date for Task A is 10 time units after the task order is issued. The representation of the data is the same for the other examples.

Through all the examples, Task A represents a harder task (i.e., longer processing times), and Task B represents an easier task (i.e., shorter processing time). Machine M1 represents a slow machine and M2 represents a fast machine.

In Examples 1 and 2, relatively moderate due dates are given for Tasks A and B, whereas in Example 3 and 4, the due date requirements for Task A are very tight but those for Task B are quite loose.

Tables 8.9–8.12 show the results of the simulations. In each example we have considered all of the possible task–machine combinations generated by three task orders; namely, four combinations for three Task As, four for three Task Bs, 12 for a combination of two Task As and one Task B, and 12 for a combination of two Task Bs and one Task A.

A notation like "B**AA*" corresponds to a possible combination of a task–machine assignment where the first three characters represent task

Case Studies Involving Information Content 315

TABLE 8.9 Example 1

Database
A, B = tasks
M1 6. 5.
M2 4. 3.
Due 10. 8.

Case	MFT	MT	MST	1/MST	INFO	PENAL
AAA***	12.00	3.33	1.33	0.750	0.00	−0.92
AA*A**	7.33	0.66	3.33	0.300	0.00	−0.28
A**AA*	6.00	0.00	4.00	0.250	0.69	—
***AAA	8.00	0.66	2.66	0.375	0.69	−0.28
BBB***	10.00	3.00	1.00	10.00	0.00	−0.99
BB*B**	6.00	0.66	2.66	0.375	0.00	−0.33
B**BB*	4.66	0.00	3.33	0.300	0.40	—
***BBB	6.00	0.33	2.33	0.428	0.40	−0.20
AA*B**	7.00	0.66	3.00	0.333	0.00	−0.28
BA*A**	6.66	0.33	3.00	0.333	0.00	−0.16
AB*A**	7.00	1.00	3.33	0.300	.000	−0.43
B**AA*	5.66	0.00	3.66	0.272	0.69	—
A**BA*	5.33	0.00	4.00	0.250	0.15	—
A**AB*	5.66	0.00	3.66	0.272	1.25	—
BAA***	11.00	2.66	1.00	1.000	0.00	−0.76
ABA***	11.33	3.33	1.33	0.750	0.00	−1.03
AAB***	11.66	3.66	1.33	0.750	0.00	−1.00
***BAA	7.00	0.33	2.66	0.375	0.15	−0.16
***ABA	7.33	0.33	2.33	0.428	1.25	−0.16
***AAB	7.66	1.00	2.66	0.375	0.69	−0.43
BB*A**	6.33	0.66	3.00	0.333	0.00	−0.33
AB*B**	6.66	1.00	3.00	0.333	0.00	−0.43
BA*B**	6.33	0.33	2.66	0.375	0.00	−0.16
A**BB*	5.00	0.00	3.66	0.272	0.40	—
B**AB*	5.33	0.00	3.33	0.300	1.25	—
B**BA*	5.00	0.00	3.66	0.272	0.15	—
ABB***	11.00	3.66	1.33	0.750	0.00	−1.12
BAB***	10.66	3.00	1.00	1.000	0.00	−0.86
BBA***	10.33	2.66	1.00	1.000	0.00	−0.89
***ABB	7.00	0.66	2.33	0.428	1.25	−0.33
***BAB	6.66	0.66	2.66	0.375	0.15	−0.33
***BBA	6.33	0.00	2.33	0.428	0.40	−0

order assignment to machine M1 and the last three, to machine M2. "*" denotes an empty assignment slot. In the case of "B**AA*," one Task B is assigned to M1 and two Task As are assigned to M2, as in Fig. 8.9.

The boxed figures in these tables show the best values in each group. For example, in Table 8.9 the candidate "A**AA*" has the minimum values for the criterion MFT, MT, 1/MST, or INFO; it should therefore be chosen out of four combinations for three Task As.

"X" in the 1/MST column represents an infinity (MST = 0). "PENAL" in the tables indicate whether or not the due date can be met. "—" in the

TABLE 8.10 Example 2

Database
A, B = tasks
M1 6. 5.
M2 4. 3.
Due 7.5 6.

Case	MFT	MT	MST	1/MST	INFO	PENAL
AAA***	12.00	5.00	0.50	2.000	0.69	−1.33
AA*A**	7.33	1.50	1.66	0.599	0.69	−0.55
A**AA*	6.00	0.16	1.66	0.599	0.69	−0.11
***AAA	8.00	1.66	1.16	0.857	0.00	−0.67
BBB***	10.00	4.33	0.33	3.000	0.91	−1.37
BB*B**	6.00	1.33	1.33	0.750	0.91	−0.58
B**BB*	4.66	0.00	1.33	0.750	0.91	−0
***BBB	6.00	1.00	1.00	1.000	0.00	−0.51
AA*B**	7.00	1.50	1.50	0.666	0.69	−0.55
BA*A**	6.66	1.16	1.50	0.666	0.91	−0.49
AB*A**	7.00	1.66	1.66	0.599	0.69	−0.64
B**AA*	5.66	0.16	1.50	0.666	0.91	−0.11
A**BA*	5.33	0.00	1.66	0.599	2.63	—
A**AB*	5.66	0.33	1.66	0.599	0.69	−0.25
BAA***	11.00	4.33	0.33	3.000	0.91	−1.24
ABA***	11.33	4.83	0.50	2.000	0.69	−1.39
AAB***	11.66	5.16	0.50	2.000	0.69	−1.38
***BAA	7.00	1.16	1.16	0.857	1.94	−0.49
***ABA	7.33	1.50	1.16	0.857	0.00	−0.74
***AAB	7.66	1.83	1.16	0.857	0.00	−0.76
BB*A**	6.33	1.33	1.50	0.666	0.91	−0.58
AB*B**	6.66	1.66	1.50	0.666	0.69	−0.64
BA*B**	6.33	1.16	1.33	0.750	0.91	−0.49
A**BB*	5.00	0.00	1.50	0.666	0.69	−0
B**AB*	5.33	0.33	1.50	0.666	0.91	−0.25
B**BA*	5.00	0.00	1.50	0.666	2.86	—
ABB***	11.00	5.00	0.50	2.000	0.69	−1.45
BAB***	10.66	4.50	0.33	3.000	0.91	−1.30
BBA***	10.33	4.16	0.33	3.000	0.91	−1.31
***ABB	7.00	1.66	1.16	0.857	0.00	−0.83
***BAB	6.66	1.33	1.16	0.857	1.94	−0.58
***BBA	6.33	0.83	1.00	1.000	0.00	−0.40

column of PENAL means that the due date can be met. If a PENAL value exists, then we can conclude the existence of an overdue. In this case, unless we are willing to assume a different probability distribution for the system range or negotiate a new due date, there is no acceptable solution.

Analysis of the Results

There are several observations that we can make based on the simulation results.

Case Studies Involving Information Content 317

TABLE 8.11 Example 3

Database
A, B = tasks
M1 4. 2.
M2 3. 1.
Due4. 5.

Case	MFT	MT	MST	1/MST	INFO	PENAL
AAA***	8.00	4.00	0.00	×	0.00	−1.54
AA*A**	5.00	1.33	0.33	3.000	0.40	−0.69
A**AA*	4.33	0.66	0.33	3.000	0.40	−0.51
***AAA	6.00	2.33	0.33	3.000	0.40	−1.25
BBB***	4.00	0.33	1.33	0.750	0.69	−0.28
BB*B**	2.33	0.00	2.66	0.375	0.69	—
B**BB*	1.66	0.00	3.33	0.300	0.00	—
***BBB	2.00	0.00	3.00	0.333	0.00	—
AA*B**	4.33	1.33	1.33	0.750	0.00	−0.69
BA*A**	3.66	0.66	1.33	0.750	0.40	−0.51
AB*A**	4.33	0.33	0.33	3.000	0.40	−0.28
B**AA*	3.66	0.66	1.33	0.750	0.40	−0.51
A**BA*	3.00	0.00	1.33	0.750	0.00	−0
A**AB*	3.66	0.00	0.66	1.500	1.09	−0
BAA***	6.00	2.66	1.00	1.000	0.00	−1.29
ABA***	6.66	2.33	0.00	×	0.00	−1.07
AAB***	7.33	3.00	0.00	×	0.00	−1.38
***BAA	4.00	1.00	1.33	0.750	0.00	−0.61
***ABA	4.66	1.00	0.66	1.500	1.09	−0.61
***AAB	5.33	1.33	0.33	3.000	0.40	−0.96
BB*A**	3.00	0.00	1.66	0.599	1.09	—
AB*B**	3.66	0.33	1.33	0.750	0.00	−0.28
BA*B**	3.00	0.66	2.33	0.428	0.00	−0.51
A**BB*	2.33	0.00	2.33	0.428	0.00	−0
B**AB*	3.00	0.00	1.66	0.599	1.09	—
B**BA*	2.33	0.00	2.33	0.428	0.00	−0
ABB***	6.00	1.33	0.00	×	0.00	−0.84
BAB***	5.33	1.66	1.00	1.000	0.00	−1.07
BBA***	4.66	1.33	1.33	0.750	0.69	−0.69
***ABB	4.00	0.00	0.66	1.500	1.09	−0
***BAB	3.33	0.00	1.33	0.750	0.00	−0
***BBA	2.66	0.33	2.33	0.428	0.00	−0.33

Observation 1. Basically Axiom 2 yields results which are similar to those of an individual or linear combination of MFT, MT, and MST. In Example 1, all of these criteria chose "A**AA*" for the group of three Task As and "B**BB*" for the group of three Task Bs. In many other task groups of other examples, these criteria chose the same candidates. This result is brought about by the fact that all of these criteria are monotonic functions of time.

Observation 2. In several cases Axiom 2 can distinguish the best candidate from those candidates other dispatching rules indicated as being equally

TABLE 8.12 Example 4

Database
A, B = tasks
M1 4.5 3.
M2 3. 1.
Due 4. 7.

Case	MFT	MT	MST	1/MST	INFO	PENAL
AAA***	9.00	5.00	0.00	×	0.00	−1.82
AA*A**	5.50	1.83	0.33	3.000	0.40	−0.94
A**AA*	4.50	0.83	0.33	3.000	0.40	−0.71
***AAA	6.00	2.33	0.33	3.000	0.40	−1.25
BBB***	6.00	0.66	1.66	0.599	1.09	−0.36
BB*B**	3.33	0.00	3.66	0.272	1.09	—
B**BB*	2.00	0.00	5.00	0.200	0.00	—
***BBB	2.00	0.00	5.00	0.200	0.00	—
AA*B**	4.83	1.83	2.00	0.500	0.00	−0.94
BA*A**	4.50	1.16	1.66	0.599	0.40	−0.65
AB*A**	5.00	0.33	0.33	3.000	0.40	−0.32
B**AA*	4.00	0.66	1.66	0.599	0.40	−0.51
A**BA*	3.16	0.16	2.00	0.500	0.00	−0.20
A**AB*	3.83	0.16	1.33	0.750	0.40	−0.20
BAA***	7.50	3.83	1.33	0.750	0.00	−1.50
ABA***	8.00	3.00	0.00	×	0.00	−1.17
AAB***	8.50	3.50	0.00	×	0.00	−1.55
***BAA	4.00	1.00	2.00	0.500	0.00	−0.61
***ABA	4.66	1.00	1.33	0.750	0.40	−0.61
***AAB	5.33	0.66	0.33	3.000	0.40	−0.51
BB*A**	4.00	0.00	2.00	0.500	1.50	—
AB*B**	4.33	0.33	2.00	0.500	0.00	−0.32
BA*B**	3.83	1.16	3.33	0.300	0.00	−0.65
A**BB*	2.50	0.16	3.66	0.272	0.00	−0.20
B**AB*	3.33	0.00	2.66	0.375	0.40	—
B**BA*	2.66	0.00	3.33	0.300	0.00	−0
ABB***	7.50	1.50	0.00	×	0.00	−0.83
BAB***	7.00	2.33	1.33	0.750	0.00	−1.17
BBA***	6.50	2.16	1.66	0.599	1.09	−0.80
***ABB	4.00	0.00	2.00	0.500	0.62	—
***BAB	3.33	0.00	2.66	0.375	0.22	−0
***BBA	2.66	0.33	3.66	0.272	0.00	−0.33

acceptable. In Example 1, shown in Fig. 8.15, "A**BB*" and "B**BA*" both have the same values for MFT, MT, and MST; therefore, both are equally good schedules by these criteria. On the other hand, Axiom 2 can choose the "B**BA*" because it has the minimum INFO values among all the qualified candidates ("A**BB*," "B**AB*," and "B**BA*"). The same argument applies to "A**BB*" and "B**BA*" in Example 2. Axiom 2 chooses those candidates in terms of the maximum probability of realization in the real world.

Case Studies Involving Information Content 319

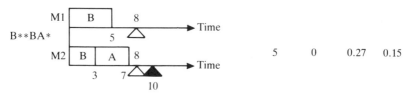

Figure 8.15. Axiom 2 distinguishes.

Observation 3. In Example 3 the choice of the best candidate and the second-best candidate depends on whether MFT, MT, and MST criterion prevails or Axiom 2 criterion does (see Fig. 8.16). If the contingency should be taken into account, we must follow the direction by Axiom 2, because the accomplishment of Task A is in marginal state in the case of "A∗∗BB∗," whereas "BB∗A∗∗" allows some contingent delays (beyond the estimated completion time, not the due date.)

Axiom 2 is more sensitive about not meeting the due dates due to delay, whereas the other criteria have the following shortcomings.

MFT: Does not evaluate the delay (lacks due date sensitivity).
MT: Does evaluate the delay, but does not take account of early completion of tasks.

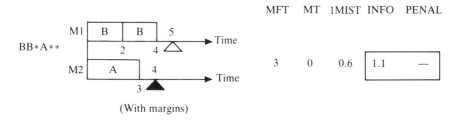

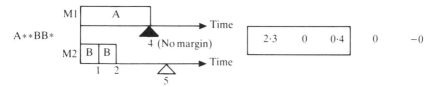

Figure 8.16. Candidates with and without margin.

MST: Does evaluate earliness of completion, but does not distinguish the degree of delay.

Observation 4. In Example 4 we can find the case where each criterion selects a different candidate (see Fig. 8.17). This phenomenon occurs when one or more due dates are tight, resulting in missed due dates (i.e., delays). Axiom 2 then assigns a large information content when the probability of not meeting the due date is large. On the other hand, the MFT, MT, or MST criterion simply ignores or assigns a relatively small value to such a critical factor. This difference in evaluation yields the difference in the results where Axiom 2 seeks the most secure candidate, whereas the other criteria often select a combination that cannot meet the deadline.

Conclusion

The Information Axiom applied to production scheduling yields quite similar results to those obtained by the conventionally used decision-making rules such as MFT, MT and MST. It can also select the best candidate in cases where these conventional methods cannot distinguish between candidates.

The axiomatic method distinguishes candidates in terms of probability. Therefore, the selected candidate is the most likely plan that can accomplish the task.

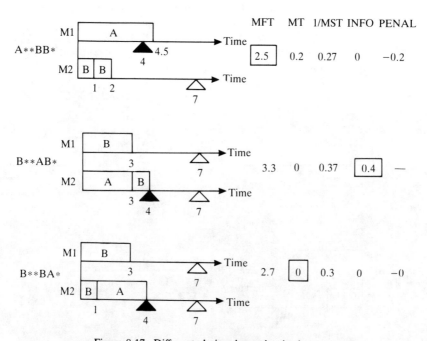

Figure 8.17. Different choices by each criterion.

8.6 Summary

The information content measure introduced as part of the axiomatic approach to design (i.e., Axiom 2, the Information Axiom) is a powerful tool in comparing various designs that satisfy Axiom 1, the Independence Axiom. The selection of the best design depends on the specification of the designer, which is subjective. Therefore, the innate ability of the designer to select a good set of specifications is a characteristic of exceptional designers.

A few points about information content need reiteration. In counting information content, only those quantities which have relevance to a given task (in terms of either FRs or DPs) should be counted. Information associated with other factors is irrelevant. Another important point is that, in many cases, the information of interest is determined by the overlap between the system specification (called the *system range*) and the specification made by the designer (called the *design range*). When the overlap is small, additional efforts must be made to satisfy the FRs. The third point that requires further reinforcement is that the information is best used on a relative basis. There is no unique solution that is best; we can only choose a better solution among those that satisfy the Independence Axiom.

One of the major strengths of the information content measure is that it provides the optimum solution when there are many variables that must be satisfied. In most other techniques used for optimization, the solution is obtained with respect to only one variable. For example, in the case of the example given in Sec. 8.4, there is no other technique that can provide an optimum solution with respect to five different variables—fuel efficiency, acceleration characteristics, noise level, trunk space, and price.

The methodology presented in this book and illustrated in this chapter can be applied to any problem that involves decision making. In this chapter three very different problems are considered: process planning, scheduling, and purchase of automobiles. The methodology was effective for all of these cases.

In order to determine the information content of a complex problem, we may need to develop computer software that can reduce the drudgery of accounting and simple calculations. To date, no general-purpose software program is available for this purpose.

Based on the observation that all information contents expressed in terms of probability are equally important, regardless of their physical origin, Theorem 16 may be stated as follows:

> Theorem 16 (Equality of Information Content)
> All information contents that are relevant to the design task are equally important, regardless of their physical origin, and no weighting factor should be applied to them.

References

Bake, K.R., *Introduction to Sequencing and Scheduling*. John Wiley, NY, 1974.
Conway, R.W., "Priority Dispatching and Job Lateness in a Job Shop," *Journal of Industrial Engineering* **16**(4):228,1965.
Nakazawa, H., "Process Planning Method by Information Concept," *Journal of Japan Society of Precision Engineering* **49**(9):1246, 1983.
Nakazawa, H., "Information Integration Method," *International Symposium on Design and Synthesis* 171–176, July 11–13, 1984, Tokyo.
Nakazawa, H., *Information Intergration Method*. Corona, Tokyo, 1987.
Nakazawa, H., and Suh, N.P., "Process Planning Based on Information Concept," *Robotics and Computer-Integrated Manufacturing* **1**(1):115–123, 1984.
Tokawa, T., "Application of the Information Axiom to Production Scheduling," Term Paper in Course 2.882, MIT, 1988.

Problems

8.1. Professor H. Nakazawa purchased a compact camera, based on the analysis of several commercially available compact cameras. The FRs used in his analysis were: the volume of the camera, its weight, its minimum distance for focus, the quality of its pictures, its apertures (i.e., f-stops), and an automatic setting for film sensitivity. Conduct a similar analysis and indicate which camera you would buy. Specify the designer range that you would choose for the camera.

8.2. Mr. Smith wishes to purchase a house either in Lexington or Sudbury. Where should he buy the house? The following are the relevant data.

	Lexington	Sudbury
Distance to work (miles)	10	20
Price of a 4-bedroom house ($)	600,000	500,000
Typical lot size (acres)	0.25	1
Nearest shopping center (miles)	5	12
Environmental quality (no. of good air-quality days)	120	300
Number of people in town	100,000	20,000
Quality of schools	Excellent	Very good

8.3. Evaluate the design (including price) features of three different desktop copiers. Which is the best buy?

8.4. The city in which you are now living wishes to expand (or construct) its subway system. Develop a strategy for establishing the best routes, based on the Information Axiom.

9
CASE STUDIES: DESIGN OF ORGANIZATIONAL AND MANUFACTURING SYSTEMS

9.1 Introduction

Not only do people design machines, processes, and software, they also design systems: organizational systems such as industrial firms, universities, and government agencies, and manufacturing systems such as factories and production lines. This chapter presents case studies involving three different kinds of systems: a government agency, a manufacturing system, and a college of engineering. Although they are very different entities, Axiom 1 is applicable to all systems.

Organizations are created to perform certain specific tasks that an individual cannot perform; they are entities with their own characteristics, goals, resources, and capabilities. In an organization, people with various expertise complement each other's talents; proprietary knowledge is accumulated to solve a similar set of problems, to produce goals, and to provide services. Certain common goals and objectives are established for everyone in the organization, in order to coordinate their activities and attain specified goals, and assets and facilities are accumulated to increase its capabilities further.

In recent years, it is well recognized that the productivity of manufacturing operations depends critically on the manufacturing system design. There are many different levels and kinds of manufacturing systems. They differ in complexity, products manufactured, and the capital equipment used, in addition to the role of human operators. The manufacturing system considered here is a cellular manufacturing system (CMS).

Organizations are also systems. The efficiency and effectiveness of an organization depend on its design; the goals of the organization must be clearly established, as they are the FRs of the organization in the functional domain. The DPs are the subunits of the organizations, which are physical units established to achieve the stated FRs. The DPs must be so chosen that the design matrix becomes a diagonal matrix. In an uncoupled organization, each subunit of the organization knows its exact mission and can pursue its objectives. When it is a coupled organization, all subunits are trying to achieve the same set of goals, thereby creating a great deal of confusion, turf fights, and inefficiency. In this case the top management must micromanage the subunits and monitor their daily actions.

There are certainly a large number of poorly designed organizations, and many which are well-organized, carefully run, and efficient. As the size of organizations grows, so the probability of finding poorly designed organizations increases, since it is more likely to find coupled designs. In organizations with coupled structures the people in the organization may not be working hard because their contributions are difficult to identify, and their initiatives become frustrated because there are no clear-cut responsibilities and authorities. However, size should not be the determining factor for the efficiency and efficacy of an organization if it is designed properly, for there are large corporations which are well-known for their corporate efficiency.

In many organizations well-defined FRs are often lacking or not completely understood by everyone in the organization, and the organizational structure does not have specific DPs to satisfy FRs. The job of the management is to define FRs and establish DPs, but this has been done ad hoc, very much as in other fields of design. In many companies they are established along product (i.e., purpose) lines, whereas in some companies, divisions are organized along their functional responsibilities, or geographic locations. What is suited for one company may not be suited for another. The organizational structure makes sense only when it is designed to serve the goals of the corporation.

There are many organizational theories taught in business schools. However, these are not based on any principles or axioms, and remain experimental. It may be interesting to analyze these organizational theories in terms of the design axioms (see Problem 9.1).

One of the case studies that we examine in this chapter is the organizational structure of a U.S. Federal government agency. The author was fortunate to have the opportunity to be in charge of the Engineering Directorate of the National Science Foundation (NSF) during the period of October 16, 1984–January 15, 1988. This was a totally unexpected transition from the academic surroundings at MIT to the political environment of Washington, D.C. As a political appointee of President Ronald Reagan, the author was given the responsibility of strengthening the NSF's policies and programs in engineering so as to serve the nation's long-term interests through engineering research and education. In this capacity he was given an opportunity to re-examine the organization of the NSF Engineering Directorate (i.e., DPs) to fulfill the FRs most effectively. The case study presented in the following section deals with the NSF's engineering activities.

9.2 Organization of a Government Agency

Introduction to NSF

The NSF was established in 1950 by the U.S. Congress as a government agency for the following purposes:

1. To ensure the progress of science and engineering.
2. To provide health, prosperity and welfare.
3. To promote national defense.

Ever since its early days, the NSF has supported basic research at universities, especially in basic sciences such as physics, chemistry, mathematics, geosciences, biology, and oceanography. The support for engineering research has been limited to about 10% of its annual budget.

Upon review of the NSF programs for engineering, it became evident that the primary emphasis had been placed on the support of ongoing research at universities, largely in well-established engineering sciences. Very little support was given to programs for those areas in which the engineering schools were notably weak, such as technology-driven areas (e.g., design, manufacturing, computer engineering, and system-integration engineering), emerging technology fields (e.g., biotechnology, lightwave technology, neuroengineering, and computational engineering), and cross-disciplinary research in areas related to large engineering systems. Also, NSF had the reputation of being an organization that supports only "safe, risk-free" research because of the peer-review process used there. The author was asked to join NSF and bolster its engineering activities.

The organization of the NSF Engineering Directorate prior to October 1984 is shown in Table 9.1. Its structure was nearly parallel to a typical engineering school structure, consisting of a Chemical and Process Engineering Division, Mechanical Engineering and Applied Mechanics Division, Civil and Environmental Engineering Division, and Electrical, Computer, and Systems Engineering Division. Each division had several programs that supported narrow disciplinary research. There was a nearly one-to-one correspondence between the NSF programs and engineering school teaching and research efforts. The similarity in the organizational structure of NSF and engineering schools is both obvious and peculiar, since the goals of NSF are very different from the primary goals of an engineering school.

The annual budget for the Engineering Directorate was about $130 million in 1984 and was increased to about $200 million in 1988, which was inadequate to meet the needs of the engineering research community. In the environment of constrained resources, it is easy to blame the lack of resources for inaction.

Problem Statement: NSF Engineering Directorate

One of the basic questions regarding the assessment of NSF activities in engineering was, "Is the NSF Engineering Directorate serving the needs of the nation, as stated in the NSF Act of 1950, which is to ensure the progress of science and engineering, to provide health, prosperity, and welfare, and to promote national defense?" The conclusion to which the

TABLE 9.1 Directorate for Engineering

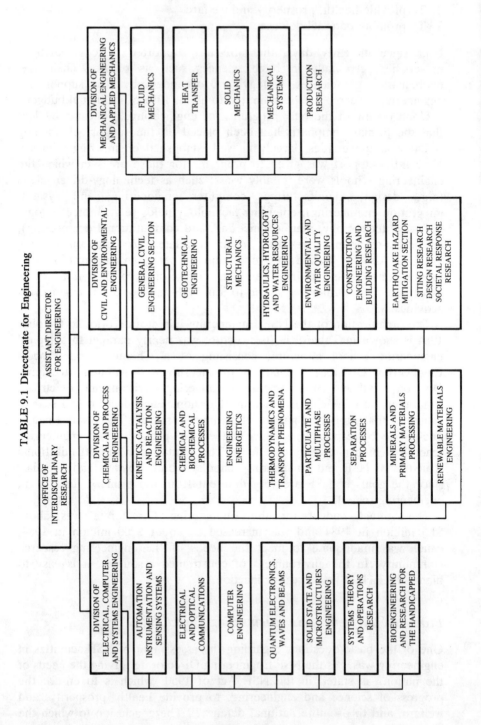

author arrived was that NSF could do better to carry out the original intent of Congress. NSF was then responding to the need of the existing engineering research community, but not necessarily to the needs of the nation.

NSF was responding to the pressure brought upon it by powerful, well-established, and vocal segments of the engineering schools which, however well-meaning, were interested in promoting their narrow research fields. In areas where the engineering schools were very weak (e.g., design and biotechnology), NSF was not receiving pressure to provide research support because these fields had no community of scholars to serve as their champions.

This situation was further aggravated by the practice of allocating the budget based on "proposal pressure"; the well-established fields with a larger number of researchers received more support than newer fields, regardless of the importance of the field in solving the future problems of the nation, because the number of proposals received by the NSF tends to be proportional to the size of the community. The budget allocation based on the proposal pressure tends to make larger research community to grow faster than smaller ones. It was not in the long-term interest of the nation to neglect the important fields of education and research, just because these fields are either recent or not popular. For example, the engineering schools could not continue to generate engineers without design skills and other new knowledge that will be important in the 1990s and the twenty-first century. Yet, many of these important fields were largely ignored by the lack of "mail-box" at NSF. The task was strengthen these areas without weakening the engineering science fields.

Organizational Solution

In view of the situation that existed at NSF prior to 1985, new goals had to be established at the NSF Engineering Directorate, to correspond to the need of the nation at the time. The FRs (i.e., goals) of the Directorate were stated as follows (NSF, 1988):

(a) To continue to strengthen the engineering science base in areas where it is well-established.
(b) To establish the science base in areas where it is absent (e.g., design, manufacturing, and computer-integrated engineering).
(c) To establish the academic infrastructure in emerging technologies.
(d) To encourage innovation in critical engineering areas (e.g., public infrastructure, mature industries, and natural hazard mitigation).
(e) To strengthen cross-disciplinary research at universities.
(f) To develop special programs for improved engineering infrastructure (e.g., undergraduate engineering education, and international cooperation).

These six goals were the primary goals of the Engineering Directorate. In order to achieve them most effectively, it was decided to create several new divisions, each of which was to be responsible for one of these goals.

The design equation may be written as

$$\begin{Bmatrix} \text{Goal a} \\ \text{Goal b} \\ \text{Goal c} \\ \text{Goal d} \\ \text{Goal e} \\ \text{Goal f} \end{Bmatrix} = \begin{bmatrix} \times & 0 & 0 & 0 & 0 & 0 \\ 0 & \times & 0 & 0 & 0 & 0 \\ 0 & 0 & \times & 0 & 0 & 0 \\ 0 & 0 & 0 & \times & 0 & 0 \\ 0 & 0 & 0 & 0 & \times & 0 \\ 0 & 0 & 0 & 0 & 0 & \times \end{bmatrix} \begin{Bmatrix} \text{Eng. Sci. Divs.} \\ \text{Design, Mfg., Comp.-Int. Eng. Div.} \\ \text{Emer. Engr. Tech. Div.} \\ \text{Critical Engr. Systems Div.} \\ \text{Cross-discip. Res. Div.} \\ \text{Engr. Inf. Dev. Office} \end{Bmatrix}$$

(9.1)

Because of the large number of engineering science programs, Goal a was subdivided into three fields:

Goal a_1 = Goal a in the field of chemical, biochemical, and thermal engineering.
Goal a_2 = Goal a in the field of mechanics, structures, and materials engineering.
Goal a_3 = Goal a in the field of electrical, communications, and systems engineering.

For each of these subfields under Goal a, three divisions were created to deal with their specific goals. The Office for Engineering Infrastructure Development was not called a Division, because of the small staff number and budget; however, it was an independent unit. The new organizational chart of the NSF Engineering Directorate is given in Table 9.2.

The functions of each division were divided into programs based on the intellectual content of the research fields. Large programs such as the Earthquake Engineering Program had four program directors by subdividing the goal of the program into four distinct but related areas; each program director was in charge of $3–7 million.

Each division administered its programs differently to achieve its goals most effectively. The Engineering Science Divisions gave nearly all of their grants to individual researchers, since the research in these fields was initiated primarily by single investigators. On the other hand, the Emerging Engineering Technology Division gave a large number of group grants. For example, in biotechnology grants may be given to a group consisting of chemical engineers, molecular biologists and chemists. The idea was to support a group of researchers so that they could generate future biotechnologists who would have all the necessary background, even though the collaborating individual professors may not have the integrated knowledge base to deal with biotechnology in its entirety; this was done to create an academic infrastructure in these emerging fields.

The Division for Design, Manufacturing, and Computer-integrated Engineering developed programs to strengthen the academic disciplines in those areas where the academic infrastructure was nearly absent. The role of the Critical Engineering Division was to encourage technological innovation in, and develop strategies for, dealing with mature technologies

TABLE 9.2 National Science Foundation Directorate for Engineering

- ASSISTANT DIRECTOR FOR ENGINEERING
  - OFFICE FOR ENGINEERING INFRASTRUCTURE DEVELOPMENT
  - CROSS DISCIPLINARY RESEARCH
    - ENGINEERING RESEARCH CENTERS
    - IUC RESEARCH PROJECTS
    - IUC RESEARCH CENTERS
  - ENGINEERING SCIENCE IN CHEMICAL, BIOCHEMICAL & THERMAL ENGINEERING
    - KINETICS & CATALYSIS
    - BIOCHEMICAL & BIOMASS ENGINEERING
    - PROCESS & REACTION ENGINEERING
    - MULTIPHASE & INTERFACIAL PHENOM
    - SEPARATION & PURIFICATION PROCESSES
    - THERMODYNAMICS & TRANSPORT PHENOM
    - THERMAL SYSTEM & ENGINEERING
  - ENGINEERING SCIENCE IN MECHANICS, STRUCTURES & MATERIALS ENGINEERING
    - SOLID & GEO-MECHANICS
    - STRUCTURES & BUILDING SYSTEMS
    - FLUID DYNAMICS & HYDRAULICS
    - TRIBOLOGY
    - DYNAMIC SYSTEMS & CONTROL
    - MATERIALS ENGINEERING & PROCESSING
  - ENGINEERING SCIENCE IN ELECTRICAL, COMMUNICATIONS & SYSTEMS ENGINEERING
    - SOLID STATE & MICROSTRUCTURES ENGINEERING
    - SYSTEMS THEORY & OPERATIONS RESEARCH
    - QUANTUM ELECTRONICS WAVES & BEAMS
    - INSTRUMENTATION SENSING & MEASUREMENT
  - SCIENCE BASE DEVELOPMENT IN DESIGN, MANUFACTURING & COMPUTER INTEGRATED ENGINEERING
    - DESIGN THEORY & METHODOLOGY
    - COMPUTER INTERGRATED ENGINEERING
    - MANUFACTURING SYSTEMS
    - AUTOMATION & SYSTEMS INTERGATION
  - FUNDAMENTAL RESEARCH IN EMERGING ENGINEERING TECHNOLOGIES
    - BIOTECHNOLOGY
    - BIOENGINEERING & RESEARCH TO AID THE HANDICAPPED
    - LIGHTWAVE TECHNOLOGY
    - COMPUTATIONAL ENGINEERING
  - FUNDAMENTAL RESEARCH IN CRITICAL ENGINEERING SYSTEMS
    - EARTHQUAKE HAZARD MITIGATION
    - ENVIRONMENTAL ENGINEERING
    - SYSTEMS ENGINEERING FOR LARGE STRUCTURES
    - NATURAL & MAN-MADE HAZARD MITIGATION

and public infrastructure problems. The Cross-disciplinary Research Division fostered interdisciplinary research by establishing Engineering Research Centers and Industry–University Cooperative Research Centers at various locations in the United States.

The Engineering Research Centers Program had the goal of establishing focal points for research to enhance the international competitiveness of the United States by enabling researchers from many disciplines to work together on a major research topic, to educate graduate and undergraduate students in a cross-disciplinary environment, to promote industry–university collaboration so as to facilitate technology transfer, to strengthen academic research through the involvement of industry in academic research, and to enable university researchers to work on large engineering systems problems (NRC, 1986). This Center's program was designed to augment research projects having single investigators and to bring about a better balance between those projects that the individual investigator is best suited to handle, and the larger projects that require teamwork.

At the time of implementing the reorganization of the NSF Engineering Directorate, the financial people at NSF advised the author that the Emerging Engineering Technology and the Critical Engineering Divisions should be housed together in one division so as to lower the overhead cost of the NSF. The attempt to comply was a failure, because the merged Division could not pursue both of these different goals effectively. The decisions made by the Division had to be monitored very carefully at the assistant director level, which was counterproductive. Therefore, they were made into separate divisions, whereupon the Emerging Engineering Technology Division became much more aggressive in promoting emerging technologies that may become the basis for new industrial enterprises in the twenty-first century, and the Critical Engineering Systems Division also became more aggresive in pursuing its goals as well. This episode is evidence that coupled organizational structure can impair the goals of the organization.

In order to implement the reorganization and support its programs, it was necessary to satisfy *constraints* in addition to FRs; that is, we had to secure the support of many people within and outside of the NSF: the director of the NSF, the Office of Management and Budget of the White House, and the Congress. Support was needed also from the engineering community, represented by engineering and professional societies, the National Academy of Engineering, and the Deans Institute (a body of the American Society of Engineering Education), as well as the National Science Board, which is the equivalent of a board of directors (except that they have no power to appoint the top-management team of the NSF; this was the prerogative of the president of the United States). These people fostered the goals of the NSF Engineering Directorate and supported the organization that was necessary to achieve its goals.

The plan was also opposed by some individual researchers, mainly those who were beneficiaries of the earlier organization. Some others also opposed the new structure, mainly because disciplinary divisions such as mechanical engineering and civil engineering disappeared from the division titles. In terms of the axiomatic approach to design, the actions undertaken to satisfy the variety of different people in the organization and other interest groups are equivalent to satisfying *constraints*.

After the reorganization was completed, it took a great deal of effort to expound the goals of the Engineering Directorate to those both inside and outside the NSF, because the tendency is to discuss the organizational structure without first agreeing what the goal of an organization should be. Once the goals are accepted, it is much easier to understand the rationale behind the structure.

In the author's opinion, the new organizational structure of the NSF Engineering Directorate served the nation well by enabling the staff, and to a lesser degree, those outside of NSF, to concentrate on the goals of NSF, rather than treating NSF as an organization that simply distributes research funds without long-term strategic views. It began to create an academic infrastructure and communities of scholars, in which engineering schools were very weak. It also promoted cross-disciplinary research among researchers and educators. NSF also began the process of improving undergraduate education through various programs created to heighten the interest in, and emphasize the importance of, undergraduate engineering education. More than 75% of the engineers in the United States have terminated their formal education after receiving degrees at B.S. level. Therefore, in engineering the B.S. degree is de facto the professional degree, whereas in pure sciences, the Ph.D. is required to practise in various fields. In spite of the importance of undergraduate education, NSF had neglected undergraduate engineering education in the past.

This case study makes the point that the basic principles of design apply even to non-technical problems, although it is more difficult to verify the outcome.

9.3 Design of a Manufacturing System (Adapted from Black and Schroer, 1988)

Introduction

As a means of illustrating how Axiom 1 can be used in designing an organization, Sec. 9.2 presents a case study involving the design of a government organization. It was a classic example of system design, in that it illustrated how different components of a system had to function to achieve common objectives of the organization.

In this section another case study involving the design of a system is presented. The system to be designed is a manned cellular manufacturing system (CMS), which has been replacing the traditional job shop–flow shop manufacturing systems (Black, 1988; Black and Schroer, 1988). The advantage of CMS over the job shop–flow shop manufacuring system is the *flexibility* of the system and the *integration* of the diverse functions of manufacturing such as quality control, inventory control and production control into the system.

Figure 9.1 shows a manned CMS with six different kinds of machines and one multifunctional worker (Black, 1988). The CMS combines processes to produce a family of parts. The worker moves around the cell, from machine to machine, performing many different functions: unloading, loading, inspecting, deburring, etc. In order to reduce inventory, Kanban and the just-in-time (1986) concept are used in CMS, which require that the worker move in the direction opposite to the part movement. With every completion of the loop, the worker completes a part. The cycle time of the cell—that is, the time to complete the loop—is not dependent on the machining time of a particular machine. The machining time for a particular machine must be less than the cycle time needed. If this is not the case, additional machines that perform the same operation must be added to the cell. The worker inspects the finished part, loads the new part into the machine, initiates the machining cycle, then moves to the next machine in sequence. After the final inspection, parts are placed in the finished parts cart. When the last part is removed from the input cart, the worker has completed the loop through the cell. Then the worker will get a production ordering Kanban (POK), which specifies the next set of parts to be made by his cell.

In order to make the CMS flexible, we need to eliminate set-up, develop multifunctional workers, design a U-shaped layout, allow small lots of parts to move between the cells, and enable one-piece part movement within the cell. The cell must accommodate changes in the design of existing parts. As the CMS cells shown in Fig. 9.1 are more computerized and automated, it loses flexibility due to coupling.

Flexibility is the primary design characteristic of a CMS. However, as CMS cells become more automated through computerization, CMS cells require a new element called a *decoupler*. Decouplers are needed to insure cell flexibility, product flow, and quality, because extensive variability in cycle time occurs in the individual processes and operations, and because it is necessary to repair and maintain machines, and control inventory. Decouplers can perform integration functions and maximize the productivity by maximizing the throughput, minimizing the inventory, and minimizing the capital investment.

In this section the decoupler, which optimizes the design of a cellular manufacturing system through decoupling of the system per Corollary 1 (decoupling of a coupled system) is described.

Functions of Decouplers

Figure 9.2 shows a manned cell with five machines, five decouplers, one worker, two carts (each with a 20-part capacity) and 10 parts in process (Black, 1988). The term "decoupler" was coined by Black from Corollary

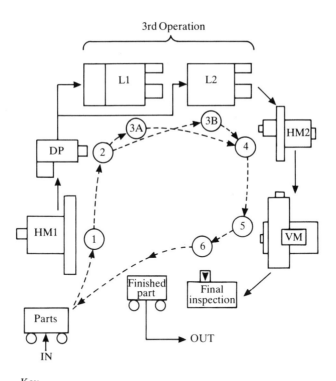

Key

DP = Drill press
L = Lathes
HM = Horizontal milling machine
VM = Vertical milling machine

-->-- = Paths of worker moving within cell
——▶ = Material movement paths
① = Operation sequence

Work sequence	Operation	Manual	Walking	Machine
1	Mill ends on work on HM 1	12"	5"	30"
2	Drill hole on DP	15"	5"	20"
3	Turn-bore on L1 or L2	13"	5", 5", 8"	180"
4	Mill flats on HM 2	12"	8", 5"	20"
5	Mill steps on VM	13"	7"	30"
6	Final inspection	10"	5"	—
		75"	35"	280"

Cycle time = 75 sec + 35 sec = 110 sec
Longest machining time = 180 sec
Total machining time = 280 sec
All machines in the cell can run untented while the operator does manual tasks (unloads, loads, inspects, deburrs) or walks from machine to machine. The time to change tools and work holders (perform setup) is not shown.

Figure 9.1. Manned cell with six machines and one multifunctional worker (lathe operation is duplicated). (From Black, 1988.)

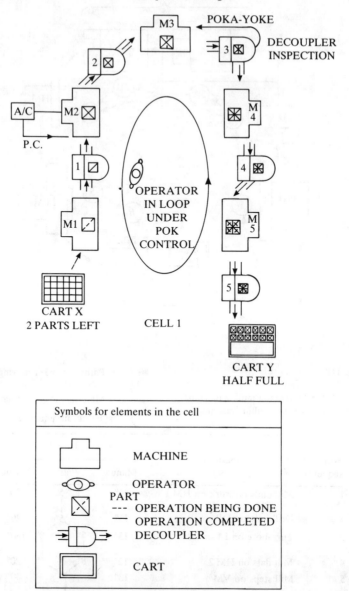

Figure 9.2. Layout of five-machine, five-decoupler, U-shaped cell to make parts one at a time. (From Black, 1988.)

1, which states: Decouple or separate parts or aspects of a solution if the FRs are coupled or become interdependent in the process proposed. "Decouplers" are devices located between the processes within the cell to enhance cell flexibility. The cell design with the decouplers relaxes the need for precise line balancing, and overcomes the coupling due to machining time variability. Decouplers reduce process dependence. A decoupler differs from a buffer, which is simply a place to store parts between the processes. Decouplers are key elements in the inventory and quality control systems within the cell, and for design of flexible cells. They are parts-holding devices with specific inputs and outputs.

Decouplers perform the following functions (Black, 1988).

1. Decouplers inspect critical dimensions of parts so that only good parts are passed onto the next station.
2. Decouplers provide flexibility in worker movement.
3. Decouplers control inventory within a cell. The number of parts within the decoupler can be zero if all of the following conditions are met: no inspection is needed, the operator moves in the same direction as the part movement, and no process delays in the process sequence are needed. For the Kanban concept to work, the operator must move in the direction opposite to the part movement. In this case the CMS will not work in the absence of decouplers, because the CMS becomes a coupled system. Similarly, if there is any variation in process times, without the decouplers, the CMS becomes a coupled system, and this functions very poorly. Ordinarily the minimum number of parts in a decoupler is one when a family of parts is being run in the cell. Then, with the decouplers, the inventory level within the cell can be doubled. If the number of parts in the decoupler is greater than one, it reflects the probability of defects, machine failure, etc.
4. Decouplers transport parts from process to process. Decouplers also branch or combine part flows within the cell.
5. Decouplers can handle families of parts. In unmanned cells utilizing robots, decouplers reduce robot handling and improve robot capability by locating, reorienting, and inspecting the part, Decouplers eliminate the need to orient the parts precisely on the input side, and allow parts to pass each other.
6. Decouplers allow process delays. When the decoupler is in place, the worker can change the process sequence, maintain machines, etc., between parts without disrupting the overall manufacturing system.

Evaluation of the Effectiveness of a Decoupler Through Simulation

Without the decouplers the CMS is a coupled system, which cannot function in a most efficient way. The machines will be underutilized, on an individual basis, by being idle and waiting for the preceding machine to complete its operation; or system-wide, in the event of a breakdown. The Kanban concept will not work. Decouplers are therefore needed to create a decoupled (or quasi-coupled) manufacturing system. The fact that the decouplers are needed to produce a CMS that satisfies the Independence Axiom can be deduced through qualitative argument without simulation or detailed analysis. Once the decision is made to use the decouplers, the efficiency of the CMS can be improved through simulations and detailed analysis programs.

Black and Schroer (1988) investigated, through simulations, the effect of the use of decouplers in a typical manned cell shown in Fig. 9.2 on the following items:

1. Cell behavior.
2. Production rates.
3. Operator utilization.
4. Machine utilization.
5. Operator delays resulting from waiting on machines and waiting for part inspectors and repairs.
6. Work-in-progress (WIP) inventory at decouplers.

The logic for the simulation model developed by Black and Schroer (1988) is given in Fig. 9.3, the key points of which may be summarized as follows.

1. Model first checks for any machine failures, and operator makes repairs.
2. Operator seizes the machine at station i when it becomes available and unloads the finished part i into automated inspection station.
3. After inspection of part i at the automated inspection station, good parts are automatically placed in decoupler i. Parts that fail are held until operator can repair them. Repaired parts are placed in the station decoupler.
4. Operator loads machine at station i with part $(i-1)$ from decoupler $(i-1)$. Operator does not start machine until just-completed part i at station i has passed inspection station at decoupler i.
5. Operator moves to station $(i-1)$ and the above sequence is repeated.

The constraints on the operation are:

1. Part flow clockwise (i.e., downstream), operating on the pull concept.
2. Operator movement is counterclockwise (upstream).
3. Inspection is automatic at each decoupler.
4. Each machine performs one operation on the part.
5. Operator unloads and loads the machines.
6. Operator does not start a machine until the previous part is inspected, and repaired if necessary.
7. Before unloading a machine, operator checks for machine failures and makes repairs immediately.

The initial conditions are:

1. Operator starts at machine 5 (because the cell operates by "pull").
2. Each decoupler has one part completed and inspected.
3. One empty cart is at final inspection.
4. One full cart is at station 1.

The material within the cell is called stock-on-hand (SOH), which is determined by the material needed to move the parts, individually,

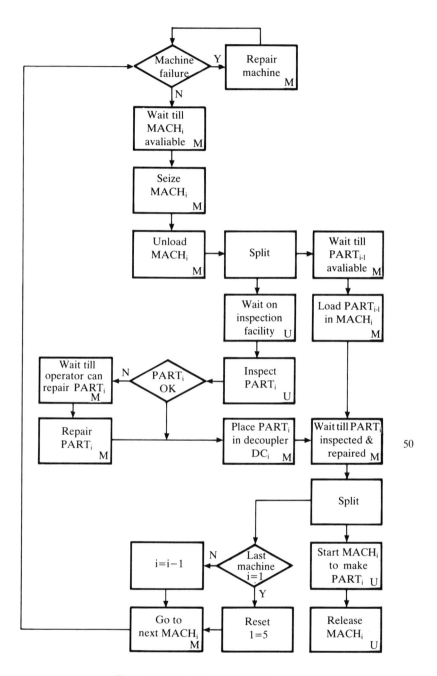

Figure 9.3. Simulation model logic.

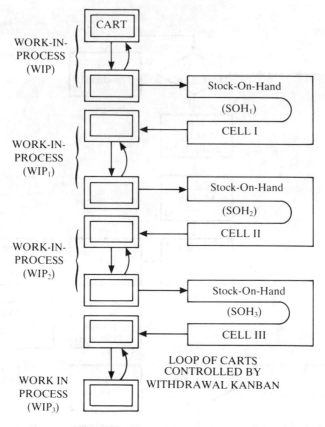

Figure 9.4. WIP and SOH differences.

through the cell satisfactorily. the WIP is defined to be the material between the cells (see Fig. 9.4).

Black and Schroer (1988) wrote the simulation model using the GPSS/PC simulation language (Minuteman Software, 1985). The model consisted of 220 blocks, 15 facilities, 26 queues, 10 logic switches and seven generate blocks. The model was run on an IBM PC/XT with an 8087 arithmetic coprocessor.

Results of Simulation

The purpose of this case study is to illustrate how the use of decouplers in a manufacturing system improves the overall efficiency of the system. However, before the effect of the decoupler is discussed, the overall validity of the model is examined by discussing three performance measures: the equilibrium, the stopping conditions, and the sample size.

The system behaves in a transient manner during the start-up phase. Therefore, it is desirable to know how the system performs when it reaches a steady-state equilibrium condition. Equilibrium is a limiting condition, but there is no single point in the simulation beyond which the system is in complete equilibrium. During simulation, the difference between the present distribution and the limiting distribution decreases with time. The acceptable error is subjective to the modeler's judgement.

Black and Schroer used Conway's (1962, 1963) subjective technique for determining equilibrium, all measurements until a measurement is neither a maximum nor a minimum are ignored. This ignored set of measurements is used as the standard set of measurements that is deleted from the collected data. GPSS/PC measurements were made after the manufacturing of 25-part lots. After each replication (each 25-part lot) any of the collected statistics can be plotted as a function of time to indicate the system's behavior.

The mean time to manufacture a part after every 25-part replication is plotted in Fig. 9.5. The fourth replication was the first measurement that was neither a maximum nor a minimum of the previous set, indicating that 100 parts were required for the system to reach equilibrium. The initial conditions selected should correspond to the system's conditions at equilibrium. However, such prior knowledge is generally unavailable. For this simulation, initial conditions were one part in each decoupler, one part in each machine ready for unloading, all queues empty and the machine idle. The operator began his work by unloading machine 5. The simulation was terminated after a given number of parts were made. In this simulation the number of parts was 500, after reaching equilibrium with 100 parts.

The sample size that provides a given confidence level is estimated using the procedure of Fishman (1968). If x_j is the mean of the jth replication after reaching the equilibrium, and N is the number of replications, then the grand mean is

$$x = \frac{\sum_{j=1}^{N} x_j}{N} \tag{9.1}$$

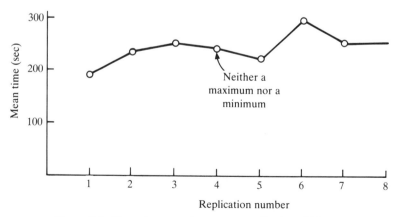

Figure 9.5. Mean time to make part per replication of 25 parts.

If the x_js are assumed to be independent and normally distributed, then the test statistic

$$c = 1 - \sum_{j=1}^{N-1} (x_j - x_{j+1})^2 \bigg/ \left(2 \sum_{j=1}^{N} (x_j - \bar{x})^2\right) \qquad (9.2)$$

is approximately normally distributed with mean of zero and variance of $(N-2)/(N^2-1)$. The two-sided hypothesis test is:

$$H_0: c \neq 0 \quad \text{(independent batches)} \qquad (9.3)$$

$$H_1: c = 1 \quad \text{(correlated batches)} \qquad (9.4)$$

TABLE 9.3 Parameters Used in Simulation runs (from Black and Schroer, 1988)

Run No.	Part Reject Rate (%)	Time to Repair Part* (sec) Mean μ	Std. dev. σ	Time to Manufacture Part† on Machine (sec) Mean μ
3	5	180	30	80
4	5	90	15	80
5	5	0	0	80
6	10	180	30	80
7	1	180	30	80
8	0	180	30	80
9	5	180	30	40
10	5	180	30	20
11	10	180	30	40
12	10	180	30	20
13	1	180	30	40
14	1	180	30	20
15	0	180	30	40
16	0	180	30	20
17	5	180	30	120
18	5	180	30	160
19	10	180	30	160
20	1	180	30	160
21	20	180	30	80
22	20	180	30	160
23	20	180	30	20
24	0	180	30	160

* Normal distribution
† Exponential distribution

Machine load time (sec) for each run follows normal distribution with $\mu = 10$ and $\sigma = 2$.
Machine unload time (sec) for each run follows normal distribution with $\mu = 10$ and $\sigma = 2$.
Decoupler inspection time (sec) for each part follows a normal distribution with $\mu = 10$ and $\sigma = 2$.
Operator movement time (sec) between machines follows normal distribution with $\mu = 10$ and $\sigma = 2$.

Case Study Design of Organizational Systems   341

The H_0 is rejected in favor of H_1 if the absolute value of $Z_{\alpha/2}$

$$Z = \frac{c}{\sqrt{(N-2)/(N^2-1)}} \tag{9.5}$$

is greater than $Z_{\alpha/2}$ where $Z_{\alpha/2}$ is the upper $\alpha/2$ point on the standard normal distribution.

The Z-statistic after five replications or a sample size of 500 was 0.696. If a 95% confidence level is assumed, the H_0 is not rejected if $|Z| < 1.645$. Since $|Z| = 0.969$, the H_0 cannot be rejected in favor of H_1. Therefore, the sample size was set at 500 after reaching equilibrium. The statistics for the first 100 parts are eliminated. The system is reset and another 500 parts are manufactured by the cell for which statistics are collected.

Twenty-two runs were made of the model, using the input parameters in Table 9.3. In Fig. 9.6 a time-bar diagram for the five-machine cell shows the relationship between the human, the machines, and the decouplers. One worker can make a part every 150 seconds, assuming the ideal condition where no part failures or delays occur due to variability of the cycle time. It was assumed that machines do not fail and that cutting-tool failure is simply one cause of part defects.

Table 9.4 summarizes the results from 22 simulation runs. Figure 9.7 gives the mean cycle time to manufacture a part in the cell as a function of part reject rate. The mean cycle time was 283 seconds with a 10% reject rate, and 217 seconds with a zero reject rate. As shown in Fig. 9.6, the ideal cycle time is 150 seconds, when no part failure or delay occurs.

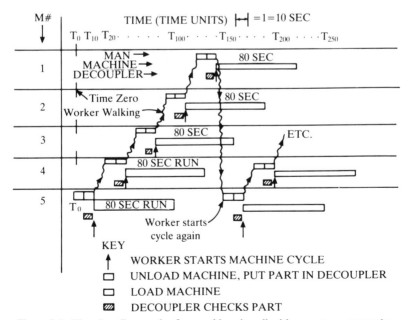

Figure 9.6. Time bar diagram for five machines in cell with operator upstreaming.

The Principles of Design

TABLE 9.4 Data Summary (from Black and Schroer, 1988)

Run No.	Mean Cycle Time (sec)	Utilization Rate	Delays Machine (sec)	Repair (sec)	Product Rate (units/hr)	Mean Cycle Time Less I Delay (sec)	Inputs Reject (%)	Machine Time (sec)
3	245	0.444	51.6	41.6	14.7	151.6	5	80
4	231	0.440	48.0	31.4	15.6	151.6	5	80
5	211	0.469	52.7	5.8	17.1	152.5	5	80
6	283	0.409	41.1	89.6	12.7	152.3	10	80
7	219	0.473	52.5	14.8	16.4	151.7	1	80
8	217	0.477	58.7	5.4	16.6	152.9	0	80
9	217	0.331	6.3	57.8	16.7	152.9	5	40
10	202	0.238	0.0	49.4	17.8	152.6	5	20
11	254	0.309	3.6	97.4	14.1	153.0	10	40
12	252	0.230	0.0	99.8	14.3	152.2	10	20
13	168	0.365	6.7	10.6	21.4	150.7	1	40
14	165	0.248	0.1	11.7	21.9	153.2	1	20
15	165	0.364	7.2	5.5	21.9	153.2	0	40
16	158	0.251	0.1	5.4	22.8	152.5	0	20
17	300	0.490	102.1	46.0	12.0	151.7	5	120
18	375	0.494	171.8	51.4	9.6	151.8	5	160
19	397	0.491	151.7	92.88	9.1	151.4	10	160
20	344	0.508	177.4	15.0	10.5	151.6	1	160
21	340	0.287	2.47	185.5	10.6	152.0	20	80
22	462	0.461	137.7	171.9	7.8	152.4	20	160
23	331	0.224	179.3	0	10.9	151.7	20	20
24	355	0.513	197.1	5.7	10.1	152.2	0	160

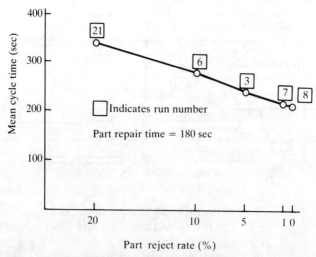

Figure 9.7. Time to manufacture parts with varying part reject rates (from Black and Schroer, 1988).

The mean manufacturing time with a zero repair time was 211 seconds, which was about equal to the mean manufacturing time for a zero reject rate. When no variability (only mean values were used with no variances and no distributions) was assumed, the time to manufacture a part was equal to the theoretical (or ideal) cycle time.

The operator can be delayed when machines are not available and (in progressing through the cell) when parts that fail inspection must be repaired. Figure 9.8 gives operator delay as a function of part repair time. It shows that operator delay caused by waiting on machine availability remained relatively constant. On the other hand, operator delay by waiting on part inspection and repair decreased with a decrease in the part repair time.

Figure 9.9 gives operator delay as a function of the part reject rate. Operator delay caused by waiting on machine availability increased with a decrease in the part reject rate, whereas the delay caused by waiting on part inspection and repair decreased with a decrease in the part reject rate. When the reject rate became very large, the operator spent much more time doing repairs. Consequently, the operator never had to wait for a machine because almost always the machining cycle had finished.

Operator delay as a function of the machining time is illustrated in Fig. 9.10. As the machining time on the individual machines decreased, the operator delay time also decreased; the delay time for repair remained about constant.

Mean machine and operator utilizations (in %) in the cell are given in Fig. 9.11 as functions of the part repair times. Operator utilization decreased with a decrease in the part repair time, but the effects on either utilization were not that large.

Figure 9.12 shows mean machine and operator utilizations (in %) in the cell as functions of the part reject rates. Operator utilization decreased

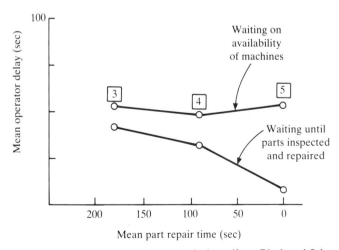

Figure 9.8. Operator delays with varying part repair times (from Black and Schroer, 1988).

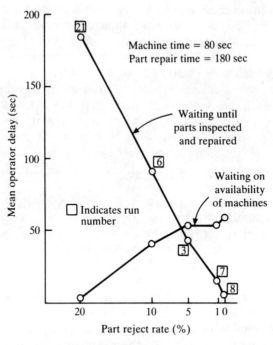

Figure 9.9. Operator delays with varying part reject rates. (From Black and Schroer, 1988).

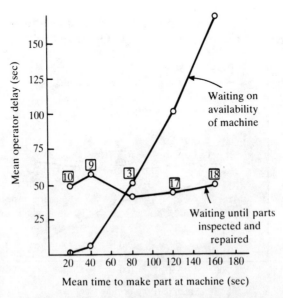

Figure 9.10. Operator delays with various times to make parts at machines. (From Black and Schroer, 1988).

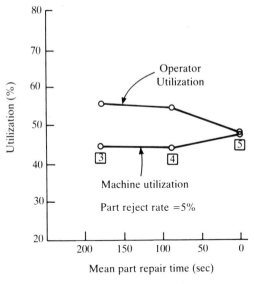

Figure 9.11. Utilizations with various part repair times. (From Black and Schroer, 1988).

with a decrease in the part reject rate. As expected, machine utilization increased when quality improved, because the operator spent less time repairing parts.

Mean machine and operator utilizations (in %) in the cell are plotted in Fig. 9.13 as functions of the machining times. Operator utilization improved with a decrease in the machining time. However, machines were

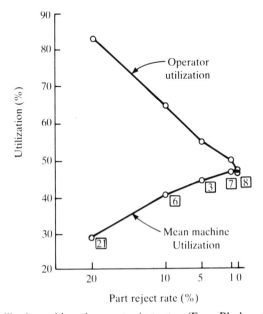

Figure 9.12. Utilizations with various part reject rates. (From Black and Schroer, 1988).

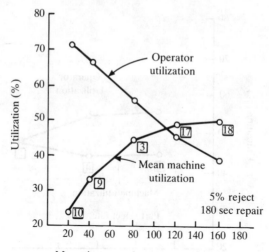

Figure 9.13. Utilizations with various times to make parts at machines. (From Black and Schroer, 1988).

idle for longer, waiting for the operator who still spent 150 seconds loading, unloading, and walking from machine to machine.

Figure 9.14 gives cell production rates, in parts per hour, with varying machine times and part reject rates. Production increased with a decrease in either the machining time or the part reject rate. The curve represented by the zero reject rate is the theoretical production limit for the cell, given the cell parameters and constraints.

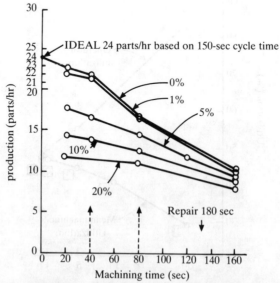

Figure 9.14. Cell production rates as function of machining times for various detect rates. (From Black and Schroer, 1988).

When the defect rate was 20%, halving the machining time had almost no effect on the production rate. At a 5% defect level, halving the machining time increased the production rate by only 2 units/hour. Increasing the machining time beyond 80 seconds had no appreciable effect on the production rate when the defect level was high (20%).

In reality, slowing the process decreases the probability of a machine tool breakdown and inproves the part quality. Therefore, the production rate can be increased and quality can be enhanced by increasing the machining time, provided that the increased machining time is less than the necessary cycle time.

Several runs were made with various SOH levels. An increase in decoupler inventory had no effect on cell performance, since an operating constraint on the cell was not to start a machine until the operator inspected and repaired the just-finished part. Additional SOH did not affect the cell performance. This simulation indicates that the ideal SOH at each decoupler is one part. However, to operate the system must have a minimum of one part at each decoupler.

The Effect of Decouplers

The main purpose of this case study was to investigate the improvements that can be made by creating decoupled systems by adding decouplers to the coupled system. Black and Schroer (1989) found that the system could not function without the decouplers when the Kanban system (pull system) was used, where the part moved clockwise and the operator moved counterclockwise. Similarly, if the part and the operator moved in the same direction without the aid of decouplers, the system will not function.

9.4 Organizational Design of an Engineering College

Universities have grown from a relatively simple organization, when they were first established, to a rather complex one, in recent years. Since universities are places where rational knowledge is pursued, one would think that the university organization is properly designed to achieve a set of goals. For the most part, that may be true, but in some cases one finds that the organizational design is less than optimum. We illustrate one of these aspects using the organization of a hypothetical engineering school as an example.

Universities consist of colleges, which in turn are divided into departments, laboratories, and centers. Many of these organizational structures of universities have evolved over the years to accommodate the specific needs of an era. Often new departments and laboratories are added to the existing structure, rather than reorganizing the entire school to meet the changing needs. It is much more difficult to reorganize universities and colleges than any other organization. The sense of loyalty and affiliation to a particular department on the part of the faculty, students, and alumni is very strong, and the external professional organizations are also structured along the traditional departmental lines, making a unilateral deviation of one school from the traditional structure difficult. Yet problems are sometimes created when new departments are established without looking at the entire organizational issue from the design point of view.

Consider a typical engineering college of a university with the following departments:

1. Civil Engineering.
2. Mechanical Engineering.
3. Electrical and Electronic engineering.
4. Chemical Engineering.
5. Materials Science and Engineering.
6. Aeronautical and Astronomical Engineering.
7. Nuclear Engineering.
8. Naval Architecture and Marine Engineering.

All of these departments are treated alike by the central administration in terms of their contributions to the university, which seems to be, at a first glance, reasonable.

Now let us examine the above hypothetical departmental structure from the Design Axiom point of view. The questions that one may ask are:

1. What are the FRs for these departments? Are they the same for all departments? Should they be different? Why?
2. What are the "structural" differences between, for example, the Department of Mechanical Engineering and Nuclear Engineering?
3. Why is there no Department of Automobile Engineering whereas there is a Department of Areonautical Engineering?

These questions may be regarded by some as, at best, irrelevant. However, if we are concerned about the efficacy of an engineering education, the intellectual development of our students, and the optimum use of the financial resources, then these questions must be taken seriously.

Since the FRs of these departments are not explicitly stated, it may be instructive to examine the physical entities of these departments in the physical domain, determine the FRs for each department, and then compare the FRs of all departments for commonality as well as for differences.

Civil engineering departments typically deal with the engineering function of solving public infrastructure problems. Mechanical engineering departments are often concerned with functional issues related to energy conversion and mechanisms. Electrical engineering departments address all issues related to electrical forces and devices. Chemical engineering departments deal with the production of materials and energy through chemical means, whereas the major concern of materials science and engineering departments is material structure, conversion, and behavior. The common characteristic of all of these departments is that they address *functions*.

This is quite a contrast from the other departments, such as nuclear engineering, aeronautical and aerospace engineering, and naval architecture and marine enginering. These departments deal mostly with products rather than with functions. Nuclear engineering departments are concerned with nuclear reactors; aeronautical engineering covers all aspects of airplanes; and the central activities of the naval architecture and marine engineering departments are related to ships. Of necessity, these product departments must cover all functions of engineering. For example, it is inconceivable to teach the design and manufacture of airplanes without covering most of the functions covered in the civil, mechanical, electrical, chemical, and materials engineering departments. Therein lies the common difficulty associated with administering such a college of engineering: when some departments are *product* departments while others are *functional* departments, product departments must *duplicate* some of the topics covered by the functional departments. Therefore, there is an inherent conflict between functional departments and product-oriented departments. Furthermore, it is more likely that the student enrollment in the product departments will fluctuate more severely than that of the functional departments, since students do not enroll in product-oriented academic departments when the products covered by the respective departments do not enjoy good sales performance (e.g., shipbuilding in the United States since the 1970s). When their enrollment is down, they are also subject to prey by other departments.

The DPs that are organizational entities in the physical domain are the departments. The FRs, as currently constituted, are either functions or products. The design equation for a highly simplified engineering school may be written as

$$\begin{Bmatrix} \text{Mechanisms} \\ \text{Chemical processes} \\ \text{Electromagnetism} \\ \text{Nuclear power plant} \\ \text{Shipbuilding} \\ \text{Airplane manufacture} \end{Bmatrix} = \begin{bmatrix} \times & 0 & 0 & \ldots & \times & \times & \times \\ 0 & \times & 0 & \ldots & \times & \times & \times \\ 0 & 0 & \times & \ldots & \times & \times & \times \\ \times & \times & \times & \ldots & \times & 0 & 0 \\ \times & 0 & \times & \ldots & 0 & \times & 0 \\ \times & 0 & \times & \ldots & 0 & 0 & \times \end{bmatrix} \times \begin{Bmatrix} \text{Mechanical Engineering Department} \\ \text{Chemical Engineering Department} \\ \text{Electrical Engineering Department} \\ \text{Nuclear Engineering Department} \\ \text{Naval Archictecture and Marine Engineering} \\ \text{Aeronautical Engineering Department} \end{Bmatrix} \quad (9.6)$$

The design matrix of Eq. 9.6 shows that the organization of this hypothetical engineering school is a highly coupled one. From this equation one can clearly see that a decoupled or an uncoupled organization for an engineering school can be designed, if we have the freedom to define the FRs and constraints for an engineering school.

9.5 Organizational Structure and Technology Transfer

The MIT Laboratory for Manufacturing and Productivity had many research contracts with industrial firms; many were large (top Fortune 500) companies and some were relatively small business firms. Although they all sponsored research at MIT for the purpose of developing new technologies and/or improving already-existing technologies, the success of these firms in transferring technology from MIT to their firms was greatly varied. The reasons for the different success rates were many, but organizational structure had profound effects on technology transfer.

Some of the large industrial firms have autonomous divisions that act as business units. The control of the divisional activities by the corporate management is done by financial means, and the executives of these divisions are rewarded for their ability to generate profits for the parent corporation. Therefore, much of their thinking centers around monthly or quarterly profit and loss (P & L) statements. Furthermore, many of these firms have subjugated research and development (R & D) to marketing organizations that dictate the funding and control the R & D agenda. Some of these firms have corporate technical staff whose function is to encourage technology innovation in divisions using corporate seed funds as matching funds, and to promote technology transfer amongst divisions by being transfer agents (somewhat like what bees are to flowers).

Many of these arrangements do not work well. There are several reasons for the difficulty. First, division technical goals are not necessarily the same as the corporate R & D goals; that is, they are trying to achieve different FRs. Secondly, there is no incentive on the part of these divisions to invest in long-term R & D and to excel technically. Thirdly, even when the corporate technical staff and the divisional staff agree on the common R & D goals, too many different DPs (e.g., the corporate technical group and the division management) are affecting the execution and the outcome. Fourthly, even when R & D generates new products or processes, there is no "champion" in the divisions who is willing to invest in the technology, and take the risk inherent in any new venture. In these firms, although the divisions have a reservoir of technical know-how and technologies, the corporate staff is unable to launch a new product by synergistic use of diverse technological capabilities residing in several divisions.

The only solution may be to create a new division with the sole task of developing the new product when the corporate technical and marketing staff identify new opportunities. Bringing in a new product that is very different from the ongoing activities of an existing division is equivalent to creating a coupled system, since the new product may disrupt all aspects of the divisional activities.

Consequently, many large corporations without *both* a central R & D group and the organizational ability to spin off new ventures have difficulties in introducing major technological innovations. These firms often acquire new technologies through the acquisition of small firms. It is best to operate these newly acquired "divisions" as separate entities even after merging with larger firms, unless the goals of these acquired small firms are the same as one or more of the existing divisions. In any case, the important thing that must be evaluated is the consistency of the goals of the acquired firm with the overall goal of the corporation. Otherwise, the acquired firm cannot function effectively and grow using the larger resources available from the parent corporation, because of the inconsistent set of FRs and constraints that the newly acquired firms have to fulfill. In this case the value of the acquired firm is less than it would be if the firm were an independent entity. In recent years many shrewd investors have realized that some of the large firms with incongruent sets of divisions would be worth more if they were spun off as independent firms. These investors began corporate takeover games for the purpose of selling the business units of these large corporations and make huge profits. This game has been the favorite pastime of certain businesspeople during the past decade.

9.6 Summary

This chapter dissects the design of organizational and manufacturing systems. Axiom 1 is applied in organizing a unit of a Federal government agency. The important point is the establishment of clear goals for the organization and the development of an uncoupled organization. Since the individual units in an organization must interact and work together, they cannot be completely independent; however, within a tolerance limit, they can be independent units and achieve a set of specific goals. The real integration of all these FRs of the operating units is done by the one-higher level organizational unit. When more than one unit is given the same task and goal, there can be conflicts amongst the units. Similarly, when one unit is given two or more independent goals (i.e., FRs), a close monitoring of the unit's performance may be necessary, reducing the effectiveness of the overall organization.

The use of *decouplers* (based on Corollary 1) in a cellular manufacturing system where the Kanban and the just-in-time systems are employed, is shown to increase the productivity by decoupling a coupled system. The coupling in the absence of decouplers occurs due to the variability in machining time, defective parts, and repair times of defective parts. Decouplers are not the same as buffers, in that, in addition to storing parts, they control inventory, monitor part quality, and rotate and align parts.

The organization of an engineering school is also analyzed in terms of the Independence Axiom. It is shown that when departments are organized along both the functional and the product lines, there is an inherent operational mismatch between these departments.

This chapter also examines the organization of large corporations from the view point of technological innovation and technology transfer. The consistency of FRs and constraints is emphasized. That is, the FRs and constraints of divisions must be derivable from those of the parent corporation through the hierarchical decomposition of the overall corporate FRs. Otherwise, Axiom 1 is likely to be violated.

References

Black, J.T., "The Design of Manufacturing Cells (Step One to Integrated Manufacturing Systems)," *Proceedings of Manufacturing International '88,* Vol. III, p. 143, Atlanta, GA, April 19, 1988. (A.S.M.E. Publication)

Black, J.T., and Schroer, B.J., "Decouplers in Integrated Cellular Manufacturing Systems," *Journal of Engineering for Industry, Transactions of A.S.M.E.* **110**:77–85, February 1988.

Black, J.T., and Schroer, B.J., Private Communication, 1989.

Conway, R.W., *Some Tactical Problems in Simulation Methods,* Memo RM-3244-PR, The Rand Corporation, Santa Monica, CA, October 1962.

Conway, R.W., "Some Tactical Problems in Digital Simulation," *Management Science* 248–253, October 1963.

Fishman, G.S., "Digital Computer Simulation: The Allocation of Computer Time in Comparing Simulation Experiments," *Operations Research* **16**:87–93, 1968.

Kanban and Just-In-Time at Toyota (Book Based on Seminars by T. Ohno et al.), Japan Management Association, ed., Lu, D.J., transl. Productivity Press, Cambridge, MA, 1986.

Minuteman Software, *GPSS/PC General Purpose Simulation.* Stowe, MA, 1985.

National Research Council, Cross-Disciplinary Engineering Research Committee, *The New Engineering Research Centers.* National Academy Press, 1986.

NSF, *Strategic Plan of the NSF Engineering Directorate.* NSF, 1988. Washington D.C.

Problems

9.1. Proctor and Gamble Co., Eastman Kodak Co., Hewlett–Packard Corporation, and IBM Corporation are known to be well-managed companies. Analyze the organizational structure of any one of these companies (or another) from the axiomatic design point of view. Suggest improvements.

9.2. There is as a continuing debate over whether or not a corporate central R & D unit is effective. Many companies have in recent years eliminated the central R & D organization. When is the divisional R & D unit more effective? Provide answers from the axiomatic design point of view.

9.3. If you have complete freedom to design a new school of engineering, what would be its organizational structure?

9.4. Design an ideal organization for an engineering school. Specify FRs and constraints.

10
MATHEMATICAL REPRESENTATION OF THE DESIGN AXIOMS AND COMPUTERIZED AXIOMATIC SYSTEM

10.1 Introduction

The design axioms are presented and used in preceding chapters in a qualitative way; we were not particularly concerned about the mathematical represention of the axioms, because the conceptual understanding of the axioms and their implication are the most difficult and important part of the axiomatic approach to design. However, once the significance of the axioms, corollaries, and theorems is understood, we must be able to express them mathematically so that they can be represented symbolically, manipulated logically, and used with precision. Furthermore, since computers are used extensively in all phases of engineering and design, the encoding of the design axioms into software is facilitated by their description in a formal representation.

As shown by the feedback diagram of Fig. 1.1, the creative process involves both analysis and synthesis, once the design problem is defined in terms of FRs and constraints. Analysis involves dissecting a given problem into manageable elements, developing models, and applying natural laws or other known postulates in order to understand the relationship among the input and output variables *quantitatively*. However, the mathematical tools used in analysis cannot be used in synthesis, because synthesis involves the *integration* of facts and knowledge for the purpose of creating a new system which is superior to, or different from, its precedents.

In this chapter symbolic logic is used to express the design axioms mathematically (Kim, 1985; Kim and Suh, 1985, 1987). This is done by introducing propositional logic and predicate logic, followed by the representation of the design axioms, using the predicate logic. The use of the computer language PROLOG (see Appendix 10A) is then introduced to develop computer software for design and for manipulation of the design axioms by computers. Using PROLOG and the design axioms, the structure of a computer-assisted design package is presented to illustrate the use of computers in axiomatic design. Finally, a concept for "Thinking Design Machine" is presented in Appendix 10B.

10.2 Introduction to Symbolic Logic

Logic is an indispensible part of reasoning or deductive thinking; in fact, most living beings possess reasoning powers and instincts that are essential to survival. Human beings have much greater capacity for reasoning than other creatures, thereby giving them supremacy in nature. Yet informal human reasoning power is very limited, sometimes inconsistent or illogical, especially as the number of variables increases. In order to deal with this problem, much scholarly effort has been directed toward formalizing the reasoning process through the development of mathematical or symbolic logic. In this section, *propositional logic* and *predicate logic* are introduced (Bittinger, 1982; Hamilton, 1978; Kneebone, 1963; Kowalski, 1979).

Propositional Logic

A *proposition* is a declarative statement or sentence. A particular proposition may be either true or false. For example, consider the following two statements: "Steel is lighter than an equal volume of water"; and "Rome, Italy, is located geographically south of Milan, Italy." The former statement is false, and the latter statement is true.

The relationship between various declarative statements can be represented symbolically, using *functors* and *connectors*. A *functor* is a function that assigns a new truth value to one or more propositions. A *connector* is a function that makes a compound statement out of simple statements. The following functors and connectors are defined as a matter of convention.

Symbol	Meaning	Functor or Connector Name
1. $\sim$	"not"	Negation functor
2. &	"and"	Conjunction connector
3. $\vee$	"or"	Disjunction connector
4. $\rightarrow$	"if–then"	Implication connector
5. $\leftarrow$	"implied by"	Contingency connector
6. $\leftrightarrow$	"if and only if" (iff)	Equivalence

Given propositions P and Q, the significance of the symbols are the following:

1. $\sim P$ The statement is true if P is false, or conversely, false if P is true.
2. $P \& Q$ The compound statement P and Q is true if both referent statements P and Q are true. If one or both of P and Q are false, then the compound statement is false.
3. $P \vee Q$ The statement is true if P or Q is true. The compound statement is false only when both P and Q are false.
4. $P \rightarrow Q$ The compound statement is false if P is true and Q is false, and true in all other cases. P is called the *antecedent* or *premise*; Q is called the *consequent* or *conclusion*.
5. $P \leftarrow Q$ The compound statement $P \leftarrow Q$ is true when $Q \leftarrow P$ is true. It is false otherwise.

Mathematical Representation of the Design Axioms 355

TABLE 10.1 Example of a Truth Table

P	Q	$(P \rightarrow Q)$	$\sim P$	$(\sim P \vee Q)$	$(P \leftrightarrow Q)$
t	t	t	f	t	t
t	f	f	f	f	f
f	t	t	t	t	f
f	f	t	t	t	t

6. $P \leftrightarrow Q$ The compound statement is true when P and Q are both true or both false. The statement is false otherwise.

Based on the definitions given above, a *truth table*, which is a table of combinations of truth assignments to propositions, may be constructed. Table 10.1 gives an example of such a truth table.

For example, the third row says that when P is false and Q is true, then the sentences $P \rightarrow Q$, $\sim P$, and $\sim P \vee Q$ are all true. The comparison of the third and fifth columns shows that they are the same. Since $P \rightarrow Q$ and $\sim P \vee Q$ take on identical truth values for all combinations of values for P and for Q, they are equivalent; i.e.

$$(P \rightarrow Q) \leftrightarrow (\sim P \vee Q)$$

The functors and connectors are not independent of each other, as the above examples illustrates.

Predicate Logic

The propositional logic is useful, but limited in scope. For example, there is no way to say, "If an object x is animal, then x is mortal" for any arbitrary object x. The concept of a proposition can be expanded to include one or more arguments.

A *predicate* is a function that assigns a truth value to some set of objects that serve as its arguments. To illustrate, let $H(x)$ be the predicate that takes on the value *true* if x is human, and is *false* otherwise. Similarly, we let $M(x)$ define the predicate "mortal." Then, $H(x) \rightarrow M(x)$ means that "if x is a human being, then it is mortal."

Quantifiers are used to qualify certain predicates. The *existential quantifier* and the *universal quantifier*, which are symbolized as $\exists$ and $\forall$ respectively, are used for this purpose. The qualifier $\exists$ signifies "there exists...." For example, $\exists x\, P(x)$ is taken to be true if there exists some object x which makes the predicate $P(x)$ true, and is false otherwise. The quantifier $\forall$ stands for "for all." For example, $\forall x\, P(x)$ is true if and only if (iff) the predicate $P(x)$ is true for all objects x.

The use of the quantifiers may be further clarified through the following examples. If we let $S(x)$ be the predicate "x is a school" and $T(y)$ "y is a teacher," then

(a) $\exists x\, S(x)$ is true iff there exists an object x, which is a school.

(b) $\exists x \, \exists y[S(x) \, \& \, T(y)]$ is true iff there exists both an object x, which is a school, and an object y, which is a teacher. Therefore, this compound statement says that "There is a teacher and a school."
(c) If we let the predicate $H(x, y)$ denote "y lives in x," the $\exists x \, \exists y \, [S(x) \, \& \, T(y) \, \& \, H(x, y)]$ states that "There is a teacher at some school who resides on campus."
(d) $\exists x[S(x) \, \& \, \forall y\{T(y) \rightarrow H(x, y)\}]$ states that "There is a school where all teachers live on campus."

10.3 Design Axioms in Predicate Logic

In order to express the Independence Axiom (Axiom 1), we need to define a few predicates. A design that satisfies the FRs and constraints is said to be feasible. Let F_s denote the set of feasible designs. We define the predicate feas(*) in the following way: for a design x, feas(x) is true if x is in F_s and is false otherwise.

Let ifm(*) denote a measure of information content defined on the set of feasible designs. Then ifm(x) < ifm(y) would imply that the information content of design x is less than that of design y. For convenience we sometimes use such infix notation to denote the associated predicate. For example, we write ifm(x) < ifm(y) to denote the predicate less than(ifm(x), ifm(y)), which is true if and only if ifm(x) is less than ifm(y).

Let coup(x) be the predicate asserting that design x is coupled, while unc(x) is defined as $\sim$coup(x). In addition, acc(x) means that design x is acceptable, and sup(x, y) holds if and only if design x is superior to design y. The first axiom may be stated in symbolic logic in the following way.

Axiom 1 *A feasible design that is uncoupled is acceptable,*

$$\forall x \, [\text{feas}(x) \, \& \, \text{unc}(x) \rightarrow \text{acc}(x)] \tag{10.1}$$

The universal quantifier is required to assert the generality of this statement. The expression in brackets states that if some object is feasible and uncoupled, then it is acceptable. The quantification ($\forall x$) stipulates, moreover, that this statement holds for *all* such objects rather than merely for *some* particular object.

This is a simplified statement that effectively classifies feasible designs into two categories: coupled and uncoupled. A more sophisticated statement would allow for degrees of functional dependence. Let idm(x) be the function that yields the independence measure of a design x. Then the general statement of the first axiom is

Axiom 1' *Of two feasible designs, the one with higher functional independence is superior,*

$$\forall x \, \forall y \, [\text{feas}(x) \, \& \, \text{feas}(y) \, \& \, \text{idm}(x) > \text{idm}(y) \rightarrow \text{sup}(x, y)] \tag{10.2}$$

This general form reduces effectively to the simpler version when a binary range is selected for the values of functional independence. For example, let idm(x) = 0 if x is coupled, and 1 otherwise. Then Axiom 1' is equivalent to Axiom 1.

The second Axiom is

> Axiom 2 *Of two acceptable designs, the one with less information is superior,*
>
> $\forall x \, \forall y \, [\text{acc}(x) \, \& \, \text{acc}(y) \, \& \, \text{ifm}(x) < \text{ifm}(y) \rightarrow \sup(x, y)]$
>
> (10.3)

When the general form of the first axiom is used, the second axiom should be modified to account for the measure of independence:

> Axiom 2' *Given two feasible designs of equal functional independence, the one with less information is superior,*
>
> $\forall x \, \forall y \, [\text{feas}(x) \, \& \, \text{feas}(y) \, \& \, \text{idm}(x) = \text{idm}(y) \, \& \, \text{ifm}(x) < \text{ifm}(y) \rightarrow \sup(x, y)]$ (10.4)

10.4 Corollaries in Predicate Logic

The corollaries given in Chapter 3, which are the intermediate consequences of the axioms, may be expressed in predicate logic. To this end some of the original statements given in Chapter 3 must be restated to be expressible in terms of the predicate logic. The following corollary relates to a coupled design that may be decoupled to yield an uncoupled one.

> Corollary 1 (Decoupling of Coupled Design)
> *Original statement:* Decouple or separate parts or aspects of a solution if FRs are coupled or become interdependent in the designs proposed.
> *Alternate statement:* An uncoupled feasible design is superior to a coupled feasible design.

This corollary is the direct consequence of Axiom 1. In terms of predicate logic, it may be stated as follows:

If a design u is feasible and decoupled, then the following facts obtain:

$$\text{feas}(u)$$
$$\text{unc}(u)$$

These facts satisfy the antecedents of Eq. 10.1, yielding the conclusion

$$\text{acc}(u)$$

Hence, a decoupled design is acceptable.

> Corollary 2 (Minimization of FRs)
> *Original statement:* Minimize the number of FRs.
> *Alternate statement:* Given two designs, u and v, where v satisfies FRs in addition to those satisfied by u, the information content associated with u is less than that associated with v.

Let $\{FR\}_s$ be a subset of $\{FR\}$. Suppose that feas(u) denotes a design that satisfies $\{FR\}_s$ and feas(v) one that satisfies the larger set $\{FR\}$. If both designs are uncoupled designs, then the corollary stipulates that the information content of u is less than that of v. This may be stated as

$$\forall x\, \forall y\, [\text{feas}(x)\, \&\, \text{feas}(y)\, \&\, \text{unc}(x)\, \&\, \text{unc}(y) \rightarrow \text{ifm}(x) < \text{ifm}(y)] \quad (10.5)$$

Let u and v be two uncoupled designs that satisfy feas(u) and feas(v), respectively. Then applying Eq. 10.5 yields

$$\text{ifm}(u) < \text{ifm}(v)$$

This may be used in Axiom 2 to conclude that u is superior to v

$$\sup(u, v)$$

> Corollary 3 (Integration of Physical Parts)
> *Original statement:* Integrate design features in a single physical part if FRs can be independently satisfied in the proposed solution.
> *Alternate statement:* The information content of uncoupled designs that have fewer components is less than those uncoupled designs that use a larger number of components to satisfy the same set of FRs and constraints.

Let x represent those uncoupled designs that use m number of parts and y those with n number of parts, where $m < n$. Then, the alternate statement of the integration corollary may be represented in predicate logic as

$$\forall x\, \forall y\, [\text{feas}(x)\, \&\, \text{feas}(y)\, \&\, \text{unc}(x)\, \&\, \text{unc}(y) \rightarrow \text{ifm}(x) < \text{ifm}(y)] \quad (10.6)$$

> Corollary 4 (Use of Standardization)
> *Original statement:* Use standard or interchangeable parts whenever possible.
> *Alternate statement:* The information content of a system that uses standard or interchangeable parts is less than that of systems that do not.
>
> In terms of predicate logic, the alternative statement of Corollary 4 may be as

$$\forall x\, \forall y\, [\text{feas}(x)\, \&\, \text{feas}(y)\, \&\, \text{unc}(x)\, \&\, \text{unc}(y)\, \&$$
$$\text{standard}(x)\, \&\, \sim\!\text{standard}(y) \rightarrow \text{ifm}(x) < \text{ifm}(y)]$$
$$(10.7)$$

where x represents designs that use standard parts and y designs that use no standard parts.

> Corollary 5 (Use of Symmetry)
> *Original statement:* Use symmetrical shapes and/or arrangements if they are consistent with the FRs.
> *Alternate statement:* The information content of symmetrical shapes or arrangements is less than that of asymmetrical parts provided that the FRs are satisfied equally well.

Let sym(x) be the predicate for a symmetric part x and asym(y) be a predicate for asymmetric part y.

$$\forall x\, \forall y\, [\text{feas}(x)\ \&\ \text{feas}(y)\ \&\ \text{unc}(x)\ \&\ \text{unc}(y)\ \&\ \text{sym}(x)\ \&\ \text{asym}(y)$$
$$\rightarrow \text{ifm}(x) < \text{ifm}(y)] \quad (10.8)$$

This statement implies that, for all parts that satisfy the FRs and Axiom 1, symmetrical design is always superior to asymmetrical design.

> Corollary 6 (Largest tolerance)
> *Original statement:* Specify the largest allowable tolerance in stating FRs.
> *Alternate statement:* Given two or more designs that satisfy the same set of FRs and constraints, and Axiom 1, the design with the largest tolerance is superior.

Let tom(x) be the predicate for a measure of tolerance. Then, this corollary can be written in predicate logic as

$$\forall x\, \forall y\, [\text{feas}(x)\ \&\ \text{feas}(y)\ \&\ \text{unc}(x)\ \&\ \text{unc}(y)\ \&\ \text{tom}(x)$$
$$> \text{tom}(y) \rightarrow \text{sup}(x, y)] \quad (10.9)$$

Equation 10.9 states that for all designs that satisfy FRs, constraints, and Axiom 1, the design that allows a larger tolerance in FR specification is superior.

> Corollary 7 (Uncoupled Design with Less Information)
> *Original statement:* Seek an uncoupled design that requires less information than coupled designs in satisfying a set of FRs.
> *Alternate statement:* Seek an uncoupled design that has less information content than a coupled design that satisfies the same set of FRs and constraints.

The alternate statement can be written in predicate logic as

$$\forall x\, \forall y\, [\text{feas}(x)\ \&\ \text{feas}(y)\ \&\ \text{unc}(x)\ \&\ \text{coup}(y) \rightarrow \text{ifm}(x) < \text{ifm}(y)] \quad (10.10)$$

Equation 10.10 states that there always exists an uncoupled design that has less information content than a coupled design that satisfies the same set of FRs and constraints.

10.5 Computerized System for Design Axioms

One of the reasons for developing symbolic logic for the design axioms is to be able to utilize the power of digital computers in making design decisions. The programming language chosen for this purpose is PROLOG (*Pro*gramming in *log*ic), which is a very high-level programming language (Kowalski, 1979; Pereira, 1984) created at Marseille (Battani and Meloni, 1973). (An introduction to PROLOG is given in Appendix 10A.) The ability to state the axioms as clauses in a logical programming language can be utilized to develop an expert system for axiomatic design. A PROLOG software driver for axiomatic design may be based on three layers or levels (Kim, 1985).

Primary Level

The primary level consists of the design axioms encoded in PROLOG. Then Axiom 1 may be written in PROLOG as (Kim and Suh, 1987)

$$acc(X):-$$
$$feas(X),$$
$$nonc(X)$$

The above states that design X is acceptable—denoted by the functor acc(x)—when it satisfies FRs and constraints and is a noncoupled (either an uncoupled or decoupled) design. In PROLOG an identifier that begins with an upper case letter (e.g., X) denotes a variable, whereas one that starts with a lower case letter (e.g., x) is a constant.

This simple encoding of the first axiom does not explicitly taken into account measures of functional independence. To this end, we define a functor idf(X, Idm) which maps a design X into its independence measure Idm. The corresponding PROLOG statement is

$$sup\ 1(X, Y):-$$
$$feas(X),$$
$$feas(Y),$$
$$idf(X, Idmx),$$
$$idf(Y, Idmy),$$
$$Idmx > Idmy$$

This statement stipulates that design X is superior to design Y if, for feasible designs X and Y, the independence measure of X exceeds that of Y.

Axiom 2 may also be written in PROLOG. Let iff(X, Ifm) be a predicate that maps a design X into its information metric Ifm. Axiom 2 may then be represented in PROLOG as (Kim, 1985)

> sup(X, Y):—
> acc(X),
> acc(Y),
> iff(X, Ifmx),
> iff(Y, Ifmy),
> Ifmx < Ifmy

The verbal translation is that design X is superior to design Y if X has less information content than Y. Again, this statement may deviate from the strict interpretation of Axiom 1, which requires that Idmx and Idmy be equal either to unity or zero. This PROLOG clause contains five predicates in its antecedent or body. The first two refer to Axiom 1, and the fifth is a simple logical test. The only substantial component pertains to the two occurrences of the iff predicate. This construct is part of the secondary level presented in the following subsection.

This encoding of Axiom 1 used the acc(x) predicate which considers functional independence only implicitly. A more general statement should take into account measures of functional independence, and decide on the basis of lower information content when two designs have equal values of independence.

The general statement of Axiom 2 is

> sup 2(X, Y):—
> feas(X),
> feas(Y),
> idf(X, Idmx)
> idf(Y, Idmy),
> Idmx = Idmy,
> iff(X, Ifmx),
> iff(Y, Ifmy),
> Ifmx < Ifmy

The verbal translation is that design X is superior to design Y if, given feasible designs of equal functional independence, X has less information content than Y.

Secondary Level

The secondary or intermediate level serves as the interface between the primary level and the data level. The first aspect of satisfying Axiom 1 relates to feasibility; that is, satisfying FRs and constraints. Design X is feasible if it satisfies the FRs of the design task

> feas(X):—meetFR(X, 1), meetFR(X, 2), – – – meetFR(X, n)

where meetFR(X, I) is satisfied if design X meets the Ith FR. The role of the meetFR(X, I) functor is to check whether the operating value of the *I*th FR for design X lies within the tolerance band associated with the *I*th FR.

The second aspect of satisfying Axiom 1 is related to the idf(X, Idm) functor, which evaluates design X in terms of the independence measures represented by Idm. A categorical measure of functional independence classifies designs into three groups, based on functional independence: coupled, decoupled, and uncoupled. As noted in Chapters 3 and 4, the design matrices for coupled, decoupled, and uncoupled designs are fully cluttered, triangular, and diagonal matrices. One possible means of assigning values to Idm would be "low" for a coupled system, "medium" for a decoupled system, and "high" for an uncoupled system. Alternatively, we may simply assign arbitrary values such as -1 for coupled design, 0 for decoupled design, and $+1$ for uncoupled design.

We may also classify the designs in terms of reangularity and semangularity, which are quantitative measures for functional independence, presented in Chapter 4. One may actually compute the values for R and S. When $R = S = 1$ the design is an uncoupled design, whereas for a coupled design $R < 1$ or $S < 1$. For a decoupled design, $R = S$. The more coupled the design is, the smaller the values of R and S, approaching $R = S = 0$ in the limit.

If we adopt the categorical classification scheme, one can assign values easily if the form of the design matrix is qualitatively known. For example, if the DM is triangular, Idm = 0; if it is diagonal, Idm = 1; and if it is neither of these of these two, Idm = -1.

Now let us consider the second-level PROLOG representation of Axiom 2. Let iff(X, Ifm) be the predicate that takes a design X and yields its information measure Ifm. The iff clause must take into account all the different types of information that are related to the set of FRs. Its general structure may be written as (Kim, 1985)

$$\begin{aligned}
&\text{iff}(X, \text{Ifm}):- \\
&\quad F1(X, I1), \\
&\quad F2(X, I2), \\
&\quad \vdots \\
&\quad Fn(X, In), \\
&\quad \text{Ifm is } (I1 + I2 + \ldots + In)
\end{aligned}$$

where Fj is the *j*th functor which takes design X and returns the information value Ij of the *j*th attribute. The last item in the body of the iff clause sums up the component information values I1–In to yield the overall measure Ifm.

A specific case for iff(X, Ifm) may be used to illustrate the points in the preceding paragraph. Suppose the FRs are hardness and electrical conductivity of dispersion-hardened copper: F1(X, I1) = hardf(X, Hardm) and

F2(X, I2) = condf(X, Condm). Then we obtain

    iff(X, Ifm):—
        hardf(X, Hardm)
        condf(X, Condm),
        Ifm is (Hardm + Condm)

The iff predicate takes design X, determines its information values due to the hardness and conductivity requirements, and returns their sum. Hardm and Condm must be computed based on the definition for information given in Chapter 5, which was defined as the logarithm to base 2 of the ratio of the system range to tolerance.

The iff clause interfaces directly with the clauses in the primary level. We also need a sublevel consisting of clauses that interface directly with the data level. This is achieved throught the Fj functor. The general stucutre of the Fj functor is

    Fj(X, Ij):—
        Dj(X, Lj),
        Oj(Lj, Ij)

the first item in the antecedent (body) finds the jth data structure Dj for design X and returns its attribute list Lj. The second item Oj is a set of predicates that takes Lj and yields the corresponding information value Ij.

Consider the data structure relating to surface tolerance. A plausible Fj clause is

    surff(X, Surfm):—
        surfd(X, List),
        infoLM(List, Surfm)

This clause corresponds to the general structure by the substitutions

    Fj ::= surff
    Ij ::= Surfm
    Dj ::= surfd
    Lj ::= List
    Oj ::= infoLM

where the compound symbol ::= means *defined as*. In other words, surff takes a design X, obtains a list of specifications and returns the corresponding information value Surfm through the functor infoLM.

The case for geometrical information is similar. Both the surff and geomf functions depend on the infoLM(L, M) predicate, which takes a list L of tolerance parameters $Ui \pm DUi$ and returns the resultant information measure M. In other words, infoLM accepts a list $L = [U1, DU1, U2, DU2, \ldots, Un, DUn]$ and calculates

$$M = lb(U1/DU1) + lb(U2/DU2) + \ldots + lb(Un/DUn),$$

where lb denotes the binary logarithm ($\log_2$).

The clauses so far described comprise the core of the Computerized Axiomatic System (CAS). These clauses are presented along with some supporting code in Fig. 10.1–10.3. Figures 10.2 and 10.3 depict code that has been tailored to the polymer processing problem described in Section 10.6.

Data Level

In order to be able to compute, for example, the information contents I_j, we need a means of representing the raw design requirements and specifications. By *data level*, we refer to the software pertaining to the raw data specification. A generalized representation of data takes the form $D_j(C, I_j)$, where D_j denotes the *j*th type of data for design X, and L_j the corresponding list of DPs.

For example, suppose the designer-specified length tolerance of a machined rod is $|\iota \pm \Delta \iota|$ the system range is U, and that the common range is DU. The appropriate data structure is then given by the pair of substitutions

$$Dj ::= \text{lengd}$$
$$Lj ::= [U, DU]$$

In other words, the resultant data structure is

$$\text{lengd}(X, [U, DU])$$

For the three-dimensional case, where the system range along the three orthogonal directions are given by U1, U2, and U3, and the common

```
/* Primary Level */

sup1(X, Y):-
   feas(X),
   feas(Y),
   idf(X, Idmx),
   idf(Y, Idmy),
   Idmx > Idmy.

sup2(X, Y):-
   feas(X),
   feas(Y),
   idf(X, Idmx),
   idf(Y, Idmy),
   Idmx = Idmy,
   iff(X, Ifmx),
   iff(Y, Ifmy),
   Ifmx < Ifmy.
```

Figure 10.1. Primary level of the Computerized Axiomatic System (CAS).

Mathematical Representation of the Design Axioms 365

/* Secondary Level*/

/** Feasibility Check**/

  feas(X):– meetFR(X, 1), meetFR(X, 2).

  meetFR(X, I):–
    fr(slab, [I, F, DF]),
    Lowval is F – DF,
    Hival is F + DF,
    dFR(X, [I, Ival]),
    Lowval = < Ival,
    Hival > = Ival.

/** Independence Measure: consists of 3 cases**/

  /** Cluttered coupling matrix
    → coupled design   **/

  idf(X, Idm):– cCoef(X, [I1, J1]), I1 < J1,
           cCoef(X, [I2, J2]), I2 > J2,
           Idm is –1.

  /** Lower or upper triangular matrix
    → decoupled design   **/

  idf(X, Idm):– cCoef(X, [I, J]), I =\= J, Idm is 0.

  /** Otherwise we have diagonal matrix
    → uncoupled design   **/

  idf(X, Idm):– Idm is 1.

Figure 10.2. Feasibility and independence components of the secondary level.

range is given by DU1, DU2, and DU3, respectively, the data structure is given by Dj ::= geomd and Lj ::= [U1, DU1, U2, DU2, U3, DU3].

When the design parameter list Lj is short, the notation is simplified by omitting square brackets that serve as delimiters for lists. For example, when Lj is empty, we use the format Dj(X) rather than Dj(X, []). When Lj is a singleton, we write Dj(X, Item), rather than Dj(X, [Item]).

An appropriate data structure for FRs may take the form

$$\text{fr}(\text{Proj}, [I, F_i, DF_i])$$

This specifies that the *I*th requirement for the design problem called Proj has the value given by Fi, with tolerance DFi. The actual values of the FRs for each design may be encoded by the data structure

$$\text{dFR}(X, [I, F_i])$$

/* Secondary Level */

/** Information Measure **/

/* Main information function */

iff(X, Ifm):–
  surff(X, Surfm),
  geomf(X, Geomm),
  Ifm is (Surfm + Geomm).

/* Information measure due to surface tolerance */

surff(X, Surfm):–
  surfd(X, List),
  infoLM(List, Surfm).

/* Information measure due to geometric tolerance */
geomf(X, Geomm):–
  geomd(X, List),
  infoLM(List, Geomm).

/* Information attributes: To convert from LIST of n pairs of (U, DU) to corresponding MEASURE */

infoLM(List, Meas):– infoLM1(List, 0, Meas).

infoLM1([], Oldm, Newm):– Newm is Oldm.
infoLM1([U, DU|L2], Oldm, Newm):–
  Ratio is U/DU,
  lb(Ratio, Mesom),
  Sum is Oldm + Mesom,
  infoLM1(L2, Sum, Newm).

Figure 10.3. Information component of the secondary level.

where Fi is the actual or intended value of the Ith FR for design X. In addition, we may use the data structure

$$cCoef(X, [F, J])$$

to mean that the $\langle I, J \rangle$th coefficient of the design matrix for design X is nonzero.

10.6 Illustrative Application (Kim, 1985)

Suppose that a polymer slab is to be fabricated whose critical parameters are its density ρ and thickness t. Then the FRs might be specified as $FR_1 = 1,600 \pm 200$ kg/m^3 and $FR_2 = 30 \pm 3$ mm.

```
/* Casebase */

/** Problem Specification **/

    fr(slab, [1, 1600, 200]).
    fr(slab, [2, 30, 3]).

/** Candidate Designs **/

    /* Design d1 */

    dFR(d1, [1, 1700]).
    dFR(d1, [2, 31]).
    cCoef(d1, [2, 1]).

    surfd(d1, [80, 10]).
    geomd(d1, [160, 5, 640, 10, 320, 5]).

    /* Design d2 */

    dFR(d2, [1, 1500]).
    dFR(d2, [2, 33]).
    cCoef(d2, [1, 2]).

    surfd(d2, [40, 10]).
    geomd(d2, [80, 5, 160, 10, 160, 5]).

    /* Design d3 */

    dFR(d3, [1, 1550]).
    dFR(d3, [2, 28]).
    cCoef(d3, [1, 2]).

    surfd(d3, [40, 10]).
    geomd(d3, [320, 10, 160, 10, 160, 5]).
```

Figure 10.4. Casebase for the polymer slab problem.

These specifications are encoded as the first component of the casebase shown in Fig. 10.4. For example,

$$\text{fr(slab, [1, 1600, 200])}$$

indicates that the first FR for the polymer slab is $1{,}600 \pm 200$ (in units of kg/m^3).

Consider the following three candidate designs for fabricating the polymer slab, as portrayed in Fig. 10.5.

(a) Design d1: Extrusion

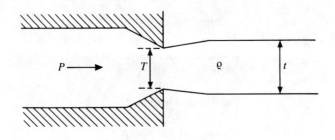

(b) Design d2: Injection Molding

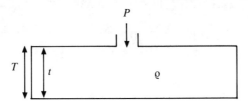

(c) Design d3: Power Liquefaction

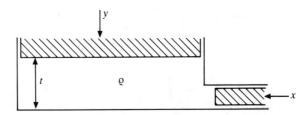

Figure 10.5. Alternative designs for fabricating a polymer slab of density $\rho \pm \Delta\rho$ and thickness $t \pm \Delta t$.

1. Design d1: extrusion

The polymer slab is extruded through a rectangular aperture, moving left to right in the diagram shown in Fig. 10.5a. Let p be the upstream pressure and T the height of the orifice. The DPs may then be defined as $DP_1 ::= P$ and $DP_2 ::= T$. This design configuration is assumed to produce a polymer slab of density $FR_1 ::= \rho = 1{,}700 \text{ kg/m}^3$ and thickness $FR_2 ::= t = 31$ mm. The density ρ of the polymer slab is to be controlled by the pressure P, and the thickness t by the aperture T. Since the slab will expand slightly after passing through the orifice, the actual thickness will be a shade greater than the gap size T. Moreover, a change in P or in T will affect both ρ and t. Since the coupling coefficients (i.e., elements of the design matrix) are given by

$$c_{ij} = \partial FR_i / \partial DP_j$$

we conclude that $c_{12} \neq 0$ and $c_{21} \neq 0$.

2. Design d2: Injection Molding.

The polymer is injected into a mold through a small aperture at the top. Let $DP_1 ::= P$, the pressure at the entrance to the aperture, and $DP_2 ::= T$, the height of the mold. The actual operating points for this design are $F_1 = 1,500 \text{ kg/m}^3$ and $F_2 = 33 \text{ mm}$. The pressure P affects the density ρ, but not the thickness t. In contrast, a change in T affects both t and ρ. We conclude that $c_{21} = 0$ and $c_{12} \neq 0$.

3. Design d3: Powder Liquefaction.

The polymer in powder form is poured into a special mold fitted with moving rods in both the vertical and horizontal axes. The apparatus yields a slab of density $F_1 = 1,550 \text{ kg/m}^3$ and thickness $F_2 = 28 \text{ mm}$. Let $DP_2 ::= y$ be the height of the vertical ram used to control density ρ and $DP_2 ::= x$ the position of the horizontal ram used to control density ρ. Then y affects both ρ and t, whereas x affects only t. Hence, $c_{21} = 0$ but $c_{12} \neq 0$.

The data pertaining to all three designs are given in Fig. 10.4. For example, cCoef(d1, [1, 2]) implies that c_{12} for design d1 does not vanish. Similarly, dFR(d1, [2, 31]) stipulates that the second FR yielded by design d1 is 31 (in units of mm). The data structure relating to surface, surfd(d1, [80, 10]), implies that the surface tolerance information for design d1 is $80 \pm 10 \ \mu m$.

In a fully fledged axiomatic system, the relationships among the various FRs would be evaluated autonomously by the software, or in collaboration with a user, through reasoning such as that described above. The system would then assert facts into the data base for subsequent use by the program.

10.7 Operating Modes of CAS

The knowledge in the preceding subsection forms the basis for a CAS. CAS may be used to address different types of queries. The three modes of operation are as follows:

1. *Basic.* Example queries are, "Is design X acceptable?," or, "Is design X superior to Y?"
2. *Information Retrieval.* An example is, "What is the information content of design X?"
3. *Enumeration.* Examples are, "Give me the list of all feasible designs," or, "Which designs are superior to which others?"

|?-sup1(d1, d2).
no
|?-sup1(d2, d1).
yes
|?-sup1(d1, d3).
no
|?-sup2(d1, d2).
no
|?-sup2(d2, d3).
yes

Figure 10.6. Basic mode of operation.

These categories are clearly related; for example, the ability to respond to questions of the first type imply the capacity to answer some of those in the third category. To render the discussion more concrete, these modes of operation are discussed in the context of the preceding example.

The basic operating mode is illustrated in Fig. 10.6. In this figure, input from the user is marked by underlining. The first query in the interaction corresponds to the question, "Is design d1 superior to design d2 according to the first axiom?" (The sup1 clause is shown in Fig. 10.1.) The system replies in the negative.

The second mode of operation relates to the acquisition of specialized information. Some examples of this type of usage are given in Fig. 10.7.

The first query corresponds to the question, "What is the independence measure of design d1?" PROLOG replies that this value is -1 (corresponding to a coupled design). The second query and its response indicate that design d2 has the Idm value of zero. The last queries and their responses show that the information contents for designs d2 and d3 are 15 and 16 bits, respectively.

The third mode of operation relates to an enumeration of objects with a

|?-idf(d1, Idm).
Idm = -1
yes
|?-idf(d2, Idm).
Idm = 0
yes
|?-iff(d2, Ifm).
Ifm = 15
yes
|?-iff(d3, Ifm).
Ifm = 16
yes

Figure 10.7. Information retrieval model

particular set of characteristics. Such enumeration may result in a partial or complete list of the relevant set.

This mode is facilitated by the *backtracking* characteristic of PROLOG. When PROLOG fails to find a solution to a query on the first pass through the data base, it retraces its steps and attempts to use alternate pathways. Even when a solution is found, PROLOG saves its pointers to the data base so that alternate solutions may be obtained if desired. Such a request is communicated to PROLOG by typing a semicolon (;). Each time PROLOG sees a semicolon in response to a generated solution, it goes back over the data base and attempts to find a new solution.

Figure 10.8 illustrates this usage. The first query, "sup1(X, Y)," corresponds to the question, "what design is superior to which other according to Axiom1?" PROLOG first replies that design d2 is superior to design d1.

The subsequent user input consisting solely of a semicolon is interpreted as a request for an alternative solution. PROLOG's response is that d3 is also superior to d1. When the semicolon is presented a second time, PROLOG is unable to find any more solutions, and therefore responds with a "no."

The query "sup2(X, Y)" is equivalent to asking "What design X is superior to which design Y according to Axiom 2?" PROLOG replies that design d2 is superior to d3. In response to the subsequent semicolon prompt, PROLOG attempts to find another such pair. However, there is no other such pair, and PROLOG replies in the negative. (A more direct way to obtain the answers in a single request is to use the bagof or setof facility. These are predicates which are usually built into the specific PROLOG implementation. However, this method may be inappropriate under certain circumstances; for example, if the expected number of solutions is large.)

|?-sup1(X, Y).
X = d2
Y = d1 :
X = d3,
y = d1 :

no
|?-sup2(X, Y).

X = d2
Y = d3 :

no

Figure 10.8. Enumeration mode.

10.8 Flexibility of Design

The modular structure of CAS allows for system expansion by simply scaling up in obvious ways. This modularity springs from two sources (Kim, 1985).

1. *Layered System Design.* The system is modular by construction. The inference engine, for example, is separate from the casebase.
2. *Hierarchical interpretation of PROLOG.* Clauses in PROLOG encourage modularity through hierarchical decomposition. Each literal in the body of the clause may be expanded into a set of literals which are evaluated in turn through a series of nested invocations.

The modularity engendered by the second characteristic may be illustrated by referring to the PROLOG statement of Axiom 2. Suppose that CAS is to be enhanced by including a new type of information, called process information. This may be achieved by adding two new predicates, called procd and procf.

The predicate procd(X, L) specifies the list L of process data pertaining to design X in the casebase. On the other hand, the function procf(X, Procm) operates on procd to yield the corresponding information measure Procm. The iff clause is modified by simply incorporating this new predicate (including the Procm variable) in the body. In other words, the new clause is given by

    iff(X, Ifm):—
        surff(X, Surfm),
        geomf(X, Geomm),
        procf(X, Procm),
        Ifm is (Surfm + Geomm + Procm)

The general structures are sufficiently flexible to accommodate extensions like these.

10.9 Architecture for a Fully Fledged Expert System

The ability to state the design axioms as clauses in a logical programming language indicates the feasibility of developing an expert system for axiomatics. This section discusses some issues relating to system capability and architecture (Kim, 1985).

System Capabilities

Expert systems may be distinguished, among other parameters, by the degree of autonomy and of capability. Autonomy relates to the amount of interaction required with a human user, whereas capability pertains to the range of system behaviors. Obviously, these two characteristics are interrelated. On the one hand, an expert system, when invoked by a user, may run to completion by itself; on the other hand, the system might be driven entirely by a series of requests from a human user.

The autonomous approach in the first type of system represents the ultimate goal of the axiomatic approach to design. However, for problems of reasonable complexity, such an expert system may be impractical until the development of subsystems that are capable of encoding knowledge in various realms of engineering. This may also have to await the development of programs that are fully capable of self-instruction.

System Architecture

A potential system architecture that is designed to support an expert system for axiomatics is shown in Fig. 10.9. The heart of the system is, of course, the axiomatics expert consisting of the primary, secondary, and data levels. These components are augmented by subsystems for explaining system behavior and for handling errors.

The user may interact with the system through a friendly interface as well as the graphics capabilities of a computer-aided design (CAD) package. The core axiomatics expert and the CAD systems communicate

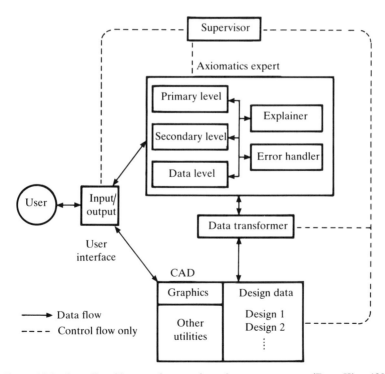

Figure 10.9. Overall architecture for an axiomatics expert system. (From Kim, 1985.)

through the data transformer, a package that converts the data for alternate designs from CAD-compatible to CAS-compatible knowledge.

10.10 Summary

In this chapter the design axioms and their corollaries are represented in terms of predicate logic. These are programmed in PROLOG to utilize computers in making design decisions. Finally, a simple software shell called the CAS is described, primarily to illustrate how an interactive program may be developed in the future. The operation of the CAS is described in the context of a simplified design example; moreover, the interactions are minimal in sophistication and in the sense that no language interpretation and generation capabilities are demonstrated. These are ancillary features which can be readily implemented by invoking PROLOG's language-generation facilities, and thereby enhance user-friendliness.

It is clear from these simple illustrations that artificial intelligence techniques can be merged with the design axioms and incorporated into a computer program. Such a program makes intelligent design decisions that embody basic principles derivable from the design axioms.

References

Battani, G., and Meloni, H., "Interpreteur du language de programmation PROLOG," Groupe de L'Intelligence Artificielle, U.E.R. de Luminy, Marseille, 1977.

Bittinger, M., *Logic, Proof and Sets*. Addison–Wesley, Reading, MA, 1982.

Hamilton, A.G., *Logic for Mathematicians*. Cambridge University Press, Cambridge, 1978.

Kim, S.H., "Mathematical Foundation of Manufacturing Science: Theory and Implications," Ph.D. Thesis, MIT, 1985.

Kim, S.H., and Suh, N.P., "Application of Symbolic Logic to the Design Axioms," *Robotics and Computer-Integrated Engineering* **2**(1):55–64, 1985.

Kim, S.H., and Suh, N.P., "Mathematical Foundations for Manufacturing," *Journal of Engineering for Industry, Transactions of A.S.M.E.* **109**:213–218, 1987.

Kneebone, G.T., *Mathematical Logic and the Foundations of Mathematics*. van Nostrand, NY, 1963.

Kowalski, R., *Logic for Problem Solving*. North-Holland, NY, 1979.

Pereira, F., ed., "C-Prolog User's Manual," Version 1.5, Department of Architecture, University of Edinburgh, February, 1984.

Problems

10.1. Assume that the following two logical statements are true.

$$\sim P \wedge Q$$
$$\sim Q \vee R$$

If P is also known to be true, can anything be said about the truth values of Q and R?

10.2. Let $P(x)$ be the predicate, "x is a person," and $M(y, x)$ the relationship, "y is the mother of x." Write the logical statement that states, "Every person has a mother." (*Hint*: If you are not careful, your logical sentence may depict the statement, "There is a person who is the mother of everyone.")

10.3. Justify Eq. 10.5 for Corollary 2 using the same type of detailed argument as shown for Corollary 1.

10.4. Consider the PROLOG statement for sup2(X, Y) in Fig. 10.1. Suppose that two designs d1 and d2 have the same values for functional independence and information. What will happen to the query sup2(d1, d2)? Modify the predicate so that both sup2(d1, d2) and sup2(d2, d1) would yield a positive response from PROLOG.

10.5. What are the relative advantages of a declarative language such as PROLOG and a procedural language such as C or FORTRAN? Under what circumstances would it be appropriate to combine both types of languages in a single software package?

10A: INTRODUCTION TO PROLOG

10A.1 Introduction

PROLOG is a programming language that is particularly suited for dealing with facts, rules and logic (see Rogers, 1986 for details). PROLOG can use rules to answer complex questions by combining facts. The knowledge is represented in descriptive rather than strictly procedural terms. PROLOG provides answers based on facts in its *data base*. In using PROLOG, we must be concerned with *syntax* (i.e., the structural organization of a sentence) as well as *semantics* (i.e., the meaning of a sentence).

The *data base* is created in PROLOG by following a set of *syntax* rules. The syntax rules for *facts* are as follows: the name that defines the relationship appears first; then the names of the objects are listed within parentheses, separated by a comma, finally, a period ends the sentence. The name of the relationship or attribute is also called a *predicate*; and its referent objects the *arguments*. For example, the PROLOG fact

$$\text{control(productivity, wage)}$$

may be used to encode the knowledge that "Productivity controls wages." The term "control" is the predicate, and "productivity" and "wage" are arguments. PROLOG facts may involve with any number of objects. To illustrate,

$$\text{student(joe)}$$

states that "Joe is a student." The statement

$$\text{design(engineer, car, people)}$$

may represent the knowledge "Engineers design cars for people."

Once the data base of facts is in place, questions can be asked about the data base. A query follows the same syntax, except that it must be specified in response to a prompt from PROLOG. This prompt is indicated by the compound symbol ?-. An example is found in the query

?- control(sun, weather)

If the data base contains the knowledge that the "Sun controls weather," PROLOG would respond

yes

If the query were

?- control(moon, weather)

then PROLOG would respond

no

if there were no fact stating that the "Moon controls weather." If the query were

?- control(weather, sun)

the PROLOG response would be

no

since the facts provided to PROLOG were in a different order. The last query would correspond to the question, Does the weather control the sum?"

PROLOG can also deal with variables. Each variable is indicated by the use of a capital letter at the beginning of its name. An example of a query is

?- control(sun, XYZ)

for which the response would be

XYZ = weather

If the query were

?- control(Variable, XYZ)

the response would be

Variable = sun,
XYZ = weather

After PROLOG provides an answer, keying in a "return" means the answer is accepted. If a semicolon (;) is typed first, followed by "return," then it means "OR" in PROLOG. In this case, the program searches its data base again to find another answer that matches the query, and provides the new answer. If there are no more matches, PROLOG responds "no." More-complex queries can be made using conjunctions (i.e., "AND"), denoted in PROLOG by a comma. If PROLOG has the following data base

$$\text{write(ink, paper)}$$
$$\text{write(pencil, paper)}$$

the following query

$$\text{?- write(ink, Medium), write(pencil, Medium)}$$

would generate the following PROLOG answer

$$\text{Medium = paper}$$

If PROLOG has exhausted all of these possibilities, or the query cannot be satisfied by the data base, PROLOG always responds "no."

Conditional knowledge in PROLOG is stated in the form of a rule. For example, the statement "Joe can fly the airplane if the weather on Sunday is fair" can be written in PROLOG as

$$\text{fly(joe, airplane):—weather(sunday, fair)}$$

The notation ":—" means "if" in this context. The fact (or predicate) written on the right-hand side of ":—" is called either the *body* or *antecedent*. The fact on the left-hand side of ":—" is the *conclusion*. When the conclusion is dependent on two or more requirements, the conditional facts are separated by commas. For example,

$$\text{pass(Student, exams):—like(professor, Student),}$$
$$\text{do(Student, homework),}$$
$$\text{get(Student, a)}$$

The above statement says that "A student passes an examination if the professor likes the student, if student does the homework, and if student gets the grade of A." If the data base contains the following

```
like(professor, joe).
do(joe, homework).
get(joe, a).
like(professor, jane).
do(jack, homework).
grade(jack, a).
athelete(kim, football).
```

PROLOG would respond to the query

$$\text{?- pass(Student, exams)}$$

as

$$\text{Student = joe;}$$

The semicolon, typed by the user, denotes "or." It is equivalent to asking "Is there anyone else?" The response would be

$$\text{no}$$

Suppose we can add the following rule to the above set of the data base:

$$\text{pass(Student, exams):— athelete(Student, football)}$$

Then, to the same query

>   ?- pass(Student, exams).

PROLOG would respond as

>   Student = joe;
>   Student = kim;
>   no

In the data base there can be rules that are based on the other rules. For example, we can assert the following rule and data to the data base given above:

>   athlete (Student, football):— height(Student, tall),
>   run(Student, marathon).
>   run(kim, marathon).
>   height(joe, tall).
>   run(jim, marathon).
>   height(jim, tall).

Now to the query

>   ?- pass(Student, exams).

PROLOG would respond

>   Student = joe;
>   Student = kim;
>   Student = jim;
>   no

Sometimes we may have rules that call themselves in the body. An example of such a *recursive* rule is defined collectively by the following pair of elementary rules:

>   influence(Master, Student):— teach(Master, Student).
>   influence(Master, Student):— teach(Master, Expert),
>   influence(Expert, Student).

The first rule says that a master influences a student if he teaches the latter. The second rule states that a master influences a student if the former teaches some expert who in turn influences the student. Note that the second rule contains the influence predicate in both its head and body.

Suppose the data base contains the additional facts

>   teach(shaw, cook).
>   teach(cook, nayak).
>   teach(nayak, smith).

To the query

>   ?- influence(Older, Younger)

the PROLOG interaction would take a form such as the following.

Older = shaw   Younger = cook;
Older = shaw   Younger = nayak;
Older = shaw   Younger = smith;
Older = cook   Younger = nayak;
Older = cook   Younger = smith;
Older = nayak Younger = smith;
no

Reference

Rogers, J.B., *A Prolog Primer*. Addison-Wesley, Reading, MA, 1986.

10B: CONCEPT FOR DESIGN OF THINKING DESIGN MACHINE

10B.1 Introduction

An ultimate goal of design automation is to develop a "thinking design machine" which can create designs or design concepts that are superior to those currently possible without the aid of such a machine. This is a difficult task since the creative processes involved in design are not yet fully understood, although it is known that "analogy," "recombination of known facts," "extrapolation," "interpolation," etc. have been identified as some of the processes used by creative people (Chapters 1 and 2). Although many computer aided design (CAD) software programs have been developed to date, they are *design aids* which deal with geometric modelling and representation rather than *creative design*. In this appendix a concept for a "thinking design machine," based on the design axioms, is presented.

A Thinking Design Machine (TDM) may be defined as an intelligent machine system that can generate creative designs. Such a machine should be able to help designers develop design concepts, particularly during the early stages of the design process. However, the TDM *concept* described in this appendix should be applicable at all stages of the design process. The concept for TDM is based on the Independence Axiom and the Information Axiom. Based on the Independence Axiom, TDM will combine a number of components (or ideas) stored in the data base to satisfy a set of FRs specified for a given design. Then, based on the Information Axiom, the TDM will select the best combinations of these components to create a design.

10B.2 Concept for Thinking Design Machine

The thinking design machine would use the following design axioms:

- Axiom 1 *The Independence Axiom*
  Maintain the independence of functional requirements.
- Axiom 2 *The Information Axiom*
  Minimize the information content.

Axiom 1 provides the criterion for good and bad design, or acceptable and unacceptable design. Axiom 2 is the criterion for selection of the best design solutions from among many acceptable proposed solutions that satisfy Axiom 1.

It was stated in Chapter 2 that the design process consists of the following four steps:

- Step 1 Definition of FRs.
- Step 2 Ideation or creation of ideas.
- Step 3 Analysis of the proposed solution.
- Step 4 Checking of the fidelity of the final solution to the original needs.

In the first step, the functional requirements (FRs) are defined to satisfy the original perceived needs. The appropriate set of DPs to satisfy the FRs must then be defined via a physical entity (i.e., product, process, software, system, or organization). The proposed solution is analyzed using the design axioms and other applicable physical laws and principles to ascertain if the solutions proposed are rational. Finally, fidelity of the final solution to the original perceived needs must also be ascertained.

Each of the steps often involves iteration, which may involve redefinition of the FRs, creation of new ideas, and modification of the proposed solutions. Ultimately, TDM should be able to execute the four steps of the design process.

10B.3 Architecture of Thinking Design Machine

The architecture of the TDM must involve at least the first three steps just discussed. However, the first step (FR definition) is highly subjective. Therefore, in this appendix only the second and third steps will be addressed in order to make the computerization of the design process possible on an objective basis.

The computer architecture consists of three software modules: the Ideation Software, the Analysis Software, and the Systems Software. The Ideation Software handles Step 2, and the Analysis Software handles Step

3. The Systems Software translates the selected physical components into a systems context to produce an overall design that satisfies the FRs defined for the design task.

10B.3.1 *The Ideation Software Concept*

The Ideation Software selects a set of plausible DPs that correspond to the set of FRs chosen as the design task. The ideation software may contain a library of data base in the form of morphological relationships between various FRs and DPs, from which appropriate DPs to satisfy a given set of FRs can be selected. Consider, for example, the design task of developing a design solution for FRs. Then, for each FR we can select a number of DPs from the data base as:

For FR_1, the corresponding plausible DPs are:
$DP_1^1, DP_1^2, DP_1^3, \ldots$

For FR_2, the corresponding plausible DPs are:
$DP_2^1, DP_2^2, DP_2^3, \ldots$

$\vdots$

For FR_n, the corresponding plausible DPs are:
$DP_n^1, DP_n^2, DP_n^3, \ldots$

If, for example, the functional requirement (FR_1) is to transfer the rotational motion, DPs may be a gear (DP_1^1), a timing belt (DP_1^2), and a chain (DP_1^3), etc. At this stage, as many DPs as possible should be considered, provided that they are within the constraints imposed upon the design. However, only a limited subset of DPs are acceptable, because the Independence Axiom does not allow functional coupling.

The Ideation Software program constructs as many plausible design matrices as possible. The relationship between FRs and DPs can be written as:

$$\{FRs\} = [DM]\{DP\} \qquad (10B.1)$$

where each element of the design matrix is given by:

$$A_{ij} = \frac{\partial FR_i}{\partial DP_j} \qquad (10B.2)$$

The element A_{ij} represents the relation between each FR_i and DP_j. If FR_i is affected by DP_j, then A_{ij} has a number larger than zero. If the FR_i is not affected by the DP_j, then A_{ij} is zero. Since there are many possible DPs for each FR, there may be many plausible design equations. However, some of these solutions must be rejected if they violate the Independence Axiom. This is done using the analysis software.

10B.3.2 *The Analysis Software Concept*

The Analysis Software analyzes the design matrix above as follows:

(i) *Uncoupled Design*
If DP is a diagonal matrix, the proposed design solution satisfies the Independence Axiom. Since each DP_i satisfies its corresponding FR_i without affecting any other FR, the independence of the functional requirements is maintained. Such an uncoupled design is the ultimate solution sought by the Analysis Software. This is accomplished by seeking DPs that yield a diagonal matrix.

(ii) *Coupled Design*
The converse of an uncoupled design is a coupled design. The design matrix of a coupled design consists of mostly non-zero elements. In this case, we cannot control each FR_i independently without changing its corresponding DP_i. Since such a change would affect other FRs, the design solution clearly violates Axiom 1. The Analysis Software discards this coupled design and begins the search process again.

(iii) *Decoupled Design*
Another acceptable solution is a decoupled design. A decoupled solution can be developed by proceeding according to the following scheme: First, a DP_1 that satisfies FR_1 must be found. Then, an FR_2 is selected and a DP_2 that satisfies FR_2 but does not affect FR_1 must be found. For FR_3, a DP_3 that affects FR_3 but not FR_1 and FR_2 should be selected. This process can continue until all the FRs are satisfied.

When it is not possible to select FRs and DPs in the specific sequence required to generate a decoupled design from the beginning, the DM must be manipulated to determine whether the DM matrix can be made triangular. The Analysis Software has the function of changing the order of the FRs and DPs so as to make the matrix triangular, if possible. The Analysis Software also has the function of detecting flaws in the proposed design, and providing suggestions for improving the design by making the DM triangular.

10B.3.3 *The System Software Concept*

The last step in the design process is the System Software. The System Software arranges the physical components chosen by the Ideation Software and the Analysis Software for each DP. The physical parts are arranged not necessarily according to the sequence of the DP control; the arrangement may be based on the sequence of flow of materials in a system, which may or may not be the same as the control sequence.

Figure 10B.1 is the schematic diagram of a Thinking Design Machine showing the decision-making steps. There can be a large number of design solutions generated by the three software types discussed. Those design solutions which satisfy Axiom 1 will then be evaluated using the Information Axiom.

Mathematical Representation of the Design Axioms

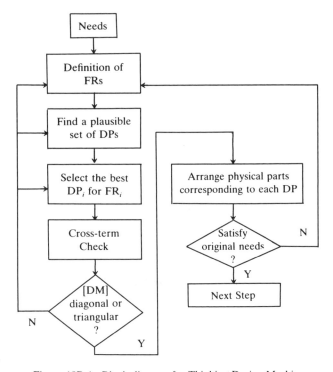

Figure 10B.1. Block diagram for Thinking Design Machine.

10B.3.4 *Hierarchical Tree in Design Process*

The design world for a product consists of an FR hierarchy in the functional space and a DP hierarchy in the physical space. The design process requires the capability of cross-over between the two hierarchical trees at a given level of the FR and DP hierarchy (Fig. 10B.2). In the design of complicated systems, the hierarchical approach simplifies the design process a great deal. The TDM can move down the hierarchical tree until the design is completed.

10B.4 Matrix Operation

10B.4.1 *Algorithm for Changing the Order of $\{FR\}$ and $\{DP\}$*

The Analysis Software has the function of changing the order of FRs and DPs to determine if the matrix can be made triangular. The algorithm is as follows:

 (i) Find the row which contains one non-zero element. Rearrange the order of {FRs} and {DPs} by putting the row and the column

Design Hierarchy

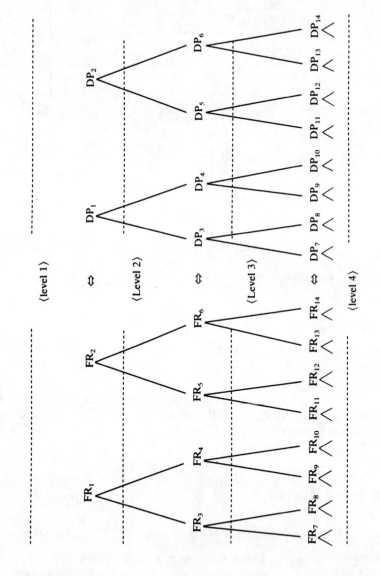

Figure 10B.2

(o) The original matrix

$$\begin{Bmatrix} FR_1 \\ FR_2 \\ FR_3 \\ FR_4 \end{Bmatrix} = \begin{bmatrix} A_{11} & A_{12} & 0 & A_{14} \\ 0 & A_{22} & A_{23} & 0 \\ 0 & 0 & A_{33} & 0 \\ 0 & A_{42} & A_{43} & A_{44} \end{bmatrix} \begin{Bmatrix} DP_1 \\ DP_2 \\ DP_3 \\ DP_4 \end{Bmatrix}$$

(i) Find a row which has one non-zero element, and put the row and the column first.

$$\begin{Bmatrix} FR_3 \\ FR_1 \\ FR_2 \\ FR_4 \end{Bmatrix} = \begin{bmatrix} A_{33} & 0 & 0 & 0 \\ 0 & A_{11} & A_{12} & A_{14} \\ A_{23} & 0 & A_{22} & 0 \\ A_{43} & 0 & A_{42} & A_{44} \end{bmatrix} \begin{Bmatrix} DP_3 \\ DP_1 \\ DP_2 \\ DP_4 \end{Bmatrix}$$

(ii) Excluding the first row and column, find a row which has one non-zero element, and put the row and the column second.

$$\begin{Bmatrix} FR_3 \\ FR_2 \\ FR_1 \\ FR_4 \end{Bmatrix} = \begin{bmatrix} A_{33} & 0 & 0 & 0 \\ A_{23} & A_{22} & 0 & 0 \\ 0 & A_{12} & A_{11} & A_{14} \\ A_{43} & A_{42} & 0 & A_{44} \end{bmatrix} \begin{Bmatrix} DP_3 \\ DP_2 \\ DP_1 \\ DP_4 \end{Bmatrix}$$

(iii) Excluding the first and second rows and columns, find a row which has one non-zero element, and put the row and the column third.

$$\begin{Bmatrix} FR_3 \\ FR_2 \\ FR_4 \\ FR_1 \end{Bmatrix} = \begin{bmatrix} A_{33} & 0 & 0 & 0 \\ A_{23} & A_{22} & 0 & 0 \\ A_{43} & A_{42} & A_{44} & 0 \\ 0 & A_{12} & A_{14} & A_{11} \end{bmatrix} \begin{Bmatrix} DP_3 \\ DP_2 \\ DP_4 \\ DP_1 \end{Bmatrix}$$

Figure 10B.3. Reordering of {FR} and {DP}

which contain the non-zero element first (i.e., if ith row contains one non-zero element at jth column, then put ith component of {FR} first and put jth component of {DP} first (Fig. 10B.3(i))).

(ii) Excluding the first row and column, find the row which contain one non-zero element. Rearrange the components of {FRs} and {DPs} by putting the row and the column which contains the non-zero element at the row and column second as shown in Fig. 10B.3(ii).

(iii) Repeat the procedure until there are no more sub-matrices to analyze (Fig. 10B.3(iii)).

At any time during the execution of the above procedure, if all of the remaining rows contain more than one non-zero element, the design is coupled. The FRs corresponding to the rows that contain no non-zero elements cannot be controlled independently, violating Axiom 1.

10B.4.2 *Improvement of the Design Solution*

Some design solutions may not be decoupled by simply reordering the FRs and DPs. In these cases the Analysis Software should provide information for improving the design solution. When the proposed design solution is coupled, the program will store the non-zero elements of the matrix and examine which element or combination of elements can be set to zero to decouple the matrix.

Figure 10B.4 illustrates how a given matrix can be decoupled by setting certain elements of the [DM] matrix equal to zero. These solutions correspond to the case when the diagonal elements cannot be made zero, because, in most cases, the diagonal elements are much more important than the off-diagonal elements. If the diagonal elements are allowed to be zero, six more cases may be obtained.

Based on these solutions for triangular design matrices, it should be

*** Original Design Matrix ***

$$\begin{matrix} FR_1 \\ FR_2 = \\ FR_3 \end{matrix} \begin{bmatrix} A_{11} & A_{12} & A_{13} \\ A_{21} & A_{22} & A_{23} \\ A_{31} & A_{32} & A_{33} \end{bmatrix} \begin{matrix} DP_1 \\ DP_2 \\ DP_3 \end{matrix}$$

*** Possible solutions ***
Permit to make the diagonal elements zero (y/n)? <u>n</u>

1) $A_{21} \Rightarrow 0$

$$\begin{matrix} FR_2 \\ FR_3 = \\ FR_1 \end{matrix} \begin{bmatrix} A_{22} & A_{23} & A_{21} \\ A_{32} & A_{33} & A_{31} \\ A_{12} & A_{13} & A_{11} \end{bmatrix} \begin{matrix} DP_2 \\ DP_3 \\ DP_1 \end{matrix}$$

2) $A_{12}, A_{32} \Rightarrow 0$

$$\begin{matrix} FR_3 \\ FR_1 = \\ FR_2 \end{matrix} \begin{bmatrix} A_{33} & A_{31} & A_{32} \\ A_{13} & A_{11} & A_{12} \\ A_{23} & A_{21} & A_{22} \end{bmatrix} \begin{matrix} DP_3 \\ DP_1 \\ DP_2 \end{matrix}$$

3) $A_{12}, A_{13} \Rightarrow 0$

$$\begin{matrix} FR_1 \\ FR_2 = \\ FR_3 \end{matrix} \begin{bmatrix} A_{11} & A_{12} & A_{13} \\ A_{21} & A_{22} & A_{23} \\ A_{31} & A_{32} & A_{33} \end{bmatrix} \begin{matrix} DP_1 \\ DP_2 \\ DP_3 \end{matrix}$$

Note: shaded elements are non-zero elements.

Figure 10B.4

possible to examine the physical significance of various DPs and the ideas proposed so as to generate improved designs that satisfy the Independence Axiom.

10B.5 Application of Thinking Design Machine to RIM System

RIM (Reaction Injection Molding) machine is a machine for producing polyurethane parts. Two resin components are mixed and a reaction is caused prior to injecting the mixture into a mold. The problem in the original design (Fig. 10B.5) is that there are fewer design parameters than functional requirements, and therefore, in violation of Axiom 1, the Independence Axiom. The FRs and DPs of the design shown are:

⟨functional requirements⟩

FR_1 = Deliver liquid at high flow rate (Q)

FR_2 = Deliver an adequately mixed liquid (X)

FR_3 = Deliver properly metered liquid (M)

⟨design parameters⟩

DP_1 = Pump speed (W)

DP_2 = Nozzle size (D)

Then, the design equation is expressed as:

$$\begin{Bmatrix} FR_1 \\ FR_2 \\ FR_3 \end{Bmatrix} = \begin{bmatrix} \times & 0 \\ \times & \times \\ \times & \times \end{bmatrix} \begin{Bmatrix} DP_1 \\ DP_2 \\ DP_3 \end{Bmatrix} \qquad (10B.3)$$

where × represents a non-zero element, and 0 represents a zero element.

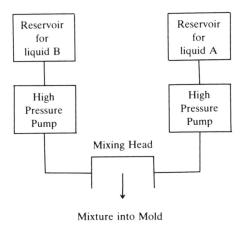

Figure 10B.5. Schematic diagram of a RIM machine

This matrix indicates that FR_1 is affected by DP_1, and the other two FRs are affected by both DP_1 and DP_2.

This design has a coupled design because there are more FRs than DPs. Therefore, FRs cannot be controlled independently. In this case a new DP to decouple the design must be found. The search for a new design parameter can also result in ideas regarding the new DP. The design matrix is expressed as:

$$\begin{Bmatrix} FR_1 \\ FR_2 \\ FR_3 \end{Bmatrix} = \begin{bmatrix} \times & 0 & ? \\ \times & \times & ? \\ \times & \times & ? \end{bmatrix} \begin{Bmatrix} DP_1 \\ DP_2 \\ DP_3 \end{Bmatrix} \qquad (10B.4)$$

Analyzing this design equation, 12 solutions are possible, including the solutions that make the old elements (A_{11}, A_{21}, A_{22}, A_{32}) zero (Fig. 10B.6). Among these solutions, the possibility of changing the old DPs, i.e., pump speed and nozzle size, is indicated. However, in this case, it is difficult to select new DPs which make the old elements zero. Therefore, it is necessary to concentrate on the solutions that make the elements of the third column of the matrix zero. The possible solutions are (5) and (7), shown in Fig. 10B.6.

The solution given by (5) of Fig. 10B.6 indicates the need for a new DP which controls only FR_3 (Deliver properly metered liquid) without affecting FR_1 (Deliver liquid at high flow rate) and FR_2 (deliver an adequately mixed liquid). The solution (7) indicates the need for a new DP which controls FR_2 (Deliver an adequately mixed liquid) without affecting FR_1 (Deliver liquid at high flow rate) and FR_3 (Deliver properly metered liquid).

[Possible Solution (1)]
** $A_{11} \to 0, A_{21} \to 0$

$$\begin{matrix} FR_1 \\ FR_2 = \\ FR_3 \end{matrix} \begin{bmatrix} A_{13} & A_{12} & A_{11} \\ A_{23} & A_{22} & A_{21} \\ A_{33} & A_{32} & A_{31} \end{bmatrix} \begin{matrix} DP_3 \\ DP_2 \\ DP_1 \end{matrix}$$

[Possible Solution (2)]
** $A_{11} \to 0, A_{22} \to 0$

$$\begin{matrix} FR_1 \\ FR_2 = \\ FR_3 \end{matrix} \begin{bmatrix} A_{13} & A_{11} & A_{12} \\ A_{23} & A_{21} & A_{22} \\ A_{33} & A_{31} & A_{32} \end{bmatrix} \begin{matrix} DP_3 \\ DP_1 \\ DP_2 \end{matrix}$$

[Possible Solution (3)]
** $A_{11} \to 0, A_{31} \to 0$

$$\begin{matrix} FR_1 \\ FR_3 = \\ FR_2 \end{matrix} \begin{bmatrix} A_{13} & A_{12} & A_{11} \\ A_{33} & A_{32} & A_{31} \\ A_{23} & A_{22} & A_{21} \end{bmatrix} \begin{matrix} DP_3 \\ DP_2 \\ DP_1 \end{matrix}$$

[Possible Solution (4)]
** $A_{11} \to 0, A_{32} \to 0$

$$\begin{matrix} FR_1 \\ FR_3 = \\ FR_2 \end{matrix} \begin{bmatrix} A_{13} & A_{11} & A_{12} \\ A_{33} & A_{31} & A_{32} \\ A_{23} & A_{21} & A_{22} \end{bmatrix} \begin{matrix} DP_3 \\ DP_1 \\ DP_2 \end{matrix}$$

[Possible Solution (5)]
** $A_{13} \to 0, A_{23} \to 0$

$$\begin{matrix} FR_1 \\ FR_2 = \\ FR_3 \end{matrix} \begin{bmatrix} A_{11} & A_{12} & A_{13} \\ A_{21} & A_{22} & A_{23} \\ A_{31} & A_{32} & A_{33} \end{bmatrix} \begin{matrix} DP_1 \\ DP_2 \\ DP_3 \end{matrix}$$

[Possible Solution (6)]
** $A_{13} \to 0, A_{22} \to 0$

$$\begin{matrix} FR_1 \\ FR_2 = \\ FR_3 \end{matrix} \begin{bmatrix} A_{11} & A_{13} & A_{12} \\ A_{21} & A_{23} & A_{22} \\ A_{31} & A_{33} & A_{32} \end{bmatrix} \begin{matrix} DP_1 \\ DP_3 \\ DP_2 \end{matrix}$$

[Possible Solution (7)]
** $A_{13} \to 0, A_{33} \to 0$

$$\begin{matrix} FR_1 \\ FR_3 = \\ FR_2 \end{matrix} \begin{bmatrix} A_{11} & A_{12} & A_{13} \\ A_{31} & A_{32} & A_{33} \\ A_{21} & A_{22} & A_{23} \end{bmatrix} \begin{matrix} DP_1 \\ DP_2 \\ DP_3 \end{matrix}$$

[Possible Solution (8)]
** $A_{13} \to 0, A_{32} \to 0$

$$\begin{matrix} FR_1 \\ FR_3 = \\ FR_2 \end{matrix} \begin{bmatrix} A_{11} & A_{13} & A_{12} \\ A_{31} & A_{33} & A_{32} \\ A_{21} & A_{23} & A_{22} \end{bmatrix} \begin{matrix} DP_1 \\ DP_3 \\ DP_2 \end{matrix}$$

[Possible Solution (9)]
** $A_{21} \to 0, A_{22} \to 0$

$$\begin{matrix} FR_2 \\ FR_1 = \\ FR_3 \end{matrix} \begin{bmatrix} A_{23} & A_{21} & A_{22} \\ A_{13} & A_{11} & A_{12} \\ A_{33} & A_{31} & A_{32} \end{bmatrix} \begin{matrix} DP_3 \\ DP_1 \\ DP_2 \end{matrix}$$

[Possible Solution (10)]
** $A_{23} \to 0, A_{22} \to 0$

$$\begin{matrix} FR_2 \\ FR_1 = \\ FR_3 \end{matrix} \begin{bmatrix} A_{21} & A_{23} & A_{22} \\ A_{11} & A_{13} & A_{12} \\ A_{31} & A_{33} & A_{32} \end{bmatrix} \begin{matrix} DP_1 \\ DP_3 \\ DP_2 \end{matrix}$$

[Possible Solution (11)]
** $A_{31} \to 0, A_{32} \to 0$

$$\begin{matrix} FR_3 \\ FR_1 = \\ FR_2 \end{matrix} \begin{bmatrix} A_{33} & A_{31} & A_{32} \\ A_{13} & A_{11} & A_{12} \\ A_{23} & A_{21} & A_{22} \end{bmatrix} \begin{matrix} DP_3 \\ DP_1 \\ DP_2 \end{matrix}$$

[Possible Solution (12)]
** $A_{33} \to 0, A_{32} \to 0$

$$\begin{matrix} FR_3 \\ FR_1 = \\ FR_2 \end{matrix} \begin{bmatrix} A_{31} & A_{33} & A_{32} \\ A_{11} & A_{13} & A_{12} \\ A_{21} & A_{23} & A_{22} \end{bmatrix} \begin{matrix} DP_1 \\ DP_3 \\ DP_2 \end{matrix}$$

Note: shaded elements are non-zero elements.

Figure 10B.6

Mathematical Representation of the Design Axioms

⟨Physical Solution⟩

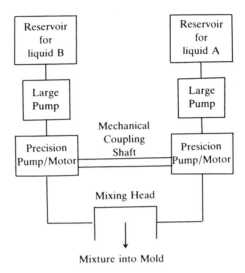

Figure 10B.7. Modified design of a RIM machine

Both of these suggested solutions are possible in terms of the matrix manipulation. However, when considering the solution in the physical domain, the solution (7) indicates that if DP_1 is intended to satisfy FR_1, then DP_2 is intended to satisfy FR_3, and finally DP_3 is intended to satisfy FR_2. However, in this sequence, it is difficult to assign DP_2 (nozzle size) to satisfy FR_3 (Deliver properly metered liquid) since a positive displacement pump is used. The final result is that it is necessary to assign a new DP which controls FR_3 (Deliver properly metered liquid) without affecting FR_1 (Deliver liquid at high flow rate) and FR_2 (Deliver properly metered liquid). Fig. 10B.7 shows one physical solution.

10B.6 Evaluation of Proposed Solutions in Terms of the Information Axiom

When a number of solutions that satisfy the Independence Axiom are presented, the information content of each must be evaluated to select the best solution among those proposed. This can be done by actually evaluating the system ranges of the proposed solutions and the common range, as per the procedure discussed in Chapters 5 and 8.

10B.7 Concluding Remarks

The ideas presented in this appendix are being pursued to develop a TDM at MIT. Much work remains to be done to make it a reality.

11
SUMMARY OF THE AXIOMATIC APPROACH

The world of design transcends the traditional academic disciplines, extending from all subdisciplines of engineering to business administration to political science. In fact, every human being designs something in his or her lifetime. A designer creates forms and/or physical entities to satisfy functions and FRs. The design process requires the integration of known facts and basic knowledge to create a solution that has not previously been in existence. This "creative process" has to be *checked* through the analysis of the fidelity of the "outputs" that are embodied in physical or organizational beings, to the "inputs" to the design process, which are in the form of FRs. Only when the proposed designs can be analyzed (i.e., checked) will the design process converge to a solution quickly. Unpromising ideas should be dispensed with early, and only promising ideas should be nurtured. Such analysis cannot be done in the absence of basic principles.

This book is a treatise on the design axioms and the axiomatic approach to design. The design axioms are extracted and abstracted from good design practice through generalization. In that sense, there should not be any surprises to those who have acquired good design practice through experience. However, in all fields of science, systematic and rational knowledge, based on logic and invariant principles, has provided humanity with a firm foundation, or platform for further intellectual advance as well as the development of practical methodologies for solving the immediate problem in hand. Expediency cannot justify the lack of knowledge, and urgency must not sidetrack an honest attempt to build the knowledge base. Rigorous discussions must prevail over quick judgements in efforts of this kind.

The design process involves four distinct aspects: (1) problem definition, which results in the definition of FRs and constraints; (2) the creative process of conceptualizing and devising a solution; (3) the analytical process of determining whether the proposed solution is a rational solution that is consistent with the problem definition; and (4) the process of checking the fidelity of the solution with the original perceived needs. These four aspects of the design process are equally important and critical. Depending on the specific situations involved, any one of these four elements may be more difficult to deal with in developing a design solution.

The basic premise of the axiomatic approach to design is that there are basic principles that govern decision making in design, just as the laws of nature govern the physics and chemistry of nature. Two design axioms, derived from the generalization of good design practice, are presented in this book. By being able to state these axioms explicitly, it is possible to derive corollaries and theorems, as well as to develop specific analytical and constructive methodologies. Just as the thermodynamic axioms have enhanced our understanding of the energy-conversion process, so the design axioms should aid the learning and comprehension of design and the design process.

Like most abstract and conceptual studies of this kind, it takes time and effort to appreciate fully the significance and implications of the design axioms. Examples and case studies are indispensible in this learning process, since they provide opportunities to examine the axioms from many different points of view. Qualitative understanding of the design axioms must precede the application of quantitative tools, but both are indispensible and mutually reinforcing.

There are only two design axioms. They are stated in their simplest form as:

Axiom 1 *The Independence Axiom*
Maintain the independence of FRs.

Axiom 2 *The Information Axiom*
Minimize the information content.

The meaning of the Independence Axiom is explored in Chapters 3 and 4, and through case studies in Chapters 6–9. The Information Axiom is discussed in Chapters 3 and 5; the case studies presented in Chapter 8 provide insight to the utility of the Information Axiom in selecting the optimum design solution from among many acceptable solutions. The technique is powerful, because it can deal realistically with multivariable design problems.

The corollaries are direct consequences of the axioms, and tend to have the flavor of design rules. They may be stated again in imperative form as:

Corollary 1 (Decoupling of Coupled Designs)
Decouple or separate parts or aspects of a solution if FRs are coupled or become interdependent in the designs proposed.

Corollary 2 (Minimization of FRs)
Minimize the number of FRs and constraints

Corollary 3 (Integration of Physical Parts)
Integrate design features in a single physical part if FRs can be independently satisfied in the proposed solution.

Corollary 4 (Use of Standardization)
Use standardized or interchangeable parts if the use of these parts is consistent with FRs and constraints.

Corollary 5 (Use of Symmetry)
Use symmetrical shapes and/or components if they are consistent with the FRs and constraints.

Corollary 6 (Largest Tolerance)
Specify the largest allowable tolerance in stating FRs.

Corollary 7 (Uncoupled Design with Less Information)
Seek an uncoupled design that requires less information than coupled designs in satisfying a set of FRs.

Corollary 8 (Effective Reangularity of a Scalar)
The effective reangularity R for a scalar coupling "matrix" or element is unity.

Good designers have used some of these corollaries intuitively. However, once the intuition is replaced with axioms or laws that state what we know intuitively with explicit statements, even inexperienced designers should quickly master the basic "tools of the trade."

Theorems are propositions that follow from axioms or other propostions that have been proven. There can be a large number of theorems, depending on the specific design problem under consideration. The theorems presented in this book are as follows.

Theorem 1 (Coupling Due to Insufficient Number of DPs)
When the number of DPs is less than the number of FRs, either a coupled design results, or the FRs cannot be satisfied.

Theorem 2 (Decoupling of Coupled Design)
When a design is coupled due to the greater number of FRs than DPs (i.e., $m > n$), it may be decoupled by the addition of new DPs so as to make the number of FRs and DPs equal to each other, if a subset of the design matrix containing $n \times n$ elements constitutes a triangular matrix.

Theorem 3 (Redundant Design)
When there are more DPs than FRs, the design is either a redundant design or a coupled design.

Theorem 4 (Ideal Design)
In an ideal design, the number of DPs is equal to the number of FRs.

Theorem 5 (Need for New Design)
When a given set of FRs is changed by the addition of a new FR, or substitution of one of the FRs with a new one, or by selection of a completely different set of FRs, the design solution given by the original DPs cannot satisfy the new set of FRs. Consequently, a new design solution must be sought.

Summary of the Axiomatic Approach   393

Theorem 6 (Path Independency of Uncoupled Design)
The information content of an uncoupled design is independent of the sequence by which the DPs are changed to satisfy the given set of FRs.

Theorem 7 (Path Dependency of Coupled and Decoupled Design)
The information contents of coupled and decoupled designs depend on the sequence by which the DPs are changed and on the specific paths of change of these DPs.

Theorem 8 (Independence and Tolerance)
A design is an uncoupled design when the designer-specified tolerance is greater than

$$\left(\sum_{\substack{j=1 \\ j \neq i}}^{n} (\partial FR_i / \partial DP_j) \Delta DP_j \right)$$

in which case the nondiagonal elements of the design matrix can be neglected from design consideration.

Theorem 9 (Design for Manufacturability)
For a product to be manufacturable, the design matrix for the product, [**A**] (which relates the **FR** vector for the product to the **DP** vector of the product) times the design matrix for the manufacturing process, [**B**] (which relates the **DP** vector to the **PV** vector of the manufacturing process) must yield either a diagonal or triangular matrix. Consequently, when any one of these design matrices; that is, either [**A**] or [**B**], represents a coupled design, the product cannot be manufactured.

Theorem 10 (Modularity of Independence Measures)
Suppose that a design matrix [**DM**] can be partitioned into square submatrices that are nonzero only along the main diagonal. Then the reangularity and semangularity for [**DM**] are equal to the products of their corresponding measures for each of the nonzero submatrices.

Theorem 11 (Invariance)
Reangularity and semangularity for a design matrix [**DM**] are invariant under alternative orderings of the FR and DP variables, as long as orderings preserve the association of each FR with its corresponding DP.

Theorem 12   (Sum of Information)
Theorem 12   The sum of information for a set of events is also information, provided that proper conditional probabilities are used when the events are not statistically independent.

Theorem 13   (Information Content of the Total System)
Theorem 13   If each DP is probablistically independent of other DPs, the information content of the total system is the sum of information of all individual events associated with the set of FRs that must be satisfied.

Theorem 14   (Information Content of Coupled versus Uncoupled Designs)
Theorem 14   When the state of FRs is changed from one state to another in the functional domain, the information required for the change is greater for a coupled process than for an uncoupled process.

Theorem 15   (Design–Manufacturing Interface)
Theorem 15   When the manufacturing system compromises the independence of the FRs of the product, either the design of the product must be modified, or a new manufacturing process must be designed and/or used to maintain the independence of the FRs of the products.

Theorem 16   (Equality of Information Content)
Theorem 16   All information contents that are relevant to the design task are equally important regardless of their physical origin, and no weighting factor should be applied to them.

A common design error that is made by designers is the creation of coupled designs. Many designers operate with a strong common intuition that physical integration of parts is a good policy, regardless of the consequent coupling of functions that may result in some cases. The qualitative use of axiom 1 can resolve many difficult design problems very easily. The Independence Axiom is a powerful concept for designers. In a complex design situation, it is desirable to measure the functional independence quantitatively to determine how far off the design is from the optimal point.

For this purpose, two quantitative measures, reangularity and semangularity, were presented. When the design is an uncoupled design, both reangularity and semangularity are equal to unity. As the degree of coupling increases, both reangularity and semangularity approach zero.

The concept of information content is closely related to the probability of fulfilling the FRs. Since the FRs specified by the designer must be satisfied by the manufacturing system, the overlap between the designer's specifications and the manufacturing system's capabilities is essential to satisfying the FRs. The information content is defined in terms of the logarithm of the ratio (system range/common range). This ratio is equal to the inverse of probability. This definition is consistent with the information measure used in information theory.

The concept of system range is important, in that it defines what the system is capable of delivering, versus what the designer must have to ensure that the entire design achieves the goals for the design task. That the system with the minimum information content is the best among the available solutions as verified from actual experiments is strong evidence in support of Axiom 2. It is a powerful tool.

Abstract concepts such as the design axioms can best be understood through examples and case studies. What is even better than reading someone else's case studies is actually to execute the entire design process using the axioms, corollaries and theorems. The problems given at the end of most chapters should be useful in that task. Some problems may take hours of work, and many of these problems may have more than one acceptable solution.

In the future, computers will improve the design process immensely. Computer-based technologies will continue to improve the physical aspects of the design process, such as those available in the form of CAD tools in the 1980s. They will also make information related to manufacturing, materials, assembly, etc., available to the designer at the most appropriate stage of design. They will also help at the conceptualization stage, as well as at the implementation stage of developing detailed design features.

Ultimately, we may be able to develop "thinking" computers, which may check the design decisions made by designers for their rationality and correctness, and which may offer alternative design solutions. Because of the extremely large data base and design possibilities, involving large numbers of variables, "thinking" design machines must be based on axioms, corollaries and theorems, some of which are presented in this book. To create software for such a "thinking" machine, the mathematical representation of these principles is essential. The use of predicate logic and languages such as PROLOG enables us to communicate with the machines and computers.

Much work must still be done to reach the goal of developing intelligent computers that can design even relatively simple things, However, we have started the journey and will eventually arrive at the destination. Even then, the ultimate designer who can create unique and beautiful things, even beyond the realm of imagination at this time, will embody the most remarkable product of nature's creation: human genius.

In all fields of human endeavor, questions lead to problem definitions, and problem definitions lead to solutions. Good solutions, in turn, should generate new questions that will raise us to a higher plateau. This book is but a chapter in this quest for knowledge—knowledge that will be both practical and enlightening.

APPENDIX

SI Units

Length	meter (m)
Mass	kilogram (kg)
Time	second (s)
Force	newton (N) = 1 kg m/s^2
Work	joule (J)
Power	watt (W)
Electrical current	ampere (A)
Electrical potential	volt (V) = 1 W/A
Electrical resistance	ohm (Ω) = 1 V/A
Pressure	pascal (Pa) = 1 N/m^2

Conversion Factors

English system			Metric System
Length	inch	25.4	millimeters (mm)
	foot	0.3048	meters (m)
	yard	0.9144	meters
	mile	1.609	kilometers (km)
Area	inch^2	645.2	millimeters^2 (mm^2)
		6.45	centimeters^2 (cm^2)
	foot^2	0.0929	meters^2 (m^2)
	yard^2	0.8361	meters^2 (m^2)
Volume	inch^3	16,387	millimeters^3 (mm^3)
		16.387	centimeters^3 (cm^3)
		0.0164	liters (L)
	quart	0.9464	liters
	gallon	3.7854	liters
	yard^3	0.7646	meters^3 (m^3)
Mass	pound mass (lbm)	0.4536	kilograms (kg)
		453.6	grams (g)
	ounce mass (ozm)	28.35	grams (g)
	ton mass (tonm)	907.18	kilograms (kg)
		0.907	tonne (t)

Appendix

Conversion Factors (Continued)

English System		Metric System	
Force	ounce force (ozf)	0.2780	newtons (N)
	pound force (lbf)	4.448	newtons
Energy	Btu	1,005	joules (J = W-s)
		252.2	calories
		6.585×10^{21}	electron volts
	foot-pound (ft-bf)	1.3558	joules
		0.1383	kgf-m
	kilowatt-hour	3.6×10^6	joules
Power	horsepower (HP)	0.746	kilowatts (kW)
Pressure or stress	pounds/square inch (p.s.i.)	0.070,31	kgf/cm^2
		6.895	kilopascals (kPa)
	atmosphere (atm)	1.033	kgf/cm^2
	inches of H_2O	0.002,54	kgf/cm^2
		0.2491	kilopascals
	(1 dyne/s cm^2 =	1.45×10^{-5}	p.s.i.)
Velocity	feet/second	30.48	cm/s
		1.097	km/h
	miles/hour	1.6093	km/h
		44.70	cm/s
Temperature	degrees Fahrenheit (°F)	(°F − 32)/1.8	degrees Celsius (°C)
	°R (= °F + 459.7)	0.556	K (= °C + 273.2)

Useful Constants

Stefan–Boltzmann number	$= 0.171 \times 10^{-8}$ Btu/ft^2 h °R^4
Avogadro's number	$= 6.022 \times 10^{23}$ mol^{-1}
Gravitational constant (G)	$= 6.670 \times 10^{-11}$ m^3/kg s^2
Gravitational acceleration	$= 9.813$ m/s^2
	$= 32.17$ ft/s^2
Universal gas constant (R)	$= 8.31434 \times 10^7$ erg (g-mole)$^{-1}$ K^{-1}
	$= 1.986$ Btu (lbm-mol)$^{-1}$ °R^{-1}
Boltzmann's constant (k)	$= 1.38 \times 10^{-16}$ erg/atom °C
	$= 6.79 \times 10^{-23}$ in.-lb/atom °R

INDEX

Alger, R. M., 11, 23
Alignment index, 140
Allen, M. S., 23
Anderson, F. H., 227, 245
Argon, A. S., 216, 245
Asada, H., 283–288, 294
Asimow, M., 11, 23
Assembly, 204
Automobiles, 307
Axiom, 19, 47, 96–131
   in predicate logic, 356
Axiom, 29, 47, 147–187
   in predicate logic, 357
Axiomatic approach:
   Historical perspective, 14
   Origin of design axioms, 17

Bake, K. R., 304, 322
Battani, G., 360, 374
Beitz, W., 11, 24
Bell, A. C., 9, 24, 47, 69
Bias, 126, 169
Bittinger, M., 354, 374
Black, J., 331–347, 352
Blackburn, J. F., 88, 95
Boothroyd, G., 11, 16, 23, 41, 44, 163, 187
Brakeforming Machines, 207
Bremmerman, H. J., 17, 23
Bremmerman's Limit, 17
Brillouin, L., 66, 69, 154, 187
Buhl, H. R., 11, 23

Can/bottle opener, 51
Carbide tools, 262
Cellular manufacturing system, 323, 331–347
Christoffersen, J., 160, 187
Clausing, D., 34, 44
Coated carbide tools, 262
Colton, J. S., 245, 253
Common range, 156
Computerized system for design axioms, 360
Constraints, 39
Conway, R. W., 304, 322, 339, 352
Conversion factors, 396, 397
Cook, N. H., 47, 69, 268
Corollaries, 47, 52, 142

Corollaries in predicate logic, 357
Creative process, 10, 29
Creativity, 9

Davis, J. T., 248, 252
Decision making in design, 5, 7
Decomposition of the design process, 36
Decoupler, 332
Definitions, 25, 37, 47, 65
Design:
   Axioms, 9, 47
   Axioms in predicate logic, 356
   Definition, 25
   Design parameter (DP), 26
   Functional requirement (FR), 26
   Mathematical representation, 54
   Non-linear design, 114
   Outputs, 40
   Process, 27
Design equation, 54
Design Helix, 42
Design for manufacture, 40, 126, 190
Design Matrix (DM), 54–57, 114
   Non-linear DM, 114
   Non-square design matrix, 56
Design of manufacturing process, 298
Design procedure, 62
Design parameter isograms, 102
Design Range, 156
Dewhurst, P., 41, 44
Dispersion strengthened metal, 216
Durill, P. L., 240, 245, 254, 255

Eder, W. E., 11, 12, 14, 24
Electroless plating, 260
Elliot, J. F., 266
Erwin, L., 201, 245
Eversheim, W., 160, 188
Expert system, 372
Extruder, 49
   Analysis of single-screw extruder, 71
   Performance of gear pump-assisted
     extruder, 78, 102

Falser, P., 160, 187
Fasal, 11, 23

Filippone, S., 171, 172, 187
Finger, S., 281, 295
Fisher, R. A., 172
Fishman, G. S., 339, 352
Four steps in design process, 6
FR isograms, 102
Functional coupling and tolerancing, 121
Functional domain (or space), 26
Functional requirements (FRs), 26, 30, 37
  FR hierarchy, 36
  Quantitative measures for independence, 115

Gallager, R. G., 154, 187
Giedt, W. H., 251, 252
Glegg, G. L., 11, 23
Gleiser, M., 266
Gogos, C. G., 49, 50, 70
Gossard, D. C., 9, 24, 47, 69
Graphical representation, 97–105
Grinding wheel, 258
Griskey, R. G., 240, 245, 254, 255

Hamilton, A. G., 354, 374
Harrisberger, L., 11, 23
Hatsopoulous, G., 15, 23
Hauser, J. R., 34, 44
Hayes, C. V., 11, 23
Hierarchy in design, 4, 36
Hildebrand, F. B., 104, 132
Hill, P. H., 11, 23
House of Quality, 34
Hubka, V., 11, 12, 14, 25

IBM, 160, 187
Ideation, 7
Independence Axiom, 9, 47, 147–187
Information Axiom, 9, 147–187
  Information Axiom and the principle of maximum entropy, 154
Information:
  Axiomatic design, 64
  Definition of information, 65, 149
  Measurement in systems context, 156
  Metric for information content, 149
  Process planning, 160
  Reduction of information content, 169
Information content, 297
Injection molding, 193
Intelligent machine, 207
Isograms, 102

Jaynes, E. T., 154, 187
Jenkins, R. V., 29, 44
Just-in-time, 352

Kanade, T., 283, 284, 294
Kanban, 335, 352
Keenan, J. M., 15
Kim, S. H., 47, 69, 140, 146, 353, 366, 372, 374

Klein, I., 72–78
Kneebone, 354, 374
Kolmogoroff, A. N., 248, 249, 252
Kowalski, R., 354, 360, 374
Kramer, B. M., 265, 267, 294
Kumar, V., 230–244, 245

Laboratory for Manufacturing and Productivity, 19
Lapidot, N., 47, 69
Logic, 353
  Predicate, 355
  Propositional, 354

MacCready, P. B., 31, 33, 44
Malguarnera, S. C., 93, 95, 227, 245
Manufacturing system, 331
Mapping process, 27, 97
Martini, J., 231, 245
Martini-Vredensky, J. E., 231, 245
Mathematical representation, 55, 353
McCree, J., 201, 245
McKelvey, J. M., 50, 69, 78–86
Meloni, H., 360, 374
Microcellular plastic, 230
Milacic, V. R., 127, 132
MIT-Industry Polymer Processing Program, 201
Mixalloy process, 214–222, 247
Mixing of powder, 222
Moon, M. G., 9, 24, 214, 245
Morphological approach, 11–14
Multi-lense plate, 190
Murphy, A. T., 89, 95

Nakazawa, H., 156, 160, 161, 187, 297, 298, 299, 302, 303, 307–310, 322
National Science Foundation, 324–331, 352
Nehru, A. K., 259, 295
Nissen, C. K., 160, 187

Oh, H. L., 289–291, 295
Organization, 324, 347
  Government, 324
  College of engineering, 347
Orthogonality index, 139
Ostrofsky, B., 11, 24

Pahl, G., 11, 24
Passive filter design, 106, 134, 175
  Analysis by Taguchi method, 175
  Functional analysis, 106
Pereira, F., 360, 374
Phadke, M. S., 125, 133, 172, 188, 293, 295
Principle of maximum entropy, 154
Problem definition:
  Existing design, 34
  Original design, 30
PROLOG, 353, 375

Quantitative measures, 115

Reangularity, 116, 142
Reethof, G., 88, 95
RIM machine, 58, 87
  Control of metering ratio, 87
  MIT RIM system, 92
Rice, W. T., 78–86
Richardson, H. H., 89, 95
Rinderle, J. R., 9, 24, 97, 104–108, 119–121, 132, 137, 139, 146, 226, 227, 245
Robot, 204, 283
Rogers, J. B., 375, 379
Roth, K., 11, 24

Saka, N., 214, 245
Scale of Frs and DPs, 119
Scheduling, 303, 311
Scheffé, H., 173, 187
Schroer, B. J., 331–347, 352
Semangularity, 116
Sensitivity analysis, 123
Shannon, C. E., 66, 69, 154, 187
Shaw, M. C., 29, 44
Shearer, J. L., 88, 89, 95
SI units, 396
Signal to Noise (SN) ratio, 125, 169
Simon, H., 5, 24
Smith, J., 281, 295
Spencer–Gilmore equation, 195
Spur, G., 160, 188
Starr, M. K., 11, 24
Statistical analysis, 124
  Taguchi method, 171
Stelson, K. A., 210, 245
Stoll, H. W., 41, 44
Su, K. Y., 216, 245
Suh, N. P., 9, 24, 41, 44, 47, 60, 61, 69, 70, 87–95, 97, 126, 132, 133, 146, 148, 156, 160, 161, 211, 214, 217, 218, 224, 226, 227, 228, 230, 231, 245, 248, 253, 259, 263, 264, 265, 267, 294, 295, 297, 322, 353
Sunraycer, 33
Suonisun, E., 160, 187
Surface groups, 301
Sutek Corporation, 221, 222
Svardson, B., 160, 187
Symbolic logic, 353
System Range, 156

Systems context:
  Information, 156
  System range, 156

Tadmor, Z., 49, 50, 70
Taguchi, G., 125, 133, 171, 172, 173, 188, 293, 295
Taguchi method, 171
Takeyama, I., 283, 284, 294, 295
Theorems 1 & 2, 57
Theorems 3 & 4, 58
Theorem 5, 62
Theorems 6 & 7, 67
Theorem 8, 122, 131
Theorem 9, 131
Theorems 10 & 11, 132, 140, 143
Theorem 12, 153, 187
Theorem 13, 154, 187
Theorem 14, 168, 187
Theorem 15, 199
Theorems, 47
Thinking Design Machine (TDM), 353, 379
Thorne, J., 231, 245
Tice, W. W., 9, 24, 268, 271–278, 295
Tokawa, T., 311, 313, 322
Toth, L. E., 265
Tribus, M., 154, 188
Tsuda, H., 231, 245
Tucker III, C. L., 60, 61, 69, 70, 72–78, 87–95, 218, 224, 227, 245, 248, 253

USM foam molding process, 210

Variance, 125, 169
Vented Compression molding, 199
Von Neumann, J., 15, 24

Waldman, F. A., 231, 245
Weaver, W., 66, 69
Weill, R., 160, 188
Wilson, D. R., 9, 24, 44, 66, 70, 153, 188
Wilson, H. G., 31, 33, 44
Woodson, T. T., 11, 24

Yasuhara, M., 11, 24
Yoshikawa, H., 11, 24
Youcef-Toumi, K., 285–288, 295

Zwicky, F., 11